# MODELING AND SIMULATION FUNDAMENTALS

# MODELING AND SIMULATION FUNDAMENTALS
## Theoretical Underpinnings and Practical Domains

Edited by

### John A. Sokolowski
### Catherine M. Banks

The Virginia Modeling Analysis and Simulation Center
Old Dominion University
Suffolk, VA

A JOHN WILEY & SONS, INC. PUBLICATION

Cover graphic: Whitney A. Sokolowski

Copyright © 2010 by John Wiley & Sons, Inc. All rights reserved.

Published by John Wiley & Sons, Inc., Hoboken, New Jersey.

Published simultaneously in Canada.

No part of this publication may be reproduced, stored in a retrieval system, or transmitted in any form or by any means, electronic, mechanical, photocopying, recording, scanning, or otherwise, except as permitted under Section 107 or 108 of the 1976 United States Copyright Act, without either the prior written permission of the Publisher, or authorization through payment of the appropriate per-copy fee to the Copyright Clearance Center, Inc., 222 Rosewood Drive, Danvers, MA 01923, (978) 750-8400, fax (978) 750-4470, or on the web at www.copyright.com. Requests to the Publisher for permission should be addressed to the Permissions Department, John Wiley & Sons, Inc., 111 River Street, Hoboken, NJ 07030, (201) 748-6011, fax (201) 748-6088, or online at http://www.wiley.com/go/ permission.

Limit of Liability/Disclaimer of Warranty: While the publisher and author have used their best efforts in preparing this book, they make no representations of warranties with respect to the accuracy or completeness of the contents of this book and specifically disclaim any implied warranties of merchantability or fitness for a particular purpose. No warranty may be created or extended by sales representatives or written sales materials. The advice and strategies contained herein may not be suitable for your situation. You should consult with a professional where appropriate. Neither the publisher nor author shall be liable for any loss of profit or any other commercial damages, including but not limited to special, incidental, consequential, or other damages.

For general information on our other products and services or for technical support, please contact our Customer Care Department within the United States at (800) 762-2974, outside the United States at (317) 572-3993 or fax (317) 572-4002.

Wiley also publishes its books in a variety of electronic formats. Some content that appears in print may not be available in electronic formats. For more information about Wiley products, visit our web site at www.wiley.com.

*Library of Congress Cataloging-in-Publication Data*

Modeling and simulation fundamentals : theoretical underpinnings and practical domains / [edited by] John A. Sokolowski, Catherine M. Banks.
    p. cm.
  Includes bibliographical references and index.
  ISBN 978-0-470-48674-0 (cloth)
  1. Mathematical models.  2. Mathematical optimization.  3. Simulation methods.
I. Sokolowski, John A., 1953–  II. Banks, Catherine M., 1960–
    QA401.M53945   2010
    511'.8–dc22
                                              2009035905

Printed in the United States of America.

10 9 8 7 6 5 4 3 2 1

This book is dedicated to

my mom, in her memory
–John A. Sokolowski

my father, who is always in my thoughts
–Catherine M. Banks

# CONTENTS

**Preface**    xi

**Contributors**    xiii

**1 Introduction to Modeling and Simulation**    1
*Catherine M. Banks*

M&S / 2
M&S Characteristics and Descriptors / 12
M&S Categories / 19
Conclusion / 22
References / 24

**2 Statistical Concepts for Discrete Event Simulation**    25
*Roland R. Mielke*

Probability / 26
Simulation Basics / 35
Input Data Modeling / 39
Output Data Analysis / 48
Conclusion / 56
References / 56

**3 Discrete-Event Simulation**    57
*Rafael Diaz and Joshua G. Behr*

Queuing System Model Components / 60
Simulation Methodology / 62
DES Example / 65
Hand Simulation—Spreadsheet Implementation / 67
Arena Simulation / 87
Conclusion / 97
References / 98

## 4 Modeling Continuous Systems — 99
*Wesley N. Colley*

System Class / 100
Modeling and Simulation (M&S) Strategy / 101
Modeling Approach / 102
Model Examples / 104
Simulating Continuous Systems / 110
Simulation Implementation / 118
Conclusion / 128
References / 129

## 5 Monte Carlo Simulation — 131
*John A. Sokolowski*

The Monte Carlo Method / 132
Sensitivity Analysis / 142
Conclusion / 145
References / 145

## 6 Systems Modeling: Analysis and Operations Research — 147
*Frederic D. McKenzie*

System Model Types / 147
Modeling Methodologies and Tools / 148
Analysis of Modeling and Simulation (M&S) / 165
OR Methods / 174
Conclusion / 179
References / 179
Further Readings / 180

## 7 Visualization — 181
*Yuzhong Shen*

Computer Graphics Fundamentals / 182
Visualization Software and Tools / 208
Case Studies / 217
Conclusion / 223
References / 224

## 8 M&S Methodologies: A Systems Approach to the Social Sciences — 227
*Barry G. Silverman, Gnana K. Bharathy, Benjamin Nye, G. Jiyun Kim, Mark Roddy, and Mjumbe Poe*

Simulating State and Substate Actors with CountrySim: Synthesizing Theories Across the Social Sciences / 229

The CountrySim Application and Sociocultural
  Game Results / 255
Conclusions and the Way Forward / 265
References / 268

## 9  Modeling Human Behavior                                            271
*Yiannis Papelis and Poornima Madhavan*

Behavioral Modeling at the Physical Level / 273
Behavioral Modeling at the Tactical and Strategic Level / 274
Techniques for Human Behavior Modeling / 277
Human Factors / 305
Human–Computer Interaction / 308
Conclusion / 320
References / 321

## 10  Verification, Validation, and Accreditation                       325
*Mikel D. Petty*

Motivation / 326
Background Definitions / 326
VV&A Definitions / 330
V&V as Comparisons / 332
Performing VV&A / 333
V&V Methods / 340
VV&A Case Studies / 354
Conclusion / 365
Acknowledgments / 368
References / 368

## 11  An Introduction to Distributed Simulation                         373
*Gabriel A. Wainer and Khaldoon Al-Zoubi*

Trends and Challenges of Distributed Simulation / 374
A Brief History of Distributed Simulation / 375
Synchronization Algorithms for Parallel and Distributed
  Simulation / 377
Distributed Simulation Middleware / 383
Conclusion / 397
References / 398

## 12  Interoperability and Composability                                403
*Andreas Tolk*

Defining Interoperability and Composability / 405
Current Interoperability Standard Solutions / 412

Engineering Methods Supporting Interoperation
and Composition / 428
Conclusion / 430
References / 431
Further Readings / 433

**Index** **435**

# PREFACE

Modeling and simulation (M&S) has evolved from tool to discipline in less than two decades. With the technology boom of the 1990s came the ability to use models and simulations in nearly every aspect of life. What was once a tool for training the military (war-gaming) is now a capability to better understand human behavior, enterprise systems, disease proliferation, and so much more. To equip developers of M&S, the theoretical underpinnings must be understood. To prepare users of M&S, practical domains must be explored. The impetus for this book is to provide students of M&S with a study of the discipline a survey at a high-level overview.

The purpose of the text is to provide a study that includes definitions, paradigms, applications, and subdisciplines as a way of orienting students to M&S as a discipline and to its body of knowledge. The text will provide general conceptual framework for further MSIM studies.

To students who will be reading this text, we offer an incisive analysis of the key concepts, body of knowledge, and application of M&S. This text is designed for graduate students with engineering, mathematical, and/or computer science undergraduate training for they must have proficiency with mathematical representations and computer programs.

The text is divided into 12 chapters that build from topic to topic to provide the foundation/theoretical underpinnings to M&S and then progress to applications/practical domains. Chapter 1, "Introduction to Modeling and Simulation," provides a brief history, terminology, and applications and domains of M&S. Chapter 2, "Statistical Concepts for Discrete Event Simulation," provides the mathematical background. Chapters 3 to 5 develop a three-part series of M&S paradigms, starting with Chapter 3, "Discrete-Event Simulation," Chapter 4, "Modeling Continuous Systems," and Chapter 5, "Monte Carlo Simulation." Chapters 6 and 7 develop two areas necessary for model development. Chapter 6, "Systems Modeling: Analysis and Operations Research," reviews model types and research methods, and Chapter 7, "Visualization," brings into the discussion the importance of graphics.

The next four chapters cover sophisticated methodologies, verification and validation, and advanced simulation techniques: Chapter 8, "M&S Methodologies: A Systems Approach to the Social Sciences," Chapter 9,

"Modeling Human Behavior," Chapter 10, "Verification, Validation, and Accreditation," and Chapter 11, "An Introduction to Distributed Simulation." The concluding chapter, "Interoperability and Composability," introduces the importance of interoperability for engaging M&S within a number of domains.

While figures in the book are not printed in color, some chapters have figures that are described using color. The color representations of these figures may be downloaded from the following site: ftp://ftp.wiley.com/public/sci_tech_med/modeling_simulation.

John A. Sokolowski
Catherine M. Banks

# CONTRIBUTORS

**Catherine M. Banks,** PhD, Virginia Modeling, Analysis, and Simulation Center, Old Dominion University, 1030 University Boulevard, Suffolk, VA 23435; Email: cmbanks@odu.edu

**Joshua G. Behr,** PhD, Department of Political Science and Geography, Old Dominion University, 5115 Hampton Boulevard, Norfolk, VA 23529; Email: jbehr@odu.edu

**Wesley N. Colley,** PhD, Senior Research Scientist, Center for Modeling, Simulation, and Analysis, University of Alabama, 301 Sparkman Drive, VBRH D-15, Huntsville, AL 35899; Email: colleyw@uah.edu

**Rafael Diaz,** PhD, Virginia Modeling, Analysis, and Simulation Center, Old Dominion University, 1030 University Boulevard, Suffolk, VA 23435; Email: rdiaz@odu.edu

**Poornima Madhavan,** PhD, Department of Psychology, Old Dominion University, 5115 Hampton Boulevard, Norfolk, VA 23529; Email: pmadhava@odu.edu

**Frederic D. McKenzie,** PhD, Department of Electrical and Computer Engineering, Old Dominion University, 5115 Hampton Boulevard, Norfolk, VA 23529; Email: rdmckenz@odu.edu

**Roland R. Mielke,** PhD, Department of Electrical and Computer Engineering, Old Dominion University, 5115 Hampton Boulevard, Norfolk, VA 23529; Email: rmielke@odu.edu

**Yiannis Papelis,** PhD, Virginia Modeling, Analysis, and Simulation Center, Old Dominion University, 1030 University Boulevard, Suffolk, VA 23435; Email: ypapelis@odu.edu

**Mikel D. Petty,** PhD, Director, Center for Modeling, Simulation, and Analysis, University of Alabama, 301 Sparkman Drive, VBRH D-14, Huntsville, AL 35899; Email: pettym@email.uah.edu

**Yuzhong Shen,** PhD, Department of Electrical and Computer Engineering, Old Dominion University, 5115 Hampton Boulevard, Norfolk, VA 23529; Email: yshen@odu.edu

**Barry G. Silverman,** PhD, Department of Systems Engineering, University of Pennsylvania, Philadelphia, PA 19104; Email: barryg@seas.upenn.edu

**John A. Sokolowski,** PhD, Virginia Modeling, Analysis, and Simulation Center, Old Dominion University, 1030 University Boulevard, Suffolk, VA 23435; Email: jsokolow@odu.edu

**Andreas Tolk,** PhD, Department of Engineering Management and Systems Engineering, Old Dominion University, 5115 Hampton Boulevard, Norfolk, VA 23529; Email: atolk@odu.edu

**Gabriel A. Wainer,** PhD, Department of Systems and Computer Engineering, Carleton University, 1125 Colonel By Drive, 3216 V-Sim, Ottawa, ON, K1S 5B6, Canada; Email: gwainer@sce.carleton.ca

**Gnana K. Bharathy,** Postdoctoral candidate, University of Pennsylvania, Philadelphia, PA 19104

**G. Jiyun Kim,** Postdoctoral candidate, University of Pennsylvania, Philadelphia, PA 19104

**Mjumbe Poe,** Research staff, University of Pennsylvania, Philadelphia, PA 19104

**Mark Roddy,** Research staff, University of Pennsylvania, Philadelphia, PA 19104

**Khaldoon Al-Zoubi,** Graduate Student, Carleton University, Ottawa, ON, K1S 5B6

**Benjamin Nye,** Graduate Student, University of Pennsylvania, Philadelphia, PA 19104

# 1

# INTRODUCTION TO MODELING AND SIMULATION

Catherine M. Banks

Modeling and simulation (M&S) is becoming an academic program of choice for science and engineering students in campuses across the country. As a discipline, it has its own body of knowledge, theory, and research methodology. Some in the M&S community consider it to be an infrastructure discipline necessary to support integration of the partial knowledge of other disciplines needed in applications. Its robust theory is based on dynamic systems, computer science, and an ontology of the domain. Theory and ontology characterize M&S as distinct in relation to other disciplines; these serve as necessary components of a body of knowledge needed to practice M&S professionally in any of its aspects.

At the core of the discipline of M&S is the fundamental notion that *models are approximations of the real world*. This is the first step in M&S, creating a model approximating an event or a system. In turn, the model can then be modified in which *simulation* allows for the repeated observation of the model. After one or many simulations of the model, *analysis* takes place to draw conclusions, verify and validate the research, and make recommendations based on various simulations of the model. As a way of representing data, *visualization* serves to interface with the model. Thus, M&S is a problem-based discipline that allows for repeated testing of a hypothesis. Significantly, M&S

*Modeling and Simulation Fundamentals: Theoretical Underpinnings and Practical Domains*, Edited by John A. Sokolowski and Catherine M. Banks
Copyright © 2010 John Wiley & Sons, Inc.

expands the capacity to analyze and communicate new research or findings. This makes M&S unique to other methods of research and development.

Accordingly, the intent of this text is to introduce students to the fundamentals, the theoretical underpinnings, and practical domains of M&S as a discipline. An understanding and application of these skills will prepare M&S professionals to engage this critical technology.

## M&S

The foundation of an M&S program of study is its curriculum built upon four precepts—modeling, simulation, visualization, and analysis. The discussion below is a detailed examination of these precepts as well as other terms integral to M&S.* A good place to start is to define some principal concepts like system, model, simulation, and M&S.

### Definition of Basic Terms and Concepts

Because system can mean different things across the disciplines, an agreed upon definition of system was developed by the International Council of Systems Engineering (INCOSE). INCOSE suggests that a *system* is a construct or collection of different elements that together produces results not obtainable by the elements alone.** The elements can include people, hardware, software, facilities, policies, documents—all things required to produce system-level qualities, properties, characteristics, functions, behavior, and performance. Importantly, the value of the system as a whole is the relationship among the parts. A system may be *physical*, something that already exists, or *notional*, a plan or concept for something physical that does not exist.

In M&S, the term system refers to the subject of model development; that is, it is the subject or thing that will be investigated or studied using M&S. When investigating a system, a quantitative assessment is of interest to the modeler—observing how the system performs with various inputs and in different environments. Of importance is a quantitative evaluation of the performance of the system with respect to some specific criteria or performance measure. There are two types of systems: (1) *discrete*, in which the state variables (variables that completely describe a system at any given moment in time) change instantaneously at separate points in time, and (2) *continuous*,

---

*Portions of this chapter are based on Banks CM. What is modeling and simulation? In *Principles of Modeling and Simulation: A Multidisciplinary Approach*. Sokolowski JA, Banks CM (Eds.). Hoboken, NJ: John Wiley & Sons; 2009; VMASC short course notes prepared by Mikel D. Petty; and course notes prepared by Roland R. Mielke, Old Dominion University.

**Additional information and definitions of system can be found at the INCOSE online glossary at http://www.incose.org/mediarelations/glossaryofseterms.aspx.

where the state variables change continuously with respect to time. There are a number of ways to study a system:

(1) the actual system versus a model of the system
(2) a physical versus mathematical representation
(3) analytic solution versus simulation solution (which exercises the simulation for inputs to observe how they affect the output measures of performance) [1].

In the study of systems, the modeler focuses on three primary concerns: (1) the quantitative analysis of the systems; (2) the techniques for system design, control, or use; and (3) the measurement or evaluation of the system performance.

The second concept, *model*, is a physical, mathematical, or otherwise logical representation of a system, entity, phenomenon, or process. Simply, models serve as representations of events and/or things that are real (such as a historic case study) or contrived (a use case). They can be representations of actual systems. This is because systems can be difficult or impossible to investigate.

As introduced above, a system might be large and complex, or it might be dangerous to impose conditions for which to study the system. Systems that are expensive or essential cannot be taken out of service; systems that are notional do not have the physical components to conduct experiments. Thus, models are developed to serve as a stand-in for systems. As a substitute, the model is what will be investigated with the goal of learning more about the system.

To produce a model, one abstracts from reality a description of the system. However, it is important to note that a model is not meant to represent all aspects of the system being studied. That would be too timely, expensive, and complex—perhaps impossible. Instead, the model should be developed as simply as possible, representing only the system aspects that affect system performance being investigated in the model. Thus, the model can depict the system at some point of abstraction or at multiple levels of the abstraction with the goal of representing the system in a reliable fashion. Often, it is challenging for the modeler to decide which aspects of a system need to be included in the model.

A model can be *physical*, such as a scale model of an airplane to study aerodynamic behavior. A physical model, such as the scale model of an airplane, can be used to study the aerodynamic behavior of the airplane through wind-tunnel tests. At times, a model consists of a set of mathematical equations or logic statements that describes the behavior of the system. These are *notional* models. Simple equations often result in analytic solutions or an analytic representation of the desired system performance characteristic under study.

Conversely, in many cases, the mathematical model is sufficiently complex that the only way to solve the equations is numerically. This process is referred

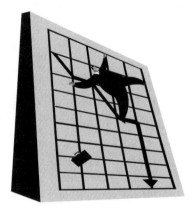

**Figure 1.1** Model example.

to as *computer simulation*. Essentially, a system is modeled using mathematical equations; then, these equations are solved numerically using a digital computer to indicate likely system behavior. There are distinct differences between the numerical and the analytic way of solving a problem: Analytic solutions are precise mathematical proofs, and as such, they cannot be conducted for all classes of models. The alternative is to solve numerically with the understanding that an amount of error may be present in the numerical solution.

Below is an example of developing a model from a mathematical equation. The goal of the model is to represent the vertical height of an object moving in one dimension under the influence of gravity (Fig. 1.1).

The model takes the form of an equation relating the object height $h$ to the time in motion $t$, the object initial height $s$, and the object initial velocity $v$, or:

$$h = \tfrac{1}{2}at^2 + vt + s,$$

where

$h$ = height (feet),
$t$ = time in motion (seconds),
$v$ = initial velocity (feet per second, + is up),
$s$ = initial height (feet),
$a$ = acceleration (feet per second per second).

This model represents a first-order approximation to the height of the object. Conversely, the model fails, however, to represent the mass of the object, the effects of air resistance, and the location of the object.

Defining the third concept, *simulation*, is not as clear-cut as defining the model. Definitions of simulation vary:

(1) a method for implementing a model over time
(2) a technique for testing, analysis, or training in which real-world systems are used, or where real-world and conceptual systems are reproduced by a model
(3) an unobtrusive scientific method of inquiry involving experiments with a model, rather than with the portion of reality that the model represents
(4) a methodology for extracting information from a model by observing the behavior of the model as it is executed
(5) a nontechnical term meaning not real, imitation

In sum, simulation is an applied methodology that can describe the behavior of that system using either a mathematical model or a symbolic model [2]. It can be the imitation of the operation of a real-world process or system over a period of time [3].

Recall, engaging a real system is not always possible because (1) it might not be accessible, (2) it might be dangerous to engage the system, (3) it might be unacceptable to engage the system, or (4) the system might simply not exist. To counter these constraints, a computer will *imitate* operations of these various real-world facilities or processes. Thus, a simulation may be used when the real system cannot be engaged.

Simulation, simulation model, or software model is also used to refer to the software implementation of a model. The mathematical model of the Model Example 1 introduced above may be represented in a software model. The example below is a *C program* that calculates the height of an object moving under gravity:

**Simulation Example 1**

```
/* Height of an object moving under gravity. */
/* Initial height v and velocity s constants. */
main()
{
   float h, v = 100.0, s = 1000.0;
   int t;
   for (t = 0, h = s; h >= 0.0; t++)
   {
      h = (-16.0 * t * t) + (v * t) + s;
      printf("Height at time %d = %f\n", t, h);
   }
}
```

This is a software implementation of the model. In an actual application, $s$ and $v$ would be identified as input variables rather than constants. The result of simulating this model, executing the software program on a computer, is a series of values for $h$ at specified times $t$.

**6**  INTRODUCTION TO MODELING AND SIMULATION

| t | v | h |
|---|---|---|
| 0 | 100 | 1000 |
| 1 | 68 | 1052 |
| 2 | 36 | 972 |
| 3 | 4 | 860 |
| 4 | −28 | 719 |
| 5 | −60 | 540 |
| 6 | −92 | 332 |
| 7 | −124 | 92 |

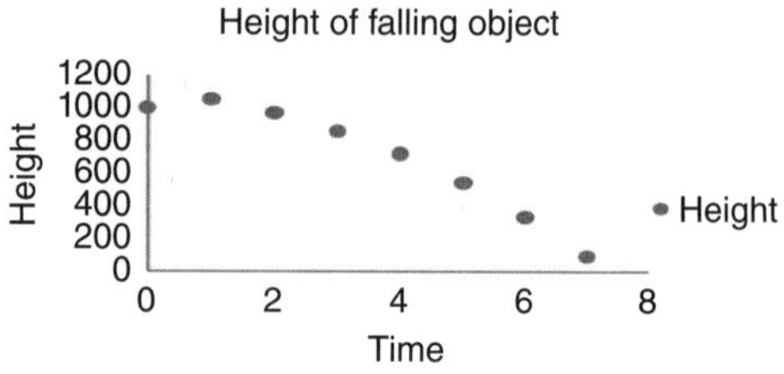

**Figure 1.2** Tabular and graphic simulation.

Below is another output of the same model showing the results of simulating or executing the model of an object moving under the influence of gravity. The simulation is conducted for an initial height of $s = 1000$ ft, and an initial velocity of $v = 100$ ft/s. Note from the example that the positive reference for velocity is up, an acceleration of −32 ft/s/s. The results of the simulation are presented in tabular and graphic forms (Fig. 1.2):

**Simulation Example 2**

Model: $h = \frac{1}{2}at^2 + vt + s\quad v = at + v_0$

Data: $v_0 = 100 \text{ ft/s}, \quad s = 1000 \text{ ft}, \quad a = -32 \text{ ft/s}^2.$

There are several terms associated with the execution of a simulation. The term *run* and/or *trial* is used to refer to a single execution of a simulation, as shown above. They may also refer to a series of related runs of a simulation as part of an analysis or experimentation process. The term *exercise* is used to refer to a series of related runs of the simulation as part of a training process. Thus, trial and exercise are similar in meaning but imply different uses of the simulation runs. Lastly, simulation also allows for virtual reality research whereby the analyst is immersed within the simulated world through the use of devices such as head-mounted display, data gloves, freedom sensors, and forced-feedback elements [2].

The fourth concept is *M&S*. M&S refers to the overall process of developing a model and then simulating that model to gather data concerning performance of a system. M&S uses models and simulations to develop data as a basis for making managerial, technical, and training decisions. For large, complex systems that have measures of uncertainty or variability, M&S might be the only feasible method of analysis of the system. M&S depends on computational science for the simulation of complex, large-scale phenomena. (Computational science is also needed to facilitate the fourth M&S precept, visualization, which serves to enhance the modeler's ability to understand or interpret that information. Visualization will be discussed in more detail below.)

In review, M&S begins with (1) developing computer simulation or a design based on a model of an actual or theoretical physical system, then (2) executing that model on a digital computer, and (3) analyzing the output. Models

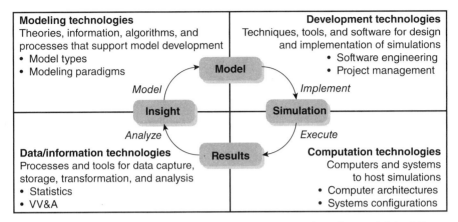

**Figure 1.3** M&S cycle and relevant technologies (adapted from Starr and Orlov [4]).

and the ability to act out with models is a credible way of understanding the complexity and particulars of a real entity or system [2].

## M&S Development Process Cycle

The process of M&S passes through four phases of a cyclic movement: model, code, execute, and analyze. Each phase depends on a different set of supporting technologies:

(1) model phase = modeling technologies
(2) code phase = development technologies
(3) execute phase = computational technologies
(4) analyze phase = data/information technologies

Figure 1.3 illustrates these phases and their related technologies [4]. The figure also depicts two processes: (1) the phases used in the development and testing of computer models and simulations and 2) the phases involved in applying M&S to the investigation of a real-world system.

***Modeling Technologies*** The construction of a model for a system requires data, knowledge, and insight about the sysyem. Different types of systems are modeled using different constructs or paradigms. The modeler must be proficient in his or her understanding of these different system classes and select the best modeling paradigm to capture or represent the system he or she is to model. As noted previuosly, modeling involves mathematics and logic to describe expected behavior; as such, only those system behaviors siginificant to the study or research question need be represented in the model.

***Development Technologies*** The development of a simulation is a software design project. Computer code must be written to algorithmically represent the mathematical statements and logical constructs of the model. This phase of the M&S cycle uses principles and tools of software engineering.

**Computational Technologies**  The simulation is next executed to produce performance data for the system. For simple simulations, this might mean implementing the simulation code on a personal computer. For complex simulations, the simulation code might be implemented in a distributed, multiprocessor or multicomputer environment where the different processing units are interconnected over a high-speed computer network. Such an implementation often requires specialized knowledge of computer architectures, computer networks, and distributed computing methodologies.

**Data/Informational Technologies**  During this phase of the M&S process, analysis of the simulation output data is conducted to produce the desired performance information that was the original focus of the M&S study. If the model contains variability and uncertainty, then techniques from probability and statistics will likely be required for the analysis. If the focus of the study is to optimize performance, then appropriate optimization techniques must be applied to analyze the simulation results.

The desired M&S process will undoubtedly take a number of iterations of the M&S cycle. The first iteration often provides information for modifying the model. It is a good practice to repeat the cycle as often as needed until the simulation team is satisfied that the results from the M&S study are close enough to the performance of the system being studied.

Figure 1.4 provides a more detailed view of the M&S cycle with the addition of details such as verification, validation, and accreditation (VV&A) activities, which serve to ensure a more correct and representative model of the system [5]. (The *dashed connectors* show how the process advances from one phase to the next. The *solid connectors* show the VV&A activities that must be integrated with the development activities.)

**Verification**  Verification ensures that M&S development is conducted correctly, while *validation* ensures that the model represents the real system and that the model is truly representative of that system. (Chapter 10 will provide a thorough discussion on the subject of VV&A.) This diagram illustrates how VV&A activities are not conducted as a phase of the M&S process, but as activities integrated throughout the M&S process.

To engage the entire M&S process, a number of related concepts and disciplines must be incorporated into the cycle.

### Related Disciplines

There are five key concepts and/or disciplines related to the M&S process: probability and statistics, analysis and operations research, computer visualization, human factors, and project management. Each will be briefly discussed.

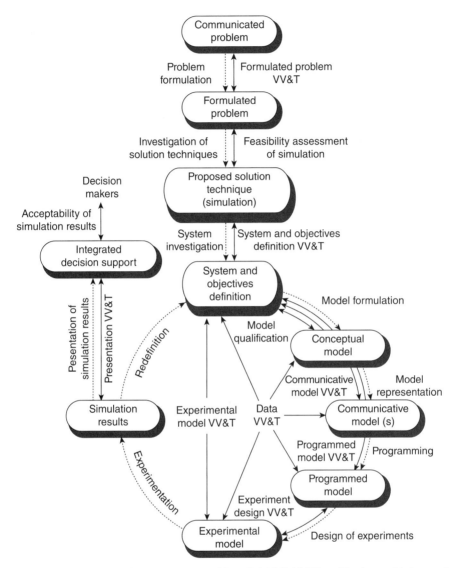

**Figure 1.4** Detailed M&S life cycle (adapted from Balci [5]). VV&T, verification, validation, and testing.

***Probability and Statistics*** Nearly all systems in the real-world display varying degrees of uncertainty. For instance, there is uncertainty in the movement of cars at a stop light:

(1) How long before the first car acknowledges the light change to green?
(2) How fast does that car take off?
(3) At what time does the second car start moving?

(4) What is the spatial interval between cars?
(5) What happens if one of the cars in the chain stalls?

In modeling a situation such as traffic movement at a stop light, one cannot ignore or attempt to average the uncertainty of response/movement because the model would then lack validity. Inclusion of uncertainty and variability requires that system parameters be represented as *random variables* or *random process*.

Working with random variables requires the use of concepts and theories from *probability and statistics*, a branch of mathematics. Probability and statistics are used with great frequency in M&S to generate random variates to model system random input variables that represent uncertainty and variability, and to anaylze the output from *stochastic models* or systems.

Stochastic models contain parameters that are described by random variables; thus, simulation of stochastic models results in outputs that are also random variables. Probability and statistics are key to analysis of these types of systems. Chapters 2 and 5 will provide further discussion of this significant branch of mathematics.

**Analysis and Operations Research** The conduct of a simulation study results in the generation of system performance data, most often in large quantities. These data are stored in a computer system as large arrays of numbers. The process of converting the data into meaningful information that describes the behavior of the system is called *analysis*. There are numerous techniques and approaches to conducting analysis. The development and use of these techniques and approaches are a function of the branch of mathematics and systems engineering called *operations research*.

M&S-based analysis has a simulation output that typically represents a dynamic response of the modeled system for a given set of conditions and inputs. Analysis is performed to transform these data when seeking answers to questions that motivated the simulation study. The simulation study can include a number of functions:

(1) *design of experiments*—the design of a set of simulation experiments suitable for addressing a specific system performance question;
(2) *performance evaluation*—the evaluation of system performance, measurement of how it approaches a desired performance level;
(3) *sensitivity analysis*—system sensitivity to a set of input parameters;
(4) *system comparison*—comparison of two or more system alternatives to derive best system performance with given conditions;
(5) *constrained optimization*—determination of optimum parameters to derive system performance objective.

Recall, analysis is one of the four precepts of M&S (along with modeling, simulation, and visualization). Simply, analysis takes place to draw conclusions, verify and validate the research, and make recommendations based on various simulations of the model. Chapter 4 delves further on the topics of queuing theory-based models, simulation methodology, and spreadsheet simulation—all functions of analysis.

***Computer Visualization***   Visualization is the ability to represent data as a way to interface with the model. (It is also one of the four precepts of M&S.) The systems that are investigated using M&S are large and complex; too often tables of data and graphs are cumbersome and do not serve to clearly understand the behavior of systems. Visualization is used to represent the data.

Computer graphics and computer visualization are used to construct two-dimensional and three-dimensional models of the system being modeled. This allows for the visual plotting and display of *system time response functions* to visualize complex data sets and to animate visual representations of systems to understand its dynamic behavior more adequately. M&S professionals who are able to engage *visualization* fully are able to provide an overview of interactive, real-time, three-dimensional computer graphics and visual simulations using high-level development tools. These tools facilitate virtual reality research, whereby the analyst is immersed within the simulated world through the use of devices such as head-mounted display, data gloves, freedom sensors, and forced-feedback elements [2]. *Computer animations* are offshoots of computational science that allow for additional variations in modeling.* Chapter 7 will provide an in-depth discussion of visualization.

***Human Factors***   Most simulations are developed to interface with a human user. These simulations place humans as system components within the model. To do this efficiently and effectively, the simulation designer must have a basic understanding of human cognition and perception. With this knowledge, the simulation designer can then create the human–computer interface to account for the strengths and weaknesses of the human user. These areas of study are called *human factors* and *human–computer interfacing*. The modeling of human factors is called *human behavior modeling*. This type of modeling focuses primarily on the computational process of human decision making. All three areas of study are typically subareas of psychology, although disciplines within the social sciences (such as history, geography, religious studies, political science) also make significant contributions to human behavior modeling.** Chapter 9 addresses human factors in M&S.

***Project Management***   The application of the M&S process to solve real-world problems is a daunting task, and, if not managed properly, it can become

---

*Computer animation* is emphasized within computer graphics, and it allows the modeler to create a more cohesive model by basing the animation on more complex model types. With the increased use of system modeling, there has been an increased use of computer animation, also called physically based modeling [4].
**For more information on human behavior modeling and case studies using systems dynamics, game theory, social network modeling, and ABM to represent human behavior, see Sokolowski JA, Banks CM. (Eds.). *Principles of Modeling and Simulation: A Multidisciplinary Approach*. New York: John Wiley & Sons; 2009; and Sokolowski JA, Banks CM. *Modeling and Simulation for Analyzing Global Events*. New York: John Wiley & Sons; 2009.

a problem in itself. For instance, there might be thousands of people and months of effort invested in a project requiring effective and efficient management tools to facilitate smooth outlay. When computer simulation is the only method available to investigate such large-scale projects, the M&S process becomes a large technical project requiring oversight and management. Thus, the M&S professional must be acquainted with project management, a subarea of engineering management.

With this introduction of M&S fundamentals, what is meant by M&S and the related areas of study that are important to the M&S process, one can progress to a more detailed discussion of M&S characteristics, paradigms, atttributes, and applications.

## M&S CHARACTERISTICS AND DESCRIPTORS

Understanding what is meant by M&S and how, as a process, it can serve a broad venue of research and development is one's initial entry into the M&S community. As M&S professionals, one must progress to understanding and engaging various simulation paradigms and modeling methods. The information below will introduce some of these characteristics and descriptors.

### Simulation Paradigms

There are different simulation paradigms that are prominent in the M&S process. First, there is the *Monte Carlo simulation* (also called the Monte Carlo method), which randomly samples values from each input variable distribution and uses that sample to calculate the model's output. This process of random sampling is repeated until there is a sense of how the output varies given the random input values. Monte Carlo simulation models system behavior using probabilities. Second is *continuous simulation* whereby the system variables are *continuous functions of time*. Time is the independent variable and the system variables evolve as time progresses. Continuous simulations systems make use of differential equations in developing the model. The third simulation paradigm is *discrete-event simulation* in which the system variables are *discrete functions in time*. These discrete functions in time result in system variables that change only at distinct instants of time. The changes are associated with an occurence of a system event. Discrete-event simulations advance time from one event to the next event. This simulation paradigm adheres to queuing theory models. Continuous and discrete-event simulations are *dynamic systems* with variables changing over time. All three of these simulation paradigms are discussed individually in Chapters 2–4.

### M&S Attributes

There are three primary descriptors applied to a *model* or *simulation* that serve as attributes or defining properties/characteristics of the model or simulation. These are fidelity, resolution, and scale.

*Fidelity* is a term used to describe how the model or the simulation closely matches reality. The model or simulation that closely matches or behaves like the real system it is representing has a high fidelity. Attaining high fidelity is not easy because models can never capture every aspect of a system. Models are built to characterize only the aspects of a system that are to be investigated. A great degree of effort is made to achieve high fidelity. A low fidelity is tolerated with regard to the components of the system that are not important to the invesigation. Similarly, different applications might call for different levels of fidelity. The simulation of the system for thesis research and development may require higher levels of fidelity than a model that is to be used for training.

Often, the term fidelity is used incorrectly with *validity* to express the accuracy of the representation. Only validity conveys three constructs of accuracy of the model: [Do not confuse validity w fidelity]

(1) *reality*—how the model closely matches reality
(2) *representation*—some aspects are represented, some are not
(3) *requirements*—different levels of fidelity required for different applications.

*Resolution* (also known as granularity) is the degree of detail with which the real world is simulated. The more detail included in the simulation, the higher the resolution. A simple illustration would be the simulation of an orange tree. A simulation that represents an entire grove would prove to have a much lower resolution of the trees than a simulation of a single tree. Simulations can go from low to high resolution. Return to the example of the tree: The model can begin with a representation of the entire forest, then a model of an individual tree, then a model of that individual tree's fruit, with a separate model of each piece of fruit in varying stages of maturity.

*Scale* is the size of the overall scenario or event the simulation represents; this is also known as level. Logically, the larger the system or scenario, the larger the scale of the simulation. Take for example a clothing factory. The simulation of a single sewing machine on the factory floor would consist of a few simulation components, and it would require the representation of only a few square feet of the entire factory. Conversely, a simulation of the entire factory would require representations of all machines, perhaps hundreds of simulation components, spread out over several hundred thousand square feet of factory space. Obviously, the simulation of the single sewing machine would have a much smaller scale than the simulation of the entire factory.

With an understanding of fidelity, resolution, and scale as individual attributes of M&S comes the ability to join these attributes to one another. The ability to relate fidelity and resolution, or fidelity and scale, or resolution and scale provides insight to the different types of simulations being used today. Table 1.1 is a comparison of *fidelity and resolution* premised on the common

**Table 1.1 Comparing fidelity and resolution**

|  | Fidelity | |
| --- | --- | --- |
| Resolution | Low | High |
| Low | Board game—*chess* | Agent-based simulation—*Swarm* |
| High | Personal computer flight simulator—*Microsoft Flight Simulator* | Platform-level training simulation—airline flight simulator |

**Table 1.2 Comparing fidelity and scale**

|  | Fidelity | |
| --- | --- | --- |
| Scale | Low | High |
| High | Board game—*Battleground* | Massive multiplayer online games—*World of Warcraft* |
| Low | Personal computer combat simulator—*Doom* | First-person shooter—*Halo* |

assumption that increasing resolution increases fidelity. This premise is not absolute because it is possible to increase the resolution of the simulation without increasing the fidelity of the simulation. Note the four combinations of fidelity–resolution.

The assumption held regarding *fidelity and scale* is that increasing scale results in decreasing fidelity. This assumption is unsound. As scale increases, it is likely that there will be an increase in the number of simulated entities. However, what if there is an aggregation of closely related entities as a single simulation entity? If the research question of the system sought to address behavior of related groups, then the increasing scale might have no effect on the fidelity of the simulation. Note the four combinations of fidelity–scale in Table 1.2.

The final comparison is that of *resolution and scale*. In general terms, more resolution leads to less scale and vice versa. Increasing scale results in decreasing resolution. This is due to the fact that the computing system hosting the simulation has a finite limit on the computing capability, especially since each simulation entity requires a specific amount of computational power for a given level of resolution. As scale increases, the number of entities increases, and these entities require additional computational capability. If the computational capability is at its limit, then increases in scale can only take place if the resolution of the simulation is lowered. As a result, high resolution, high-scale simulations are constrained by computing requirements. Note the four combinations of resolution–scale in Table 1.3.

Once a model has been developed with the correct simulation paradigm engaged and a full appreciation of fidelity, resolution, and scale as attributes

**Table 1.3 Comparing resolution and scale**

| Resolution | Scale | |
|---|---|---|
| | Low | High |
| Low | Not interesting | Operational-level training simulation—*WarSim* |
| High | Urban warfare personal computer game—*Shrapnel: Urban Warfare 2025* | Not practical |

of the simulation acknowledged, the modeler must then consider, *is the model correct and usable*? This is done through the process of verification and validation (V&V).

## VV&A Process

No discussion of M&S characteristics and descriptors would be complete without addressing the importance of the VV&A process. VV&A is the process of determining if the model and/or simulation is correct and usable for the purpose of which it has been designed. Simply, one might ask, was the model *built correctly* and was it the *correct model*? It is also the process of developing and delimiting confidence that a model can be used for a specific purpose. The first phase, *verification*, is the process of determining if a model accurately represents the conceptual description and specifications of the model. Verification requires a check on the coding by determining if the simulation is coded correctly. Asking, *does the simulation code correctly implement the model* is the way verification tests the software quality. A number of software engineering tests and techniques that are part of the verification process will be introduced in Chapter 10.

The process of determining the degree to which a model is an accurate representation of the real-world system from the perspective of the model's intended use is *validation*. Validity answers the question, *is the right thing coded* or *how well does the model match reality in the context of purpose of the model*? Validity speaks to modeling quality. In essence, validity is a measure of model fidelity in a specific application of the model. Validation methods or techniques exist to verify and validate a model and its simulation. These techniques go from *informal* to *inspection-like* to *formal* with the use of logic to prove correctness. Table 1.4 is a small representation of some of these techniques [5].

The third aspect of VV&A is the process of accreditation as the official certification by a responsible authority that a model is acceptable for a specific use [6]. The authority is an agency or person responsible for the results of using the model. As such, the authority should be separate from the developer of the model or simulation. This is not a general-purpose approval as each model is accredited for a specific purpose or use.

**Table 1.4 Verification and validation techniques**

| Informal | Static | Dynamic | Formal |
|---|---|---|---|
| Audit | Cause–effect graphing | Acceptance testing | Induction |
| Desk checking | Control analysis | Alpha testing | Inductive assertions |
| Documentation checking | Data analysis | Assertion checking | Inference |
| Face validation | Fault/failure analysis | Beta testing | Logical deduction |
| Inspections | Interface analysis | Bottom-up testing | Lambda calculus |
| Reviews | Semantic analysis | Comparison testing | Predicate calculus |
| Turing test | Structural analysis | Statistical techniques | Predicate transformation |
| Walkthroughs | Symbolic analysis | Structural testing | Proof of correctness |
| | Syntax analysis | Submodel/ module testing | |
| | Traceability assessment | Visualization/ animation | |

## Model Types

The final subtopic under M&S characteristics and descriptors is *model types*. The following modeling types are common in M&S, and this listing will serve to define these modeling methods succinctly. Discussion of many of the model types will be developed in greater detail in the following chapters.

***Physics-Based Modeling***  Physics-based modeling is solidly grounded in mathematics. A physics-based model is a mathematical model where the model equations are derived from basic physical principles. Model Example 1 is a physics-based model. Unique to physics-based models is the fact that the physics equations are models themselves in that many physics-based models are not truly things—they are intangibles; hence, they are representations of phenomena. Another example is Newton's law of gravity, which describes the gravitational attraction between bodies with mass. His idea was first published in 1687; in contemporary text it reads:

Every point mass attracts every other point mass by a force pointing along the line intersecting both points. The force is directly proportional to the product of the two masses and inversely proportional to the square of the distance between the point masses:

$$F = G\frac{m_1 m_2}{r^2},$$

where

$F$ is the magnitude of the gravitational force between the two point masses
$G$ is the gravitational constant
$m_1$ is the mass of the first point mass
$m_2$ is the mass of the second point mass
$r$ is the distance between the two point masses.

Physics-based models are based on *first principles* (as such, they may be referred to as first principle models). These principles, however, do not guarantee fidelity as this type of model may not represent all aspects of a system, or it might be based on assumptions that constrain the use of the model so that it is suitable only under certain conditions. There may be assumptions and omissions that affect the fidelity. Going back to Model Example 1, the height of the building is recognized but not the air resistance—and the model assumes the location will be near the surface of the Earth.

Physics-based models may also suffer *invalid composition*. This occurs when many simulations combine multiple physics-based models. Combining multiple models usually takes place with the development of a large-scale model; in essence, larger-scale models are the combination of smaller-scale models. When this combination takes place, changes to one or many of the models' components result. When this occurs, there might be invalid composition.

**Finite Element Modeling (FEM)** FEM is the method used for modeling large or complicated objects by decomposing these elements into a set of small elements and then modeling the small elements. This type of modeling is widely used for engineering simulation, particularly mechanical and aerospace engineering. These subdisciplines conduct research that requires structural analysis or fluid dynamics problems. FEM facilitates the decomposition of a large object into a set of smaller objects labeled *elements*.

These individual elements and the neighbor relationships that occur with elements in proximity are represented by a mesh of nodes. The state of the nodes is modeled using physics-based equations that take into account the current state of the node, the previous state, the state of the nearest neighboring node, and any knowledge of interactions between the neighbors. These computations of the state of the nodes are iterated over simulation time.

**Data-Based Modeling** Data-based modeling results from models based on data describing represented aspects of the subject of the model. Model development begins with advanced research or data collection, which is used in simulations. Data sources for this type of modeling can include actual field experience via the real-world or real system, operational testing and evaluation of a real system, other simulations of the system, and qualitative and quantitative research, as well as best guesses from subject matter experts. The model is developed with the view that the system is exercised under varying conditions with varying inputs. As the outputs unfold, their results are recorded and tabulated so as to review appropriate responses whenever similar conditions and inputs are present in the model.

Data-based modeling is often used when the real system cannot be engaged or when the subject of the model is notional. When the physics of the model subject is not understood or computations costs are high, data-based modeling can substitute. This modeling relies on data availability—it functions at its best when the data are accurate and reliable.

***Agent-Based Modeling (ABM)*** ABM is an important modeling paradigm for investigating many types of human and social phenomena [7]. The important idea here is that of a computer being able to create a complex system on its own by following a set of rules or directions and not having the complex system defined beforehand by a human. ABMs consist of *agents* that are defined as *autonomous software entities that interact with their environment or other agents to achieve some goal or accomplish some task*. This definition has several important elements to recognize. Probably the most important of these elements is the concept of *autonomy*. This characteristic is what sets agents apart from other object-oriented constructs in computer science. Agents act in their own self-interest independent of the control of other agents in the system. That is not to say that they are not influenced by other agents. They do not take direction from other agents. Because of this autonomy, each agent decides for itself what it will do, when it will do it, and how it will be done. These decisions are based on behaviors incorporated into the agent by its designer.

An agent's *environment* and the existence of other agents in that environment also play a key role on how an agent may behave. An agent is embodied with the ability to sense its environment, which includes everything it is aware of external to itself except other agents. As it senses this environment, it may respond to changes in it or it may just observe the changes waiting for a specific event to take place. It is also aware of the other agents. It may monitor what they are doing and may communicate with them to request they accomplish some task or it may respond to a request it has received. This closely represents how a human interacts with its surroundings and the other persons in it.

Finally, an agent acts to achieve some *goal* or accomplish some task. A task may be to retrieve a piece of data from a specific source or move to a certain location in virtual space. Task accomplishment is generally reactive in nature and does not require some complex set of reasoning to carry out. The use of artificial intelligence techniques can be incorporated into the model to modify the behavior of agents and rules of interaction among them. As such, ABMs vary widely in implementation and level of sophistication. (For a detailed discussion, see Sokolowski and Banks [8].)

***Aggregate Modeling*** This modeling method facilitates a number of smaller objects and actions represented in a combined, or aggregated, manner. Aggregate models are used most commonly when the focus of the M&S study is on aggregate performance. The model can also scale and number represented entities that are large and can compromise the time required to conduct a simulation. These models are most often used in constructive models; they are not physics-based models.

***Hybrid Modeling*** Hybrid modeling entails combining more than one modeling paradigm. This type of modeling is becoming a common practice among model developers. Hybrid modeling makes use of several modeling methods; however, they are disadvantaged in that composing several different types of models correctly is a difficult process.

There are numerous other model types. Some of these include Markov chains, finite-state automata, particle systems, queuing models, bond graphs, and Petri nets. The challenge for the modeler is to choose the best modeling paradigm that represents the designated system and answer specific research questions or training needs.

## M&S CATEGORIES

Categorizing models and simulations into different groupings or assemblages is a useful exercise in that it facilitates a clearer understanding of what makes some models and simulations similar and some different. This grouping or partitioning of models and simulations is grounded on shared, common characteristics. Categorizing models and simulations establishes what can be considered *coordinates of M&S space*. Where a model or simulation is placed in this space identifies its individual properties. There are four category dimensions: type, application, randomness, and domain.

### Type

M&S is classed into three types: live, virtual, and constructive. These types vary in operator and environment. For example, the model and simulation can include real people doing real things, or real people operating in unreal or simulated environments, and real people making inputs into simulations that execute those inputs by simulated people.

A *live simulation* involves real people operating real systems. This simulation strives to be as close as possible to real use, and it often involves real equipment or systems. The military train using live simulation when they conduct war games that place real soldiers and real platforms in an engagement situation in which actual weapon firings or impacts have been replaced with instrumentation. The purpose of live simulation training is to provide a meaningful and useful experience for the trainee.

A *virtual simulation* is different from live simulation in that it involves real people operating in simulated systems. These systems are recreated with

**Table 1.5  M&S types**

| Category | Participants | Systems |
|---|---|---|
| Live | Real | Real |
| Virtual | Real | Simulated |
| Constructive | Simulated | Simulated |

simulators, and they are designed to immerse the user in a realistic environment. A good example of virtual simulation training is the cockpit simulator used to train aircraft pilots. This simulator uses a physical representation of the actual cockpit with computer models to generate flight dynamics, out-of-window visuals, and various environmental/atmospheric changes to which the pilot must respond. This type of training is designed to provide useful piloting experience without leaving the ground.

The third type of M&S is *constructive simulation*. This simulation involves real people making inputs into a simulation that carry out those inputs by simulated people operating in simulated systems. As real people provide directives or inputs, activity begins within the simulation. There are no virtual environments or simulators, and the systems are operated by nonparticipants. The expected result of constructive simulation is that it will provide a useful result. The military has made use of constructive simulation via the modular semiautomated forces (ModSAF). ModSAF is a constructive combat model designed to train doctrine and rules of engagement. Table 1.5 delineates the types of simulation and the nature of the participants and systems.

In sum, the three categories are distingushed by the nature of the participants and the systems. One of the challenges for modelers, in both engineering and sciences, is to combine the live, virtual, and constructive simulations into a single training environment. That environment is well constructed when the participants are unable to indentify if they are contending with real, virtual, or constructive threats.

**Applications**

The purposes for developing models and simulations vary. These purposes, also called *applications*, include training, analysis, experimentation, engineering, and acquisition. As discussed above, training is key to model development and simulation categorization. Relative to applications, the intent of *training* is to produce learning in the user (or participant). The training environment or the training activity must be realistic to the point that it produces effective, useful skills and/or knowledge. Training in a simulated environment or a simulation is safe, reliable, and less costly. Because this environment is reproducible, it can outlast a live-training environment. One drawback to training is that it can also produce negative learning or habits. Care must be taken so that the experiences in which the participant is learning are present in the real environment, and any scenarios or situations in which the learned responses are not present in the real environment should be removed.

*Analysis* as an application is the process of conducting a detailed study of a system to prepare for the design, testing, performance, evaluation, and/or prediction of behavior in different environments. The system can be real or notional. Simulation is often used for analysis; however, these simulations require a higher degree of fidelity than would simulations developed for training. There must also be a carefully crafted experimental design in that trials planned in advance will cover many cases and a sufficient number of trials are conducted to achieve statistical significance.

A third application or purpose for using M&S is *experimentation*. The intent of experimentation is used to explore design or solution spaces; it also serves to gain insight into an incompletely understood situation [9]. Experimentaion is likened to analysis; however, it lacks some of the structure and control found in analysis. Simply, experimentation allows for the *what-if* questions; it explores possibilities and varying outcomes. Experimentation is an iterative process of collecting, developing, and exploring concepts to identify and recommend value-added solutions for change.

In conjunction with M&S, *engineering applications* are used to design systems. These designs can be tested or changed in the simulation. Validity is the desired end with an engineering application. Engineering applications begin at the undergraduate level where students are taught the development of a model and ways to execute—simulate—the model. Simulation tools used in this application include finite element M&S tools, MATLAB (for modeling continuous systems), and ARENA (for modeling discrete-event systems).

The *acquisition application* entails the process of specifying, designing, developing, and implementing new systems. The process includes the entire life cycle of a system from concept to disposal. The intent of this application is to use the simulation to evaluate cost-effectiveness and correctness before committing funds for an acquisition.

## Randomness

The concept of randomness is simple: Does an M&S process include uncertainty and variability? Randomness is comprised of two types of simulations: deterministic and stochastic. *Deterministic simulation* takes place when a given set of inputs produce a determined, unique set of outputs. Thus, these simulations include *no* uncertainty and *no* variability. Physics-based simulations and engineering simulations can be deterministic simulations. For both deterministic and stochastic simulations, output is determined by input. Conversely, *stochastic simulation* accepts random variables as inputs, which logically lead to random outputs. This type of simulation is more difficult to represent and analyze because appropriate statistical techniques must be used. Thus, these simulations *do* include uncertainty and variability. Stochastic simulations are common in models for discrete-event systems.

## Domains

The *domain* of an M&S process refers to the subject area of the process. There are numerous domains, and as M&S becomes fluent in the user community, more domains will be engaged. Of course, the military has been using M&S for many years and so military simulation as a domain has a long association with research, development, and the education components of M&S. Within the past decade, a number of other domains have made inroads in the M&S community: transportation, decision support, training and education (aka game-based learning), medical simulation, homeland security simulation, M&S for the social sciences, and virtual environments.

M&S as a discipline is expanding the body of knowledge in an effort to explain the theory and ontology of M&S. This will no doubt spawn professionals who will develop models that can faciliate investigation in the various domains. As research and development continue, the user community (non-M&S academics or professionals) will be able to make use of M&S as a tool for representing their findings, predicting outcomes, and proffering solutions.

## CONCLUSION

This chapter introduced three fundamental precepts of M&S: the basic notion that *models* are approximations for the real world; a well-developed model can then be followed by *simulation*, which allows for the repeated observation of the model; and that *analysis* facilitates drawing conclusions, V&V, and recommendations based on various iterations/simulations of the model. These three principles coupled with *visualization*, the ability to represent data as a way to interface with the model, make M&S a problem-based discipline that allows for repeated testing of a hypothesis.

Those who chose to engage M&S are aware of its useful attributes:

(1) It allows for precise abstraction of reality.
(2) It hosts a methodology to master complexity.
(3) It requires techniques and tools.
(4) It is validated by solid mathematical foundations.

These attributes lend themselves to better research and analyses. Note the advantages to using M&S as determined by the Institute of Industrial Engineers (IIE) [10]. In 1998, IIE published the following list:

(1) The ability to *choose correctly* by testing every aspect of a proposed change without committing additional resources
(2) *Compress and expand time* to allow the user to speed up or slow down behavior or phenomena to facilitate in-depth research

(3) *Understand why* by reconstructing the scenario and examining the scenario closely by controlling the system
(4) *Explore possibilities* in the context of policies, operating procedures, and methods without disrupting the actual or real system
(5) *Diagnose problems* by understanding the interaction among variables that make up complex systems
(6) *Identify constraints* by reviewing delays on process, information, and materials to ascertain whether or not the constraint is the effect or cause
(7) *Develop understanding* by observing how a system operates rather than predictions about how it will operate
(8) *Visualize the plan* with the use of animation to observe the system or organization actually operating
(9) *Build consensus* for an objective opinion because M&S can avoid inferences
(10) *Prepare for change* by answering the "what if" in the design or modification of the system
(11) *Invest wisely* because a simulated study costs much less than the cost of changing or modifying a system
(12) *Better training* can be done less expensively and with less disruption than on-the-job training
(13) *Specify requirements* for a system design that can be modified to reach the desired goal.*

The chapter also introduced related disciplines and concepts such as probability and statistics, analysis and operations research, computer visualization, human factors, and project management—all integral to the M&S process. The discussion on modeling paradigms and types introduced three simulations: Monte Carlo, continuous, and discrete event. A review of attributes, fidelity, resolution, and scale, as well as their distribution emphasized the importance of understanding the intent of the model so that the modeler can give attention to the appropriate attribute. The importance of V&V in model creation must be stressed for any model (and modeler) to retain credibility.

Understanding the various model types, physics-based, finite element, data-based, agent-based, and so on, is important for all students of M&S. Whether the simulation is live, virtual, or constructive; whether it is used for training, analysis, experimentation, acquisition, or engineering; and whether

---

*The IIE also made a noticeably shorter list of the disadvantages: *special training* needed for building models; *difficulty in interpreting results* when the observation may be the result of system interrelationships or randomness; *cost in money and time* due to the fact that simulation modeling and analysis can be time consuming and expensive; and *inappropriate use* of M&S when an analytic solution is best.

is it deterministic or stochastic, the M&S professional must understand all of these concepts and capabilities and how they come into play in model development. Lastly, the chapter listed some of the domains in which M&S is leading in research and development: military, homeland security, medical, transportation, education and training, decision support, M&S for the social sciences, and virtual environments.

## REFERENCES

[1] Law AM, Kelton WD. *Simulation, Modeling, and Analysis*. 4th ed. New York: McGraw-Hill; 2006.

[2] Fishwick PA. *Simulation Model Design and Execution: Building Digital Worlds*. Upper Saddle River, NJ: Prentice Hall; 1995.

[3] Banks J (Ed.). *Handbook of Simulation: Principles, Methodology, Advances, Applications, and Practice*. New York: John Wiley & Sons; 1998.

[4] Starr SH, Orlov RD. Simulation Technology 2007 (SIMTECH 2007). *Phalanx*, September 1999, pp. 26–35.

[5] Balci O. Verification, validation, and testing. In *Handbook of Simulation: Principles, Advances, Applications, and Practice*. Banks J (Ed.). New York: John Wiley & Sons; 1998, pp. 335–393.

[6] Department of Defense. *Instruction 5000.61*, M&S VV&A, 1996.

[7] Sokolowski JA, Banks CM. *Modeling and Simulation for Analyzing Global Events*. Hoboken, NJ: John Wiley Publishers; 2009.

[8] Sokolowski JA, Banks CM. Agent-based modeling and social networks. In *Modeling and Simulation for Analyzing Global Events*. Sokolowski JA, Banks CM (Eds.). Hoboken, NJ: John Wiley & Sons; 2009, pp. 63–79.

[9] Ceranowicz A, Torpey M, Helfinstine B, Bakeman D, McCarthy J, Messerschmidt L, McGarry S, Moore S. J9901: Federation development for joint experimentation. Proceedings of the Fall 1999 Simulation Interoperability Workshop, Paper 99F-SIW-120, 1999.

[10] Colwell RR. Complexity and connectivity: A new cartography for science and engineering. Remarks from the American Geophysical Union's fall meeting. San Francisco, CA, 1999.

# 2

# STATISTICAL CONCEPTS FOR DISCRETE EVENT SIMULATION

Roland R. Mielke

One of the most important characteristics of discrete event simulation is the apparent ease with which uncertainty and variability can be included. Virtually all systems contain some uncertainty or variability, that is, randomness. In many systems, it is absolutely essential to include this randomness as part of the system model if reasonable fidelity is to be achieved. The price that is paid for including randomness in system models and simulations is that analysis becomes somewhat more difficult. Fundamental concepts from probability and statistics must be used, both to characterize the randomness that is included as part of the model and to understand the data result of simulating the model. The focus of this chapter is to introduce several statistical concepts necessary for developing and using discrete event simulation.

This chapter has two main objectives. The first objective is to investigate some of the statistical methods associated with including randomness as part of a system model. Sources of randomness must be represented analytically, and then this representation must be used to generate the streams of random variates required to drive a simulation. These methods are collectively referred to as input data modeling. The second objective is to investigate some of the methods associated with processing the simulation output to assess system performance. For a model that includes randomness, each simulation run produces just one sample value of a performance measure from a population of

---

*Modeling and Simulation Fundamentals: Theoretical Underpinnings and Practical Domains*,
Edited by John A. Sokolowski and Catherine M. Banks
Copyright © 2010 John Wiley & Sons, Inc.

all possible performance measure values. A number of sample values must be generated and then used to estimate the population performance parameter of interest. These methods are collectively referred to as output data analysis. It is clear that the modeling and simulation (M&S) professional must have an understanding of input data modeling and output data analysis to conduct discrete event simulation experiments.

The intent of this chapter is to present an overview of the methods and issues associated with input data modeling and output data analysis. Additional reading and study will be necessary to become knowledgeable and proficient in the exercise of these methods. However, there is sufficient depth in the presentation to allow the reader to undertake input data modeling and output data analysis for simple situations that occur quite frequently in practice.

The chapter is organized in four main sections. First, important concepts from the probability theory are reviewed. Properties of sets and functions are stated and used to define a probability space. Then, the notion of a random variable defined on the probability space is presented. Random variables and probability spaces are the underlying foundation for representing randomness. Second, several basic simulation techniques are introduced. Common theoretical statistical distributions, often used in M&S to characterize randomness, are identified. Then, algorithmic procedures for generating random samples from these distributions are explained. Third, the input data modeling problem is addressed. Methods for representing sample data using empirical distributions and theoretical distributions are explained. Some of the more difficult input data modeling issues are identified. And fourth, the output data analysis problem is investigated. The confidence interval estimate of the mean is introduced as an important technique for interpreting simulation output data. Some of the more complex output data analysis problems are described. The chapter ends with a brief conclusion and references.

## PROBABILITY

The purpose of this section is to provide a brief overview of selected concepts from the probability theory. As will be observed as this chapter develops, probability is a fundamental building block of statistics. Statistics in turn is the source of several very powerful techniques used to represent randomness, to generate the random variate streams that drive a stochastic simulation, and to interpret the output data obtained from conducting a stochastic simulation experiment. Probability then is the foundation upon which this chapter is developed.

### Sets and Functions

Throughout this chapter, the concepts of set and function will be used to define other important quantities. Thus, this section begins with a brief review of these constructs. A more detailed presentation of these topics is found in Hein [1].

A *set* is a collection of objects. The objects are called *elements* of the set and may be anything. If $S$ is a set and $x$ is an element of $S$, then we write $x \in S$. The set consisting of elements $x$, $y$, and $z$ is denoted as $S = \{x, y, z\}$. There is no particular significance given to the order or arrangement of the elements of a set. Thus, the set $S$ can also be expressed as $S = \{z, x, y\}$. There are no repeated occurrences of an element in the definition of a set; an object is either an element of a set or it is not. A set having no elements is called the *null set* and is denoted by $\Phi$.

Several important set characteristics are defined in the following statements:

(1) A set is said to be *countable* if the elements of the set can be put in a one-to-one correspondence with the positive integers.
(2) A set is called *finite* if it is empty or if it has elements that can be counted with the counting process terminating. A set that is not finite is called *infinite*.
(3) If every element of set $A$ is also an element of set $B$, then $A$ is said to be contained in $B$ and $A$ is called a *subset* of $B$ denoted $A \subseteq B$.
(4) Two sets are said to be *disjoint* if they have no common elements.
(5) The largest or all-encompassing set of elements under discussion in a given situation is called the *universal set* denoted $\Omega$.
(6) Two sets $A$ and $B$ are said to be *equal*, denoted $A = B$, if set $A$ and set $B$ contain exactly the same elements.

Finally, we will define the set operations of union, intersection, difference, and complement. Each of these operations combines sets to form new sets:

(1) The *union* of two sets $A$ and $B$, denoted $A \cup B$, is the set consisting of all elements that are contained either in set $A$ or in set $B$.
(2) The *intersection* of two sets $A$ and $B$, denoted $A \cap B$, is the set consisting of all elements that are contained in both set $A$ and set $B$.
(3) The *difference* of two sets $A$ and $B$, denoted $A - B$, is the set consisting of the elements contained in set $A$ that are not contained in set $B$. It should be observed that, in general, $A - B \neq B - A$.
(4) Let $\Omega$ be the universal set and let $A$ be a subset of the universal set. The *complement* of set $A$, denoted $\bar{A}$, is the set $\Omega - A$.

Consider the following example to illustrate some of these definitions concerning sets. A universal set is given as $\Omega = \{2, 4, 6, 8, 10, 12\}$ and two subsets are defined as $A = \{2, 4, 10\}$ and $B = \{4, 6, 8, 10\}$. Then, $A \cup B = \{2, 4, 6, 8, 10\}$, $A \cap B = \{4, 10\}$, $A - B = \{2\}$, $B - A = \{6, 8\}$, and $\bar{A} = \{6, 8, 12\}$.

Having first defined sets, we are now able to define functions. A *function* is a mathematical construct consisting of three components: (1) a set of elements called the *domain* set $X$, (2) a set of elements called the *codomain* set $Y$, and (3) a *rule of correspondence* $f$ that associates each element of the domain set with exactly one element of the codomain set. For $x \in X$ and $y \in Y$, a function is denoted by $f:X \to Y$ or $f(x) = y$. Two functions are equal if they have the same domain set, the same codomain set, and the same rule of correspondence. A change in any component of a function results in the definition of a different function.

Several very important properties of functions are defined in the following statements:

(1) The *range* of a function $f$ is the set of elements in the codomain $Y$ that are associated with some element of the domain $X$ under the rule of correspondence $f$.
(2) If the codomain of a function and the range of the function are equal, then the function is said to *onto*.
(3) If each element of the domain $X$ maps to a unique element of the codomain $Y$, then the function is said to be *one-to-one*.
(4) A function $f$ has an *inverse*, denoted $f^{-1}$, if and only if $f$ is one-to-one and onto.
(5) For function $f:X \to Y$ one-to-one and onto, the inverse is a function $f^{-1}:Y \to X$ such that $f^{-1}[f(x)] = x$ for all $x \in X$.

An example is presented to illustrate several of the definitions concerning functions. Let $R$ denote the set of real numbers and define a function having domain $X = R$, codomain $Y = R$, and rule of correspondence $y = f(x) = mx + b$, where $m$ and $b$ are real-valued constants. This function is shown graphically in Figure 2.1. It is clear that the function represents the graph of a straight line in the $x$–$y$ plane; the line has slope $m$ and $y$-axis intercept $b$. For $m \neq 0$, the function is one-to-one and onto. In this case, the inverse function exists and is given as $x = f^{-1}(y) = (1/m)y - (b/m)$. For $m = 0$, the function becomes $y = f(x) = b$. Now, all points in the domain map to a single point in the codomain. Thus, the range of the function is the set $\{b\}$ so the function is neither one-to-one nor onto, and no inverse function exists.

## Probability Space

Next, we define a mathematical construct called probability space. Problems in probability are often formulated or stated in the form of a random experiment. A *random experiment* is a well-defined experiment in which the experimental outcome cannot be predicted before conducting the experiment. The use of the word random in random experiment does not mean that the condi-

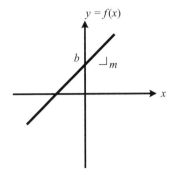

**Figure 2.1** Graph of function y = mx + b.

tions of the experiment are loosely or incompletely defined, but rather that the outcome is not known beforehand. All probability problems expressed as a random experiment have an underlying probability space. It is usually beneficial to identify this probability space explicitly even though it is rarely requested as part of a problem solution. The probability space construct is essential to developing a conceptual understanding of problems involving probability [2].

The probability space concept is developed through the following set of definitions. These definitions are followed by two examples, a discrete example and a continuous example, that illustrate the construction of a probability space:

(1) The *probability space* for a random experiment is a mathematical construct consisting of three components: the sample space $S$, the sigma algebra of events $A$, and the probability measure $P$.
(2) The *sample space* $S$ is a set consisting of the possible outcomes for a random experiment.
(3) An *event* $E$ is a subset of the sample space from a random experiment.
(4) A *sigma algebra of events* $A$ is a set of events from a random experiment that satisfies the following properties: (1) if $E \in A$, then $\bar{E} \in A$; (2) if $E_1 \in A$ and $E_2 \in A$, then $E_1 \cup E_2 \in A$; (3) if $E_1 \in A$ and $E_2 \in A$, then $E_1 \cap E_2 \in A$; and (4) $S \in A$.
(4) A *probability measure* $P$ is a function having domain $A$, codomain $[0, 1] \subset R$, and rule of correspondence $P$ such that the following properties hold: (1) $P(E) \geq 0$ for all $E \in A$; (2) $P(S) = 1$; and (3) $P(E_1 \cup E_2) = P(E_1) + P(E_2)$ when $E_1 \cap E_2 = \Phi$.

As a first example, consider the discrete random experiment consisting of rolling a three-sided die (like an ordinary six-sided die but having only three sides). The example is said to be discrete because the sample space and the event space are defined using discrete sets. The sides of the die are labeled $B$,

**Table 2.1   Probability measure for discrete example**

| Event | Probability |
|---|---|
| $\Phi$ | 0 |
| $\{B\}$ | 1/6 |
| $\{C\}$ | 1/3 |
| $\{D\}$ | 1/2 |
| $\{B, C\}$ | 1/2 |
| $\{B, D\}$ | 2/3 |
| $\{C, D\}$ | 5/6 |
| $S$ | 1 |

C, and D. When the die is rolled many times, it is found that side B shows one-sixth of the time, side C shows one-third of the time, and side D shows one-half of the time. It is desired to construct a probability space for this random experiment. From the statement defining the random experiment, it is clear that the experiment has three possible outcomes: side B showing; side C showing; or side D showing. Thus, the sample space S is defined as $S = \{B, C, D\}$. Next, we construct the sigma algebra of events A. Set A contains the events corresponding to the individual random experiment outcomes, events $\{B\}, \{C\},$ and $\{D\}$. Using the rules for a sigma algebra, the complement of each of these singleton sets, $\{C, D\}, \{B, D\},$ and $\{B, C\}$, must be elements in A. Then, the union, intersection and complement of each pair of these sets must be added to A. It quickly becomes apparent that the sigma algebra of events includes all possible subsets of S, including S itself and the null set $\Phi$. For this example, $A = \{\Phi, \{B\}, \{C\}, \{D\}, \{B, C\}, \{B, D\}, \{C, D\}, S\}$. In general, if a discrete sample space S contains k outcomes, the associated sigma algebra of events A will contain $2^k$ events. Finally, the probability measure component of the probability space is constructed. Most often, the interpretation of probability as indicating the relative frequency of occurrence of the experiment outcomes suggests the probability values for the singleton event sets. Here, it is desirable to assign $P(\{B\}) = \frac{1}{6}$, $P(\{C\}) = \frac{1}{3}$, and $P(\{D\}) = \frac{1}{2}$. The rules of probability measure then indicate how to assign probability values to the other events in set A. In this example, the probability measure is defined by enumerating the probability value assigned to each event. The probability measure is given in Table 2.1.

As a second example, consider the continuous random experiment of spinning the pointer on a wheel of chance. This example is said to be continuous because the sample space and the event space are defined using continuous variables. In this experiment, the perimeter of the wheel of chance is marked in degrees and an experimental outcome is indicated by the arc of rotation of the pointer from the zero degrees mark to the stopping point $\theta$; that is, the set $(0, \theta]$ where $0 < \theta \le 360$. The sample space S for this experiment is $S = (0, 360]$. An event corresponding to a possible outcome will have the form $(0, \theta]$.

When all possible unions, intersections, and complements of these events are considered, the result is an event having the form of a union of disjoint sets of the form $E = (\theta_1, \theta_2]$, where $0 \leq \theta_1 \leq \theta_2 \leq 360$. Thus, the sigma algebra of events $A$ consists of the null set $\Phi$, the sample space $S$, and an infinite number of unions of disjoint sets of the form of set $E$. The probability measure for this experiment, assuming a fair wheel of chance, is given as $P(E) = (\theta_2 - \theta_1)/360$ for each event $E \in A$.

The purpose of defining the sigma algebra of events $A$ and associating probabilities with events in $A$ now becomes clear. The events corresponding to experimental outcomes form only a subset of all events that one might wish to associate with a probability measure. By enumerating the sigma algebra of events, all possible events are identified. By associating probability values with events, rather than outcomes, we are ensuring that a probability value is assigned to every possible event.

Sometimes it is possible that two events may not be mutually exclusive because the events contain common elements from the sample space. For two such events $B$ and $C$, the common elements form the event $B \cap C$ and the probability for this event, $P(B \cap C)$, is called the *joint probability*. In this case, knowledge that event $C$ occurred can alter the probability that event $B$ also occurred. The *conditional probability* of event $B$ given event $C$ is defined as $P(B|C) = P(B \cap C)/P(C)$ for $P(C) \neq 0$. Events $B$ and $C$ are said to be *independent* if $P(B \cap C) = P(B) P(C)$. If events $B$ and $C$ are independent, then $P(B|C) = P(B)$ and $P(C|B) = P(C)$. In this case, knowledge of the occurrence of one of the two events conveys no knowledge concerning the occurrence of the other event. Independence will be an important prerequisite to several statistical analysis procedures to be considered later in this chapter.

## Random Variables

Outcomes are elements of the sample space that represent the result of conducting a random experiment. Outcomes can be complex objects that may be difficult or inconvenient to use directly. Often, it is desirable to associate a number with an outcome and then to work with that number rather than the outcome. In this section, we define a random variable as a means of associating real numbers with experimental outcomes. The concepts of cumulative distribution functions and probability density functions are also introduced.

Let $\mathcal{P} = (S, A, P)$ be a probability space with sample space $S$, sigma algebra of events $A$, and probability measure $P$. A *random variable* $X$ defined on this probability space is a function with domain set $S$, codomain set $R$, and rule of correspondence $X:S \to R$ such that the set $\{X \leq x\} = \{s | s \in S \text{ and } X(s) \leq x\}$ is an event in $A$ for all $x$ in $R$. Thus, the random variable $X$ is simply a function that assigns to each experimental outcome $s \in S$ a real number $X(s)$. However, not just any function will qualify as a random variable. To qualify, the set of outcomes $s$ that satisfy $X(s) \leq x$ must be an event in the sigma algebra of events for all values of $x \in R$.

Associated with a random variable $X$ is a function called the *cumulative probability distribution function*, denoted $F(x)$. The name of this function is often abbreviated and called the cumulative distribution function or even just the distribution function. The distribution function is a function having domain $R$, codomain $[0, 1] \subset R$, and rule of correspondence $F:R \to [0, 1]$ such that $F(x) = P(\{X(s) \leq x\})$. It has already been noted that the set $\{X(s) \leq x\}$ is an event in $A$ for all $x \in R$. Since each event in $A$ is assigned a probability via the definition of probability measure, this definition of distribution function is consistent and has meaning within the context of the underlying probability space.

Distribution functions have a number of properties that occur as the result of the distribution function being a probability. These properties are presented in the following statements:

(1) $F(-\infty) = 0$ and $F(+\infty) = 1$.
(2) $0 \leq F(x) \leq 1$ for all $x \in R$.
(3) $F(x)$ is a nondecreasing function of $x$; that is, $F(x_1) \leq F(x_2)$ for $x_1 < x_2$.
(4) $F(x)$ is continuous from the right; that is, $F(x^+) = F(x)$.
(5) $P(\{x_1 < X \leq x_2\}) = F(x_2) - F(x_1)$.

Another function associated with a random variable is the *probability density function* $p(x)$. This function is often called the density function. The density function is a function having domain $R$, codomain $R$, and rule of correspondence $p:R \to R$ such that $p(x) = dF(x)/dx$. When $F(x)$ is discontinuous, as occurs with discrete random variables, it is necessary to call upon the impulse function from the theory of generalized functions to facilitate the calculation of the derivative operation [2].

Density functions also have several important properties that are presented in the following statements:

(1) $p(x) \geq 0$ for all $x \in R$.

(2) The total area under the $p(x)$ curve is one; that is, $\int_{-\infty}^{\infty} p(x)\,dx = 1$.

(3) The distribution function is related to the density function via integration; that is, $F(x) = \int_{-\infty}^{x} p(\sigma)\,d\sigma$.

(4) $P(\{x_1 < X \leq x_2\}) = \int_{x_1}^{x_2} p(x)\,dx$.

To illustrate the concepts of this section, random variables are defined on the probability spaces defined in the previous section. Then, distribution and density functions are constructed for these random variables. Consider first the probability space for the three-sided die. The following function, shown in tabular form in Table 2.2, is used to define a random variable $X$.

PROBABILITY 33

Table 2.2  Random variable definition for example

| Outcome $s_k$ | Random Variable Value $X(s_k) = x_k$ |
|---|---|
| B | 1 |
| C | 2 |
| D | 3 |

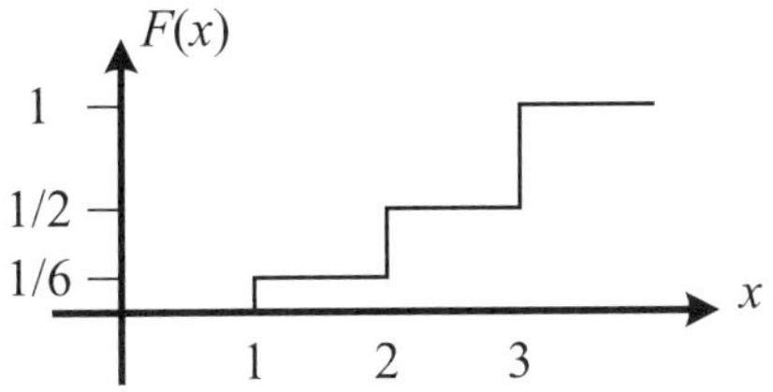

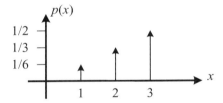

Figure 2.2  Cumulative distribution function for the three-sided die example.

Figure 2.3  Density function for the three-sided die example.

Then, the events corresponding to the different ranges of random variable values are computed:

(1) $\{X(s) < 1\} = \Phi$
(2) $\{X(s) < 2\} = \Phi \cup \{B\} = \{B\}$
(3) $\{X(s) < 3\} = \Phi \cup \{B\} \cup \{C\} \cup \{B,C\} = \{B, C\}$
(4) $\{X(s) < \infty\} = \Phi \cup \{B\} \cup \{C\} \cup \{D\} = S$

The resulting plot of the cumulative distribution function $F(x)$ for this example is shown in Figure 2.2. The density function $p(x)$, obtained by differentiating $F(x)$, is shown in Figure 2.3. Expressions for the distribution function and the density function can be written as follows:

$$F(x) = \sum_{k=1}^{3} P(x_k) u(x - x_k),$$

$$p(x) = \sum_{k=1}^{3} P(x_k) \delta(x - x_k).$$

The notation $u(x)$ represents the unit step function defined as

$$u(x) = \begin{cases} 1 & x > 0 \\ 0 & x < 0, \end{cases}$$

and $\delta(x)$ represents the unit impulse function.

Now consider the example of spinning the pointer on a wheel of chance. An outcome for this random experiment is the arc of rotation from zero to the point at which the pointer comes to rest, that is, the rotation interval $(0, \theta]$ with $0 < \theta \leq 360$. The probability assigned to the event corresponding to this outcome is $P((0, \theta]) = \theta/360$. We now define a random variable $X$ by superimposing a standard clock face on the wheel of chance. This is equivalent to defining the random variable as $X(\theta) = \theta/30$. Then,

$$F(x) = P(X(\theta) \leq x) = P\left(\frac{\theta}{30} \leq x\right) = P(\theta \leq 30x) = \frac{x}{12}$$

for $0 < x \leq 12$. The cumulative distribution function is shown graphically in Figure 2.4. The density function $p(x)$ is obtained from the distribution function as $p(x) = dF(x)/dx$.

The density function is shown in Figure 2.5. The distribution function and the density function are piecewise continuous functions because the underlying probability space is continuous.

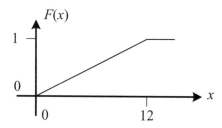

**Figure 2.4** Cumulative distribution function for the wheel of chance example.

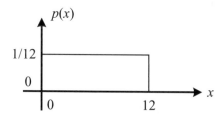

**Figure 2.5** Density function for the wheel of chance example.

## SIMULATION BASICS

Now that we have completed a brief review of probability, it is time to investigate how concepts from probability and statistics are utilized in M&S. One of the most important characteristics of discrete event simulation is the ability to include uncertainty and variability that occurs naturally in many systems. For example, in queuing system models there is often uncertainty concerning the time of arrival of entities entering the system. Additionally, there is often variability in the times required to provide service as an entity passes through the system. This randomness is modeled by drawing random variates from distribution functions that are representative of the entity interarrival times and service times. In this section, we first review some of the theoretical distributions that are used commonly in M&S. Then, we investigate methods for generating samples from these distributions for use as the random variates needed to drive our simulations.

### Common Theoretical Distribution Functions

There are numerous theoretical distributions that are often used to model uncertainty and variability in M&S. In this section, only four of these distribution functions are reviewed. The reader is encouraged to consult the references at the end of this chapter, especially Law [3], for information on additional distribution functions. Theoretical distributions are often referred to as parametric distributions because each distribution has one or more associated parameters. The parameters are normally selected to control the physical characteristics of the distribution, such as distribution location, scale, and shape. Specification of the name and parameters of a theoretical distribution is sufficient to define the distribution.

The *uniform distribution*, denoted UNIFORM $(a, b)$, is often used when a random variable having equally likely values over a finite range is needed. This distribution has two parameters: "$a$" specifies the minimum value of the random variable and "$b$" specifies the maximum value of the random variable, where $a,b \in R$ and $-\infty < a < b < \infty$. The probability density function for this distribution is

$$p(x) = \begin{cases} \left(\dfrac{1}{b-a}\right) & a \leq x \leq b \\ 0 & \text{otherwise.} \end{cases}$$

The density function for the uniform distribution is shown in Figure 2.6.

The *triangular distribution*, denoted TRIANGULAR $(a, m, b)$, is used in situations in which the exact form of the distribution is not known, but approximate values are known for the minimum, maximum, and most likely values. This distribution has three parameters: "$a$" specifies the minimum value of the distribution, "$b$" specifies the maximum value of the distribution,

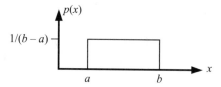

**Figure 2.6** Uniform probability density function.

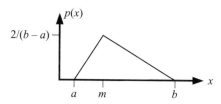

**Figure 2.7** Triangular probability density function.

and "m" specifies the mode or most likely value of the distribution, where $a$, $m, b \in R$ and $-\infty < a < m < b < \infty$. The probability density function for this distribution is

$$p(x) = \begin{cases} \dfrac{2}{(b-a)(m-a)}(x-a) & a \le x \le m \\ \dfrac{2}{(b-a)(b-m)}(b-x) & m \le x \le b \\ 0 & \text{otherwise.} \end{cases}$$

The probability density function for the triangular distribution is shown in Figure 2.7.

The *exponential distribution*, denoted EXPONENTIAL $(m)$, is often used to model interarrival times for entities arriving at a system. This distribution has a single parameter "m" that specifies the mean of the distribution, where $m \in R$ and $0 < m < \infty$. The probability density function for this distribution is

$$p(x) = \begin{cases} \dfrac{1}{m}e^{-x/m} & x > 0 \\ 0 & \text{otherwise.} \end{cases}$$

The probability density function for the exponential distribution is shown in Figure 2.8.

The *normal distribution*, denoted NORMAL $(m, \sigma)$, is often used to represent variability of quantities about some average value. By the central limit theorem, this distribution also represents the distribution of a random variable formed as the sum of a large number of other random variables, even when

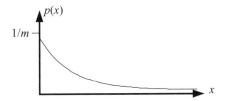

**Figure 2.8** Exponential probability density function.

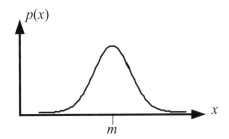

**Figure 2.9** Normal probability density function.

the other random variables are not normally distributed. The normal distribution has two parameters: "$m$" specifies the mean of the distribution, and "$\sigma$" specifies the standard deviation of the distribution, where $m, \sigma \in R, -\infty < m < \infty$, and $0 < \sigma < \infty$. The probability density function for this distribution is

$$p(x) = \left(\frac{1}{\sqrt{2\pi}\sigma}\right) e^{-(x-m)^2/2\sigma^2} \quad -\infty \leq x \leq \infty.$$

The probability density function for the normal distribution is shown in Figure 2.9.

## Generation of Random Variates

When developing a model, it is often possible to characterize a particular source of uncertainty or variability as being represented by a random variable having a theoretical distribution function. For example, the random times between entity arrivals at a queuing system are often modeled by selecting independent samples from a random variable having an exponential distribution. The random service times required to process entities as they pass through the server in a queuing system are often modeled by selecting independent samples from a random variable having a triangular distribution. The independent samples selected from a random variable having a theoretical distribution are called *random variates*. The capability to quickly and efficiently generate random variates is clearly an essential requirement for all discrete event simulation software systems.

The generation of random variates is usually implemented as a two-step process. In the first step, a random number is generated. In the second step, the random number is transformed to a random variate. This process of generating random variates is described in this section. The reader is referred to Law and Leemis for more detailed presentations [3,4].

A *random number* is defined as a sample selected randomly from the uniform distribution UNIFORM (0, 1). Thus, a random number is a real number between zero and one; at each random number draw, each real number in the range [0, 1] has an equally likely chance of being selected.

In simulation, it is necessary to generate random numbers on a computer using a recursive algorithm. Thus, the very best that can be done is to simulate or approximate the generation of random numbers. Historically, the most common algorithms used to simulate random number generation are called *linear congruential generators* (LCG). Using an LCG, a sequence of integers $Z_1, Z_2, Z_3, \ldots$ is generated by the recursive formula

$$Z_k = (aZ_{k-1} + c) \bmod(m).$$

In this formula, the constant "$a$" is called the multiplier, the constant "$c$" is called the increment, and the constant "$m$" is called the modulus; $a$, $c$, and $m$ must be nonnegative integers. Generation of an integer sequence is initiated by specifying a starting integer Z-value, $Z_0$, called the *seed*. Each successive integer $Z_k$ is obtained from the previous value by dividing the quantity $(aZ_{k-1} + c)$ by $m$ and retaining the integer remainder. It can be shown that the resulting values $Z_k$ are limited to the range $0 \leq Z_k \leq m - 1$ and are uniformly distributed. The output of the LCG, $U_k$, is obtained by dividing each integer $Z_k$ by $m$.

For given values of $a$, $c$, and $m$, the value of $Z_k$ depends only on $Z_{k-1}$; in addition, the algorithm is capable of producing at most $m$ distinct integers. Thus, each seed $Z_0$ will produce a sequence of $p$ distinct integers, $p \leq m$, and then the sequence will begin to repeat. Generators with $p = m$ are called full-cycle or full-period generators. Thus, the useful output of an LCG is a sequence of $p \leq m$ distinct rational numbers $U_1, U_2, \ldots, U_p$ uniformly distributed over (0, 1). While these numbers approximate samples from UNIFORM (0, 1), it is clear that they are not random at all. For this reason, the LCG output values are usually called *pseudorandom numbers*. To be useful in simulation, the constants in the LCG and the initial seed must be selected to produce a very long cycle length. It is important to note that the LCG algorithm is elegantly simple and very efficient when implemented in computer code. In addition, the LCG output stream is reproducible. This is often important when repeating an experiment or for debugging a simulation implementation.

As an example, the discrete event simulation tool ProModel uses an LCG with $a$ = 630,360,016; $c$ = 0; and $m = 2^{31} - 1$ [5]. It is capable of producing a cycle length of more than 2.1 billion pseudorandom numbers. The LCG output is divided into 100 unique random number streams each of length greater than 21 million numbers.

The objective of this section is to generate random variates, that is, samples from random variables having distributions other than UNIFORM (0, 1). Conceptually, the most basic method for generating a random variate from a random number is to utilize the *inverse transform theorem*. This theorem states that if $X$ is a random variable with continuous cumulative distribution function $F(x)$, then the random variable $Y = F(X)$ has distribution UNIFORM (0, 1). This theorem is proved in Goldsman [6]. Let $X$ be a random variable with distribution function $F(x)$ and suppose we wish to generate random variates from this distribution. If $U$ is the random variable with density function UNIFORM (0, 1), then by the inverse transform theorem, we know that $F(X) = U$. Therefore, if the function $F(\cdot)$ has an inverse, then $X = F^{-1}(U)$. For most theoretical distributions of interest, the function $F(\cdot)$ is strictly monotone increasing. This property is sufficient to guarantee the existence of $F^{-1}(\cdot)$. For obvious reasons, this method of generating random variates is called the inverse transform method.

As an example, suppose we desire to generate random variates from the distribution EXPONENTIAL (2). The cumulative distribution function for the random variable $X$ having density function EXPONENTIAL (2) is $F(x) = 1 - e^{-2x}$. Setting $F(X) = U = 1 - e^{-2X}$ and solving for $X$, we obtain $X = F^{-1}(U) = \left(\frac{1}{2}\right)\ln(1-U)$. This equation transforms each random number $U$ into a random variate $X$ from the distribution EXPONENTIAL (2).

The inverse transform method can also be applied to discrete distributions, but a pseudoinverse must be utilized because the cumulative distribution function inverse does not exist in this case. Conceptually, the inverse transform method seems easy to apply; however, in practice, it may be difficult or impossible to calculate the required inverse function in closed form. For this reason, there are a number of additional methods that are also used to convert random numbers to random variates [4]. Virtually all commercial discrete event simulation tools have automated the process of generating random numbers and random variates. The user simply specifies a distribution, and possibly a desired random number stream, and the simulation does the rest.

## INPUT DATA MODELING

In this section, we address one of the two main objectives of this chapter, *input data modeling*. As described in the previous chapter, the development of a system model has two distinct but closely related components. The first component is often referred to as structural modeling. In this component, the entities, locations, resources, and processes that describe the structure and operation of the system are defined. The result is a mathematical or logical representation, such as a queuing system model, that describes how the system behaves when executed. The second component is often referred to as data modeling. In this component, descriptive data required to execute the structural model are developed. These data include entity interarrival times, process service times, resource schedules and failure rates, entity travel times, and all of the other system data needed to describe the operation of the system quantitatively. The data model is used to generate the many random variate streams that must be supplied during the simulation of the system model.

Our focus in this section is on the second model component, development of the data model. In many M&S projects, input data modeling is the most difficult, time-consuming, and expensive part of the overall project activity.

We will introduce the topic of input data modeling by considering the most simple model development situation. First, we assume that real examples of the modeled system exist and are accessible. Thus, it is possible to monitor these systems to make measurements of each required data component. Abstractly, our data gathering for each data component results in the collection of a sample from an underlying data component population represented by a population probability distribution. Second, we assume that our data samples are independent and identically distributed (IID). Independent means that there is no relationship or influence between successive samples and identically distributed means that each sample comes from the same underlying distribution. Finally, we assume that the underlying distribution is simple (unimodal) and stationary (not time varying). More complex situations are described at the end of this section and will be the focus of study in several future M&S courses that you will likely take.

Once a set of samples representing a system data input has been acquired, there are three choices for generating the random variate stream required for simulation. First, the sample values can be used directly to form the random variate stream. Second, the set of samples can be used to generate an empirical distribution and the empirical distribution in turn can be used to generate the random variate stream. Third, a theoretic distribution function that fits the set of samples can be identified, and this theoretical distribution can be used to generate the random variate stream. These three approaches are described in this section.

**Direct Use of Sample Data**

When sample data are to be used directly, the sample values are stored in a data file. Each time a random variate is needed during simulation execution, the next sample value is read from this data file. This approach has the obvious advantage of generating only legal and realistic random variates. This is guaranteed because each sample value was observed in the operation of the actual system. However, this approach also has several substantial disadvantages. First, in a limited set of sample values, legal and realistic values that occur infrequently may not be present. In this case, these values will not be present in the random variate stream, and their absence in the simulation may change computed system performance. Second, making repeated calls to data memory to retrieve stored sample values is computationally very slow. Thus, this approach results in substantially greater computational times for simulation execution. Third, and most significant, each simulation execution may require a large number of random variate draws. As will be shown in the next section, it is necessary to conduct repeated simulation executions to obtain sufficient data to statistically estimate desired system performance measures. Thus, it is common for a simulation study to require extremely large numbers of sample values. In practice, it is almost never possible to gather enough sample values to provide the number of random variates needed to conduct a simulation study.

Therefore, while at first glance direct use of sample data appears to be a good approach to generating random variate streams for simulation, it is used only infrequently because of practical limitations.

## Use of Empirical Distributions

An *empirical distribution* is a nonparametric distribution constructed to exactly represent the available set of sample values. The use of an empirical distribution has the advantage of facilitating the generation a random variate stream of unlimited length. The empirical distribution can be used to transform a stream of random numbers of arbitrary length to the desired random variate stream. In addition, each random variate will be a legal and realistic value. However, as with the direct use of sample data, this approach cannot generate outlying values that may actually occur but are only infrequently observed. The empirical distribution is generally used when a good fit to the sample data using a theoretical distribution cannot be found.

Somewhat different construction approaches are required for discrete empirical distributions and continuous empirical distributions. The construction process for a discrete empirical distribution is similar to the construction for any discrete distribution, except that the process is conducted with sample values rather than with all population values. Let $Y = (y_1, y_2, \ldots, y_m)$ be the discrete data sample. In general, $Y$ is not a set because some of the sample values may be repeated. Let $X = \{x_1, x_2, \ldots, x_n\}$ be the set of distinct values in $Y$. Typically, the number of sample values $m$ should be selected so that the number of distinct sample values $n$ is much smaller than $m$. Define $p(x_k)$ as the number of appearances of $x_k$ in the data sample $Y$ divided by the total number of sample values; that is, $p(x_k)$ is the fraction of sample values having the value $x_k$. Then the empirical density function is defined as

$$p(x) = \sum_{k=1}^{n} p(x_k) \delta(x - x_k),$$

where $\delta(\cdot)$ is the unit impulse function. The empirical distribution function is defined as

$$F(x) = \sum_{k=1}^{n} p(x_k) u(x - x_k),$$

where $u(\cdot)$ is the unit step function.

The construction process for a continuous empirical distribution differs from the construction process for a discrete empirical distribution in two ways. First, a *data histogram* must be constructed by grouping the sample values by size into equal width bins. This is done in order to utilize the relative frequency definition of probability. Second, linear interpolation is normally used to smooth the cumulative distribution function. This construction process is first explained and then illustrated by example in the following.

Let $Y = \{y_1, y_2, \ldots, y_m\}$ be the real-valued data sample. Choose real numbers "$a$" and "$b$" to be upper and lower bounds, respectively, for the elements of $Y$. That is, $a$ and $b$ are selected so that $a \leq y_k < b$ for all $k$. The semi-open interval $[a, b)$ is then divided into $n$ equal width intervals $[a_q, b_q)$ where $a_q = a + (q-1)h$, $b_q = a + (q)h$, and $h = (b-a)/n$. Finally, let $p_q$ denote the number of data samples in $Y$ that are also contained in the $q$th data bin $[a_q, b_q)$ divided by the total number of sample values $m$. Then, the data histogram for data sample $Y$ is defined as a bar plot of $p_q$ versus $q$ for $q = 1, 2, \ldots, n$. The continuous empirical distribution function for data sample $Y$ is constructed from the data histogram for $Y$. The empirical distribution function $F(x)$ is assigned value zero for $x < a$ and value one for $x \geq b$. Over the interval $[a, b)$, $F(x)$ is defined by the piecewise continuous union of $n$ straight-line segments. The $q$th line segment, corresponding to the $q$th data bin $[a_q, b_q)$ of the histogram, has end points $(a_q, p_{q-1})$ and $(b_q, p_q)$. This empirical distribution function is shown in Figure 2.10.

Since this empirical distribution consists of the piecewise continuous union of straight-line segments and each line segment is monotonic increasing, an inverse for $F(x)$ can be calculated over domain set $(0, 1)$. Thus, this empirical distribution can be used to transform a random number stream into a random variate stream representative of the sample data.

An example is presented to illustrate the development of a continuous empirical distribution function. As part of the development of a continuous-time, discrete event model of some manufacturing system, it is necessary to

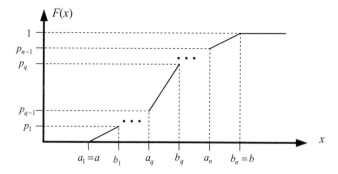

**Figure 2.10** Construction of the continuous empirical distribution function.

Table 2.3 Listing of sample values and distribution into bins

| Sample Number | Sample Value | Bin Number |
|---|---|---|
| 1 | 6.25 | 4 |
| 2 | 5.58 | 3 |
| 3 | 4.65 | 2 |
| 4 | 4.94 | 2 |
| 5 | 7.72 | 5 |
| 6 | 6.49 | 4 |
| 7 | 5.11 | 3 |
| 8 | 3.30 | 1 |
| 9 | 5.54 | 3 |
| 10 | 5.39 | 3 |
| 11 | 6.19 | 4 |
| 12 | 4.29 | 2 |
| 13 | 5.97 | 3 |
| 14 | 4.03 | 2 |
| 15 | 5.26 | 3 |
| 16 | 5.40 | 3 |
| 17 | 4.88 | 2 |
| 18 | 4.15 | 2 |
| 19 | 6.18 | 4 |
| 20 | 4.43 | 2 |
| 21 | 4.84 | 2 |
| 22 | 4.76 | 2 |
| 23 | 4.54 | 2 |
| 24 | 5.34 | 3 |
| 25 | 5.43 | 3 |
| 26 | 5.82 | 3 |
| 27 | 4.86 | 2 |
| 28 | 3.79 | 1 |
| 29 | 4.78 | 2 |
| 30 | 7.06 | 5 |

model the service time for a value-added process. The service times are thought to be stationary, independent, and identically distributed. A set of 30 sample points for this service time is gathered and used to develop a continuous empirical distribution. The smallest sample value is 3.30 and the largest sample value is 7.72. For convenience, $a = 3.00$ and $b = 8.00$ are selected as the lower bound and upper bound, respectively. The interval [3.00, 8.00) is divided into five subintervals or bins that collectively cover the range of sample times: [3.00, 4.00), [4.00, 5.00), [5.00, 6.00), [6.00, 7.00), and [7.00, 8.00). The sample values are listed in Table 2.3. This table also identifies the bin to which each sample value is assigned. The relative frequency of occurrence for each bin is computed as $p_1 = 0.0667$, $p_2 = 0.4000$, $p_3 = 0.3333$, $p_4 = 0.1333$, and

**44** STATISTICAL CONCEPTS FOR DISCRETE EVENT SIMULATION

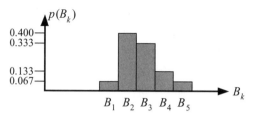

**Figure 2.11** Histogram of sample service times for the example.

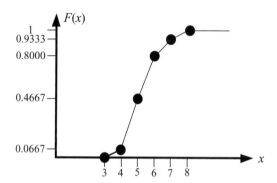

**Figure 2.12** Continuous empirical distribution function for the example.

$p_5 = 0.0667$. These values are used to construct the histogram for the sample service times; this histogram is shown in Figure 2.11.

The final continuous empirical distribution function is shown in Figure 2.12. The inverse for this distribution function, when driven by a stream of random numbers, generates the random variate stream of service times required to simulate the manufacturing system.

## Use of Theoretical Distributions

A *theoretical distribution* is a distribution function that is defined mathematically in terms of distribution parameters. Parameters are often related to the position, scale, or shape of the theoretical distribution. The exponential distribution, the normal distribution, and the triangular distribution are examples of theoretical distributions having one, two, and three parameters, respectively. Generally, if a histogram of the sample data has a single mode and there are no large gaps lacking sample points, then it is likely that one or more theoretical distributions can be fit to the sample data.

The use of theoretical distributions to represent simulation input data has several advantages compared with the use of empirical distributions or direct use of sample data. First, the theoretical distribution tends to smooth irregularities resulting from using a limited number of sample values. The theoretical distribution will generate values outside the range of observed values, often capturing values that might actually occur but that were missed in the sampling process because they occur only infrequently. Second, as we have already observed, the use of a theoretical distribution to generate random variate streams is computational efficient. This approach results in the smallest computational overhead during runtime of the three input data modeling approaches. The major disadvantage of using a theoretical distribution is the computational effort required to fit a distribution to sample data and, for some sample data sets, the difficulty in finding a good fit.

The process of fitting a theoretical distribution to sample data normally consists of three steps. In step one, a histogram for the sample data is constructed. The overall shape of the histogram is compared with probability density functions from theoretical distributions. The purpose of this step is to identify several candidate theoretical distributions for further processing. Sometimes comparing summary statistics for the sample data to the same statistics for the candidate theoretical distributions is helpful. It is important to include all likely theoretical distribution candidates at this step; those that do not fit well will be identified and eliminated in step three.

The second step is to determine the theoretical distribution parameters to obtain the best fit of the theoretical distribution to the sample data. This should be done for each candidate theoretical distribution identified in step one. There are two approaches that are often used in this step: the *maximum likelihood estimation method* and the *method of moments*. The maximum likelihood method identifies the theoretical distribution parameters that make the resulting distribution the most likely to have produced the sample data. The method of moments equates the first $q$ population moments with the first $q$ sample moments, where $q$ is equal to the number of theoretical distribution parameters. Both approaches use concepts from statistics that are beyond those presented in this chapter [7]. These concepts are normally investigated in first-level graduate courses in statistics. Fortunately, many of the commercially available discrete event simulation software environments have built in tools to automate this step. In Arena, this step is done in the Input Analyzer, while in ProModel this step is done in Stat::Fit [5,8]. For each candidate theoretical distribution, the result of this step is a set of distribution parameters that make the distribution fit the sample data as closely as possible.

The third and final step is to determine the best theoretical distribution, from among the candidate distributions, to represent the sample data. This step often relies on the use of statistical *goodness-of-fit tests* to help identify the best theoretical distribution. Two goodness-of-fit tests are commonly used for this purpose: the *chi-square test* and the *Kolmogorov–Smirnov (K-S) test* [3]. The chi-square test statistic is a measure of the squared distance between the sample data histogram and the fitted theoretical probability density function. The K-S test statistic is a measure of the largest vertical distance between

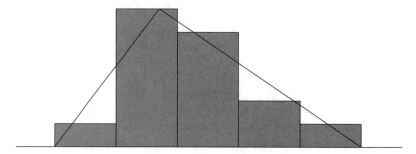

**Figure 2.13** Triangular distribution fitted to the data sample.

an empirical distribution and the fitted theoretical cumulative distribution function. Both of these tests are statistical hypothesis tests. Rather than report the test statistics, it is common to report the test results as a *p*-value. The *p*-value is a measure of the probability that another data sample will compare the same as the present data sample given that the theoretical distribution fit is appropriate. A small *p*-value indicates that another data sample probably would not compare the same and the fit should be rejected. A high *p*-value indicates that another data sample probably would compare the same and the fit should not be rejected. When comparing candidate theoretical distributions for goodness-of-fit, the distribution with the highest *p*-values is likely to provide the best fit. These goodness-of-fit tests are automated in many discrete event simulation tools. Generally, a *p*-value greater than 0.05 is considered acceptable, while a *p*-value less than 0.05 indicates a poor fit.

To illustrate this process, we use the Arena Input Analyzer to fit a theoretical distribution to the data sample presented in Table 2.3 [8]. The histogram for this data sample is shown in Figure 2.11. Viewing the histogram, it appears that both the triangular distribution and the normal distribution might be viable candidates for a fit. The result of fitting a triangular distribution to the data sample is shown in Figure 2.13. The report that accompanies this figure indicates that the best fit for a triangular distribution is TRIANGULAR (3.00, 4.70, 8.00). For this distribution, the chi-square test yields a *p*-value of 0.427, while the K-S test yields a *p*-value of greater than 0.15. The goodness-of-fit tests indicate that the triangular distribution remains a strong candidate for representing the data sample.

The result of fitting a normal distribution to the data sample is shown in Figure 2.14. The report indicates that the best fit for a normal distribution is NORMAL (5.23, 0.947). For this distribution, the chi-square test yields a *p*-value of less than 0.005, while the K-S test yields a *p*-value of greater than 0.15. The low *p*-value from the chi-square test indicates that the normal distribution is not a good candidate to fit the data sample.

In practice, a number of other continuous theoretical distributions could be considered candidates for fitting the data sample. These theoretical distribu-

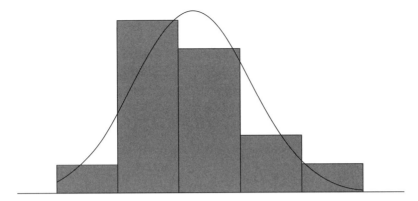

**Figure 2.14** Normal distribution fitted to the data sample.

tions include the beta distribution, the Erlang distribution, the gamma distribution, the lognormal distribution, and the Weibull distribution. Each of these distributions was also fitted to the data sample. Each produced a K-S test $p$-value of greater than 0.15 but a chi-square test $p$-value of less than 0.005. Therefore, based on the results of our tests, we would likely choose the triangular distribution to model our data sample.

## Other Input Data Modeling Situations

In this section, we have investigated only the simple cases of modeling input data. However, for many simulation projects, the task of developing good input data models can be considerably more complex. The diagram shown in Figure 2.15 categorizes some of these more complex situations.

For some simulation projects, it is either very difficult or even impossible to collect sample data. This occurs when simulating a new system design before an actual prototype system is developed. In other situations, sample data can be obtained, but the data have more complex statistical properties. Sample data sometimes appear to exhibit multimodal behavior; that is, the data histogram displays several peaks separated by intervals containing few sample values. Sample data are sometimes time varying. For example, the interarrival times for customers at a fast-food restaurant likely vary with time of day and day of the week. Sometimes different random parameters in a system are correlated. Entities that require larger service times in one manufacturing area may also require larger service times in other manufacturing areas. These complex input data modeling situations occur frequently in practice and an M&S professional must learn approaches and techniques for dealing with these cases. Fortunately, we will not attempt to address these issues in this chapter. Rather, these are some of the many topics that will be treated in later courses in probability and statistics and courses in discrete event simulation.

**48** STATISTICAL CONCEPTS FOR DISCRETE EVENT SIMULATION

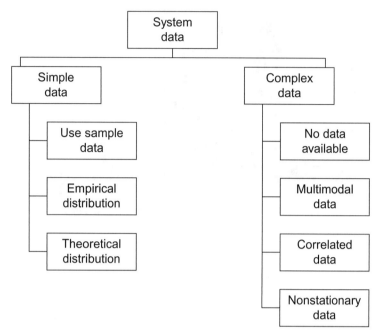

**Figure 2.15** Categorization of input data modeling situations.

## OUTPUT DATA ANALYSIS

In this section, we address the second of the two main objectives of this chapter: output data analysis. *Output data analysis* is the process of analyzing simulation output data to produce useful information concerning system performance. First, a simple queuing system example is presented and used to illustrate why output data analysis is necessary. Then, the confidence interval estimate of the mean is described. This statistical technique is used to estimate the population mean for a performance measure from a set of sample values from that population. The confidence interval estimate yields information about both the accuracy and the correctness of the estimate. Finally, some of the more complex output data analysis situations are identified, and approaches for addressing these situations are briefly discussed.

### A Motivational Example

We begin by describing a simulation experiment performed on an M/M/1 queuing system Z. Let Z consist of a first-in, first-out (FIFO) queue having infinite capacity followed by a server having a single resource. The interarrival times for entities arriving at Z are assumed to be IID; that is, each interarrival time is independent of all other interarrival times, and all interarrival times are obtained by selecting sample values from the same distribution. The inter-

**Table 2.4  Results of experimental runs for Z with n = 5**

|       | Run  |      |       |      |      |
|-------|------|------|-------|------|------|
|       | One  | Two  | Three | Four | Five |
| Delay | 1.83 | 2.95 | 0.49  | 1.16 | 1.56 |

*[handwritten annotation: Avg of avgs ≈ 1.60]*

arrival times are exponentially distributed with an average interarrival time of 1 time unit. The entity service times at the server are also IID and are exponentially distributed with an average service time of 0.8 time units. The system starts empty and idle, the first entity arrives at $t = 0$, and the system processes entities for 50 time units. The system performance measure of interest is the average entity wait time in queue. For each simulation run of 50 time units, the wait time of each entity that enters and then leaves the queue is measured. These wait times are averaged to produce an average wait time for that simulation run. The complete experiment consists of making $n$ simulation runs. After each simulation run, the system is reinitialized, and another simulation run is started using new independent sets of random variate draws for interarrival times and service times.

The experiment is conducted for $n = 5$. Each experimental run produces a new and different average wait time. These average wait times vary in size from a low of 0.49 time units to a high of 2.95 time units; the average of these five average wait times is 1.60 time units. The results from each of the five individual runs are shown in Table 2.4.

When the experiment is repeated with $n = 100$, a new set of 100 different average wait times is obtained. In this case, the average wait times range in size from a low of 0.20 time units to a high of 6.72 time units; the average of the 100 average wait times is 1.98 time units.

This simple example raises several very significant issues. First, each time a simulation run is conducted, a different average wait time is generated. Thus, it is not clear which result, if any, is the average wait time that is sought. Second, when a number of simulation runs are conducted and the wait times resulting from each run are averaged, the average of the average wait times appears to be a function of the number of simulation runs conducted. It is not apparent how many simulation runs should be conducted or how increasing the number of simulation runs enhances the experimental result. Third, there is no indication of the accuracy or the correctness of the result obtained from the simulation experiment. What is needed is a way to estimate the average wait time that conveys some measure of both the accuracy and the correctness of the result.

Generalizing from the results of this simple simulation experiment, the presence of uncertainty (random interarrival times) and variability (random service times) in the system $Z$ results in a system output (average wait time) that is also random. If it was feasible to conduct all possible simulation experiments, the resulting output set would include the entire population of all

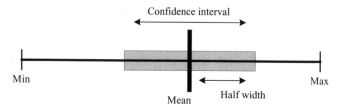

**Figure 2.16** Graphic display of confidence interval.

possible system outputs. It is convenient to think of this population as being a random variable having some underlying distribution that governs the behavior of the population. Our objective is to determine the population mean, denoted as $\mu$. In almost all cases, the best that can be done using simulation is to generate only a relatively few samples of the many possible outputs that are present in the output population. It is then necessary to use statistical analysis to estimate a population parameter by performing calculations on samples from that population.

It should be clear from this discussion that a single simulation run normally results in the generation of a single sample from the output population. It is never advisable to reach a conclusion about system performance, that is, the output population, using a single sample from that population. At best, the sample is just one of many possible values that could have occurred. At worst, the sample may be an outlier, an output value that occurs with very low probability, and has little to do with the desired population parameter.

## Confidence Interval Estimate of the Mean

*Statistical estimation* is the process of estimating a population parameter based on knowledge of a sample statistic. The sample statistic used in estimating a population parameter is called an estimator. Perhaps the most popular form of an estimator is the *confidence interval estimate*. The focus of this section is the confidence interval estimate of the mean for some population.

A typical graphic display of a confidence interval is shown in Figure 2.16. The confidence interval estimate of the mean is computed using $n$ samples obtained randomly from the underlying population and is used to estimate the population mean. The "min" and "max" values indicate the minimum and the maximum sample values, respectively, while "mean" indicates the average of all sample values. The *confidence interval* is centered about the sample mean and extends a distance equal to the *half width* on both sides of the mean. The procedure for calculating the confidence interval specifies how to compute the sample mean and the confidence interval half width.

*Precision* refers to the accuracy of an estimate. In a confidence interval, the precision of the estimate is indicated by the width of the confidence interval. The smaller the confidence interval, the higher is the precision of the estimate.

*Reliability* refers to the probability that the estimate is correct. Reliability depends on the choice of a parameter α used in the calculation of the half width. Thus, the confidence interval estimate includes information concerning both the precision and the reliability of the estimate. These are essential pieces of information when deciding the significance to attribute to the estimate. Unfortunately, precision and reliability are competing quantities; one may be traded for the other in any estimate. To demonstrate this, consider two estimates of the mean height of all males in the U.S. population. The confidence interval estimate 5′9″ ± 0.01″ is very precise but not very reliable. On the other hand, the confidence interval estimate 5′9″ ± 1.00′ is not very precise but is highly reliable. The objective is to obtain an estimate having acceptable reliability and precision.

Additional descriptors used to characterize estimators are the terms unbiased, consistent, and efficient. An estimator is said to be *unbiased* if the expected value of the estimator is equal to the parameter being estimated. An estimator is said to be *consistent* if the precision and reliability of the estimator improve as the sample size is increased. Finally, an estimator is said to be more *efficient* than another if for the same sample size it produces greater precision and reliability. Thus, it is advantageous to utilize an estimator that is known to be unbiased, consistent, and efficient.

Let $\{X_1, X_2, \ldots, X_N\}$ be IID random variables (observations) obtained from a population having mean $\mu$ and variance $\sigma^2$. The *sample mean*, defined as

$$\bar{X}(n) = \frac{\sum_{k=1}^{n} X_k}{n}, \qquad (2.1)$$

is an unbiased estimator for $\mu$. Similarly, the *sample variance*, defined as

$$S^2(n) = \frac{\sum_{k=1}^{n} \left[ X_k - \bar{X}(n) \right]^2}{n-1}, \qquad (2.2)$$

is an unbiased estimator for $\sigma^2$. Assuming the samples $X_k$ are samples from a normal distribution, then the random variable defined as

$$t = \frac{\left[ \bar{X}(n) - \mu \right]}{\sqrt{\frac{S^2(n)}{n}}} \qquad (2.3)$$

has a *Student t distribution* with degrees of freedom df = $n - 1$. The probability density function for the Student $t$ random variable is shown in Figure 2.17. The value $t = -t_{1-\alpha/2}$ is selected so $P\{-\infty < t \leq -t_{1-\alpha/2}\} = \alpha/2$; similarly,

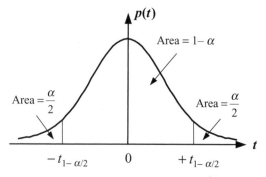

**Figure 2.17** Student $t$ probability density function.

since the Student $t$ density function is symmetric about the $t = 0$ point, $P\{+t_{1-\alpha/2} < t < \infty\} = \alpha/2$. Since the total area under a probability density function is one, it follows that

$$P\{-t_{1-\alpha/2} \leq t \leq +t_{1-\alpha/2}\} = 1 - \alpha. \tag{2.4}$$

Substituting for $t$ in Equation (2.4) using Equation (2.3), and then rearranging the inequality to indicate a bound for the population mean $\mu$ results in

$$P\left\{\bar{X}(n) - t_{1-\alpha/2}\sqrt{\frac{S^2(n)}{n}} \leq \mu \leq \bar{X}(n) + t_{1-\alpha/2}\sqrt{\frac{S^2(n)}{n}}\right\} = 1 - \alpha. \tag{2.5}$$

Equation (2.5) indicates that the confidence interval estimate for the population mean $\mu$ is centered about the sample mean $\bar{X}(n)$ and has a half width of $t_{1-\alpha/2}\sqrt{S^2(n)/n}$. The reliability associated with the confidence interval is $1 - \alpha$. The meaning of the reliability of the confidence interval is as follows: If this simulation experiment is conducted repeatedly and each experiment is used to form a confidence interval estimate for $\mu$, then exactly $(1 - \alpha)100\%$ of the confidence intervals will contain $\mu$.

As an example, consider the five sample points listed in Table 2.4 that represent the average wait time for entities passing through a simple M/M/1 queuing system over a time interval of $T = 50$ time units. These data points are listed again for convenience:

$$\{X_1, X_2, X_3, X_4, X_5\} = \{1.83, 2.95, 0.49, 1.16, 1.56\}.$$

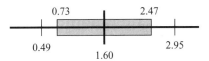

**Figure 2.18** Ninety percent confidence interval for the M/M/1 queuing system example.

A 90 percent confidence interval is calculated for these sample points. First, the sample mean $\bar{X}$ is computed using Equation (2.1):

$$\bar{X} = \{1.83 + 2.95 + 0.49 + 1.16 + 1.56\}/5 = 1.60.$$

Next, the sample variance $S^2$ is computed using Equation (2.2):

$$S^2 = \{(1.83 - 1.60)^2 + (2.95 - 1.60)^2 + (0.49 - 1.60)^2 + (1.16 - 1.60)^2 + (1.56 - 1.60)^2\}/4 = 0.83.$$

The values of the Student $t$ parameter $t_{n-1, 1-\alpha/2}$ are usually tabulated. For this example, the degrees of freedom df $= n - 1 = 4$ and the quantity $1 - \alpha/2 = 0.95$, so $t_{4, 0.95}$ is read from a table as 2.13 [3]. Thus, the 90 percent confidence interval estimate for the average wait time is given as $\bar{X} \pm t_{4, 0.95} \sqrt{S^2/5} = 1.60 \pm 0.87 = [0.73, 2.47]$. This confidence interval is shown graphically in Figure 2.18.

Controlling the reliability of a confidence interval estimate is accomplished through the choice of the parameter $\alpha$. If in the previous example it is desired to calculate a 95 percent confidence interval estimate, the only adjustment required is to change the value of $\alpha$ from 0.1 to 0.05. This in turn changes the Student $t$ parameter; the calculation now requires $t_{n-1, 1-\alpha/2} = t_{4, 0.975} = 2.78$. The 95 percent confidence interval estimate, based on the same five sample values, is $1.60 \pm 1.13$. As expected, increasing the reliability for the same set of sample values results in a decrease of the estimate precision, that is, an increase in the width of the confidence interval. This occurs because reliability and precision are competing factors and one can be traded for the other. However, for a given reliability, the precision of a confidence interval estimate can be improved by increasing the number of sample values. Suppose it is desired to improve the precision of the confidence interval estimate by a factor of $k$. Comparing the ratio of confidence interval estimates for different numbers of samples shows that the improvement in precision is approximately proportional to $k^2$. Thus, if the precision is to be increased by a factor of two, it is necessary to increase the number of sample values by $(2)^2 = 4$. Repeating the experiment with the M/M/1 queuing system, this time with 20 repetitions, results in a 95 percent confidence interval estimate of average wait time of $1.93 \pm 0.44$.

The confidence interval estimate of the mean provides an approach to obtain meaningful results from discrete event simulation experiments containing random quantities. The confidence interval provides insight to both the reliability and the precision of the measurement. The reliability of the confidence interval estimate is controlled by choice of the parameter $\alpha$, while the precision is controlled through the number of sample values used in the computation.

## More Complex Output Data Analysis Situations

So far in this section, we have discussed one of the most important output data analysis problems of discrete event simulation. We have investigated the calculation of a confidence interval estimate of the mean for some system performance measure. The calculation was made using sample data obtained by conducting multiple simulation runs of the system model. While this is certainly the most common output data analysis problem, it is not the only output data analysis problem. In the remainder of this section, some of the other more complex output data analysis situations are described. Students of M&S will meet many of these situations, along with appropriate statistical methods for addressing the situations, in future studies. The purpose here is to simply identify that there are other output data analysis problems that occur in discrete event simulation.

Discrete event simulations are often categorized as being terminating simulations or nonterminating simulations. A *terminating simulation* is a simulation for which there is an event that naturally defines the end of a simulation run. The simple queuing system simulation described at the beginning of this section is an example of a terminating simulation. The terminating event for that simulation is the event of the simulation time variable reaching the value $T = 50$ time units. Output analysis of a terminating simulation is usually straightforward because there is no ambiguity about when or how to measure the desired output quantity. A *nonterminating simulation* is a simulation for which there is no natural terminating event. This situation often occurs when it is desired to investigate the long-term or steady-state behavior of a system. Nonterminating simulations present several new problems not encountered in terminating simulations. First, the analyst must determine when the output actually reaches steady state. A graphic procedure known as Welch's procedure is one method that has been developed for this purpose [3]. The time interval that occurs from simulation start to achieving steady state is called the warm-up period. To accurately measure steady-state performance, a nonterminating simulation is started at simulation time zero, but performance data are not collected until after the warm-up period has elapsed.

This process of deleting performance data collected during warm-up is called the replication/deletion procedure [3]. Welch's procedure and the replication/deletion procedure represent additional overhead in conducting output data analysis for a nonterminating simulation. Fortunately, the output analysis components of most discrete event simulation tools provide support for these two procedures.

In some system studies, information beyond just average system performance is required. Statistical descriptors of performance typically fall into three categories: (1) descriptors that help locate the center of a distribution, such as the mean; (2) descriptors that measure the spread of a distribution, such as the variance; and (3) descriptors that indicate relative standing within a population, such as the probability of occurrence or quartile location. Statistical procedures, similar to the confidence interval estimate of mean, exist for estimating population variance, probability of occurrence, and quartile location [4].

Simulation is often used to identify the better of two alternatives or the best of several alternatives. In such studies, the alternatives are simulated to determine estimated performance, and then the different estimated performances are compared. Since performance is described with samples from a distribution, comparison of performance is not as simple as comparing two numbers. Very sophisticated statistical procedures exist for comparing two systems and for comparing multiple systems [3].

The precision of performance estimates is related to the sample variance of the simulation output data. Sample variance is generally reduced by increasing the sample size. However, in many situations, practical considerations place limits on the number of simulation runs, and thus the sample size, that can be obtained. A number of techniques, called variance reduction techniques, have been developed to help reduce sample variance without increasing sample size [3]. These variance reduction techniques are especially useful when comparing system alternatives and when conducting nonterminating simulations having long warm-up times.

Finally, simulation is often used to optimize system performance. In these studies, system performance is determined by a set of system parameters. For a particular choice of parameters, system performance is estimated using simulation. An iterative process is defined that selects a new set of parameters, estimates system performance for these parameters, and then compares the performance with these parameters to that achieved with the previous set of parameters. It is especially important to orchestrate an efficient search process in the parameter space so that an acceptable solution is found with as little simulating as possible. Many commercial discrete event simulation tools have separate components for conducting simulation-based optimization. Arena uses a component called OptQuest, ProModel uses a component called SimRunner, and AutoMod uses a component called AutoStat [5,8,9]. Each of these optimization components uses a different approach for conducting the optimization search.

## CONCLUSION

In this chapter, we have been introduced to some of the statistical concepts required to conduct a discrete event simulation experiment. Our focus has been on two important components of discrete event simulation: input data modeling and output data analysis. Input data modeling is the process of selecting an appropriate distribution to represent each random input quantity, and then generating random variates from these distributions as required during simulation runtime. Output data analysis consists of recognizing that each simulation run produces one sample value from a performance measure population. Repeated simulation runs must be conducted to estimate the desired population statistic. Confidence interval estimates are often used because they provide information about the reliability and precision of the performance statistic measurement.

The study of statistical techniques for discrete event simulation is a very rich and exciting area. While a large and impressive body of knowledge in this area has been developed, there remain many important unanswered questions and research challenges for students of M&S.

## REFERENCES

[1] Hein JL. *Discrete Structures, Logic, and Computability*. 2nd ed. Sudbury, MA: Jones and Bartlett; 2002.

[2] Peebles PZ. *Probability, Random Variables, and Random Signal Principles*. New York: McGraw-Hill; 1980.

[3] Law AM. *Simulation Modeling and Analysis*. 4th ed. New York: McGraw-Hill; 2007.

[4] Leemis LM, Park SK. *Discrete Event Simulation: A First Course*. Upper Saddle River, NJ: Pearson Prentice Hall; 2006.

[5] Harrell C, Ghosh BK, Bowden RO. *Simulation Using ProModel*. 2nd ed. New York: McGraw-Hill; 2004.

[6] Goldsman D. Introduction to simulation. In *Proceedings of the Winter Simulation Conference*. Henderson S, et al. (Eds.), 2007, pp. 26–37.

[7] Mendenhall W, Sincich T. *Statistics for Engineering and the Sciences*. 4th ed. Upper Saddle River, NJ: Prentice Hall; 1995.

[8] Kelton WD, Sadowski RP, Swets NB. *Simulation with Arena*. 5th ed. New York: McGraw-Hill; 2010.

[9] Banks J, Carson JS, Nelson BL, Nicol DM. *Discrete-Event Simulation Systems*. 4th ed. Upper Saddle River, NJ: Prentice Hall; 2005.

# 3

# DISCRETE-EVENT SIMULATION

Rafael Diaz and Joshua G. Behr

A queue is a line of either people or objects waiting for service or handling. *Queuing* is a generic term used to refer to the process of people or objects forming a line in preparation to receive service or handling. Queues and queuing are an integral part of the ordered, normal environment that defines our modern world. Part of our social, cultural, and civil norms is defined by the unwritten rules that guide where and how we queue. In fact, our familiarity and frequent application of these unwritten rules in our daily lives has allowed them to become almost natural. For example, we form lines and wait to be serviced by the next available customer representative at the bank, to use a restroom during intermission, and for the traffic light to return to green. The rules that govern queuing are more formalized in controlled environments such as manufacturing, warehousing, and distribution. In a production setting, for example, semifinished products are queued awaiting to receive the next treatment in an assembly line, followed by packaging and shipping to either a final destination or a distribution center. At a congested distribution center, trucks form queues awaiting loading.

As you can see, queues and queuing are quite common, and since the terms, by definition, entail "waiting," the evaluation of a queuing system's performance, often a point of interest for consumers and operations managers alike,

*Modeling and Simulation Fundamentals: Theoretical Underpinnings and Practical Domains*,
Edited by John A. Sokolowski and Catherine M. Banks
Copyright © 2010 John Wiley & Sons, Inc.

may be measured by some form of "wait time." When evaluating the performance of a real-world queuing system, there are a number of measures one may consider: the average and maximum wait times, the average and maximum number of persons or objects in a queue, service utilization time, total time (entry to exit) in the system, and so forth. From the perspective of the consumer, when seeking a service, if the wait time is too lengthy, then the evaluation of the business tends to be rather poor. From the operations side, it is quite helpful to understand what configuration of a queuing system's components, vis-à-vis the number of scheduled servers and number of customer lines, will yield wait times that are consistent with the organization's business model. For example, given the pace and arrival times of patients within an emergency department, what deployments of registration representatives and medical staff will yield an acceptable wait time for patient treatment? Or, in a production environment with finite space, given the nature and quantity of orders placed by retail stores, what is the number of assembly queues that will yield a given turnaround time between order and delivery?

One approach to answering these questions requires the following: first, the modeling of the environment where queuing takes place; second, a process where the model is executed or simulated to allow it to play out over time, revealing the dynamic behavior of the system; and third, evaluation where performance measures resulting from the modeling and simulation process are analyzed, potential adjustments made to the model, and the simulation potentially run again.

Modeling simply means making a logical representation, usually a simplified one, of the real-world queuing environment or system. When modeling the queuing system, not all the information detailing the environment is included; some component parts of a system are more important than others relative to the measured performance of the system. Thus, the model is not an identical replication of a real-world system; rather, it is a plausible representation of the parts of a system that matter the most to the overall performance of that system. If the real-world system evidences a single queue line and four service stations (essential components relative to the performance of the system), then the initial model ought to replicate this arrangement. On the other hand, representing the colors on a facility's walls ought not to be considered for model inclusion since its relation to a system's performance is likely tenuous.

The building of a valid model requires awareness not only on the number of queues and servers present in the real-world system, but also on the foundational information about patterns of behavior for entities and servers. These behaviors can be derived from our understanding of historical data such as those associated with past performance measures of the system as well as historical data that describe, as distributions, the past behavior of entities (e.g., probabilistic or deterministic arrival times) and servers (e.g., probabilistic or deterministic service times) within the system. This information informs the modeler in the building of the model. When historical data are not available, the researcher has the options of either engaging the real-world system expressly to gather these foundational data or relying on subject matter experts to express the behavior of entities and servers.

The arrival and service times within a real-world system may be stochastic. The articulation of mathematical equations, complemented by probability theory, has been used as a close approximation for the real-world behaviors of entities and servers within queuing systems. Thus, probabilistic distributions under proper assumptions may be employed to represent the stochastic behavior of entities within the system.

Simulation is the execution of the model and is commonly a two-phase process. The first phase entails initializing the simulation, generating either random or deterministic numbers that describe the behavior of entities and servers, entering of an entity into the system, updating the state data and statistical accumulators, repeating the two preceding operations for a predetermined number of time steps, and terminating the simulation. In most cases, a time step is occasioned by an event such as the entry into, or departure from, the system by an entity. While the real-world queuing system within an emergency department, for example, may experience thousands of time steps associated with the arrival and departure of patients unfolding over a period of many days, a digital computer can execute these time-stepped events in a matter of seconds. The arrival times of entities entered at each step as well as the service times can be derived from probability distributions, which, in turn, have been informed by our knowledge of the arrival and service patterns evidenced in the historical data. The state variables are updated and the statistical accumulators expanded at each time step. Termination occurs after numerous iterations through this process loop and the summation of accumulated time steps reaches a predetermined threshold or termination criteria. Final performance measures are then summarized for this initial simulation run.

The second phase of the simulation process, if necessary, entails the replication of the first phase. Since the arrival and service utilization times for a particular simulation run are potentially derived from a set of random numbers unique to that run, the resultant queue sizes and wait times are also unique to that run. Thus, the performance measures yielded from each replication will differ because of the generation and input into the simulation of unique random numbers. The number of simulation replications depends upon the sample mean and the sample standard deviation. Since each replication is independent and identically distributed (IID), one can build a confidence interval for measuring the expected performance of the system. Thus, through a process of evaluation, the investigator learns how the model behaves under prescribed conditions. The utility of modeling and simulation is found in the derived inferences; the investigator has learned about the model's performance under various sets of given conditions, and, since the model is a representation of a real-world queuing system, inferences may be made about the behavior of the real-world system. Simply, the investigator upon completion asserts that he or she is able to predict, within a certain degree of confidence, how the system will perform when confronted with those simulated conditions.

In this chapter, we discuss the construction and performance of a queuing system model. We illustrate a simulation methodology by describing a general discrete-event simulation (DES) framework and further illustrate the mechanics of the simulation. Next, we introduce the main components and aspects of Arena® simulation software, and, finally, a simple simulation model is developed and executed using this simulation software.

## QUEUING SYSTEM MODEL COMPONENTS

A *queuing system model* can be defined as a representation that captures and quantifies the phenomenon of waiting in lines. The three basic elements within a queuing system are entities, servers, and queues. *Entities* can represent either customers or objects, servers can represent persons or production stations that treat or interact with the entity, and queues are the holding or waiting position of entities. The queue size may be assumed to be either finite or infinite. For example, the physical confines of a buffer area (i.e., queue) between two workstations along a production line may behave as a cap on the potential number of entities waiting for service; since historically we have knowledge that the maximal number of entities has filled on occasion this queuing area, we say the queue size is finite. In contrast, a call center may have the capacity to queue, or place on hold, up to 1000 incoming calls unable to be serviced by the three service representatives; since there is no history of the call center approaching its holding queue capacity, for all intents and purposes this queue can be assumed to be infinite.

The arrival process is characterized by the interarrival time, or the interval of time between successive customers or objects entering the system in preparation to receive service. As mentioned before, the interarrival time may be described in terms of probabilistic behavior. Often, an assumption is made that the arrival time is a random variable that is IID, meaning that the arrival times of the entities are independent, or autonomous, of each other and that the probability distribution is identical, or alike, for all entities. Upon arrival, an entity enters into the system at which point the entity proceeds directly either to a server to receive service or to a queue if it is the case that all servers are busy. Any particular server may be either busy, as is the case when servicing an entity, or idle, as is the case when there are no entities in queue. If more than one server is modeled, then the design of the service may involve series, parallel, or networked servers. Similar to interarrival time, the service time may also be expressed in terms of probabilistic behavior. The customer may also engage in strategic behavior known as jockeying or balking; assuming there is more than one queue, a customer may choose to remove himself or herself from one queue and enter another queue.

A relevant component of the queuing system is the queue discipline that rules the model. *Queue discipline* is the order that either customers or objects are selected from a queue to advance to receive service. Two common disciplines are first come, first served (FCFS) and shortest process time (SPT), but the researcher may also customize a heuristic that prioritizes the customer or object according to unique conditions. For example, within an emergency department, customers arriving with predetermined conditions such as chest pain or head injury are advanced to the front of the queue. Or, within a processing and distribution center, customer orders tagged for expedited service may be directed to the front of the packaging and shipping queue.

We have introduced above several "components" that define a queuing system model. A queuing system is characterized by the specification of a combination of these components. For example, the M/M/1 queuing model is characterized by a Markovian arrival process in which entities arrive independently and distributed identically from an exponential distribution (first M); a Markovian service time (second M), again from an exponential distribution; and finally with one single server (the third component, 1). The M/M/1 queuing model, like other queuing models, uses mathematical expressions to describe performance. Because certain assumptions are employed when deriving these expressions, it may be said that the model has been constrained.

## State Variables, Events, and Attributes

*State variables* are those measures that characterize a system at a particular moment or state. Thus, the state of the queue at time $t$ (i.e., $Q(t)$) may be characterized by the number of entities in the queue, while the state of the server at time $t$ (i.e., $Q(t)$) may be characterized as either busy or idle. *Events* are the arrival to, or departure from, the system by entities or objects. The termination, or end, of the simulation is also a form of an event, sometimes referred to as a pseudoevent since it is not marked by either the arrival or the departure of an entity.

Entities and servers are described by attributes. *Attributes* are components of the system state. A server can have an attribute that describes its state as either busy or idle. In addition, when the server is busy, an attribute may describe the type of activity the server is performing. Entities can have an attribute that indicates its amount of time required to service as well as an attribute that describes the type of service sought. An example is a service center that fields customers' calls about a particular product. The types of service sought by a calling customer may include the attributes "product complaint," "product praise," and "product technical question." The attributes associated with a server (assuming the server is engaged) may be "proficient" and "slow." A newly hired service representative who is still early in the learning curve may be slow at managing a product complaint but proficient at registering a product praise. If there is more than one customer who would like to register a product complaint, then these customers have a shared attribute and may be grouped into a set. Likewise, if there are several servers that share the attribute "proficient," then these servers may also be grouped into a set.

## SIMULATION METHODOLOGY

In most cases, systems can be classified into two types: discrete and continuous. The system is called discrete when the state variables are updated instantaneously at specified times (e.g., $t_1$, $t_2$, $t_3$ ...). For example, envision a lunch queue and server utilization at a local fast-food restaurant throughout the peak midday business hours. One could rigidly take a snapshot of the state of customers precisely every 60 seconds and update the state variables to capture the progression of the lunch rush. However, a more meaningful approach, and the one advanced here, is the taking of a snapshot each time a new customer either arrives or departs the restaurant. With each snapshot, or event, the number and position of customers in the restaurant adjusts relative to the previous snapshot; new customers have entered the system and joined one of the several queues, previously queued customers have stepped up to the counter and are now being served, others have jockeyed from one queue to another, and still yet others—hopefully satisfied and well fed—have departed the system. Continuous systems, on the other hand, denote the situation where the state variables change continuously with the progression of time. For example, the state variable "volume" exhibits continuous change over time as a tank is filled with fluid.

The modeling and simulation approach considered in this chapter is the DES model. DES models a system as it evolves over time by a representation in which the state variables change instantaneously at separate, countable points in time [1]. As enumerated on the event list, the advancement, or action, from one event to the next is sequential, but does not necessarily occur at a precise interval [2]. These instantaneous actions are associated with updates to the state of the system. Each event produces a large amount of data that characterizes the entire system at that particular moment. Analysis of these data can be cumbersome and, thus, digital computers may be necessary to perform more comprehensive DESs.

### Time-Advance Mechanism

There are two well-known approaches for advancing the simulation clock: fixed-increment advance and next-event time advance. The fixed-increment advance, uncommon in simulation software, initializes at time zero and then advances at fixed time increments. In the fixed-increment advance approach, events such as an entity's arrival to, or departure from, the system, as generated by the probability distribution and recorded in the event list, may fall between successive moments that demarcate the time interval. Thus, there is disjuncture, or asynchronicity, between the moment of the interval and the moment of the event. The updating of the state variables takes place not with the occurrence of an event, but with the passing of a time increment moment. In the fixed-increment approach, for the purpose of recording data and updating state variables, the event is artificially forced in sync with the interval. Another characteristic of the fixed-increment advance is that the size of the time increment is subjective. The choice of increment has implications for the performance measures of the system. There are caveats both for an interval that is relatively large and for an interval that is relatively small. Any number of arrival or departure events may occur within a particular interval if the interval is relatively large. By syncing these events with the singular most proximate future interval moment, this approach treats the time differences among the various events as nonconsequential, even when the differences may have very real meaning. On the other hand, a time interval that is relatively small may advance many times without the occurrence of a single event. Thus, settling on an appropriate time interval that balances these two competing potentialities—too many events occurring within a single interval and no events occurring within an extended series of intervals—is as much art as science.

In contrast, the versatile next-event time-advance approach, employed within most simulation software, initializes the simulation clock at zero, progresses to the most proximate, forthcoming event as tendered by the event list, and then updates the state variables. Once this is complete, the simulation clock is again progressed to the moment of the next most forthcoming event, and the state of the system is again revised. While there is much variability in the timing of events, the updating of the state variables is in sync with the occurrence of events. Thus, the performance measures that are yielded from the state variables avoid some of the potential distortion found in the fixed-increment advance. Since all state changes take place exclusively at event times, the next-event time-advance approach omits periods of inactivity. Figure 3.1 illustrates a hypothetical situation in which a series of events, denoted by $E_0$, $E_1$ …, is in sync with the arrivals and departures of entities. The state variables are updated at each time $t$. Notice that $t_1$, $t_2$ … represents the arrival and departure times while and $I_1$, $I_2$ … corresponds to the interarrival intervals.

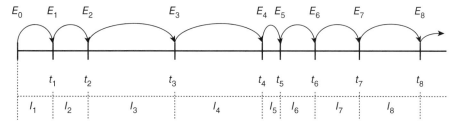

**Figure 3.1** Time advance mechanisms.

## Main Components

The following provides an enumeration of some common terminology associated with DES:

(1) *System State.* A characterization of the state of the system at a particular moment; expressed as state variables.
(2) *Simulation Clock.* A tool that provides the elapsed real-world time.
(3) *Next-Event List.* An ordered list of times that each event will take place.
(4) *Statistical Counter or Accumulator.* A tool that both records and expresses a system's evolving performance; it accumulates information as the simulation unfolds.
(5) *Initialization Subprogram.* A protocol utilized in the initialization of the simulation, usually setting the start time to zero.
(6) *Timing Subprogram.* A protocol that, drawing from a next-event list, sets the next event and progresses the simulation clock to the moment when an event is to happen.
(7) *Event Subprogram.* A protocol that launches a routine that updates the state of the system with the occurrence of each event.
(8) *Library Subprogram.* A protocol used to produce random observations drawn generally from predetermined probability distributions.
(9) *Report Generator.* A tool that calculates and reports statistics that describe the performance of the system.
(10) *Main Program.* A routine that coordinates the concert of subordinate routines, executing these in the correct sequence. It initializes the timing subprogram that determines the subsequent event, passes control to the related event subprogram, and updates the system state. This routine verifies for termination and triggers the report generator once the simulation ends.

## Simulation Flowchart

Figure 3.2 presents the workings of a general DES model. At time zero, the main program invokes the initialization subprogram, which sets the simulation clock to zero and initializes the system state, the statistical counter, and the event list. Next, the main program brings into play the timing subprogram. The timing subprogram verifies the forthcoming event and progresses the simulation clock. After this, the main program invokes the event and calls upon the event subprogram. This subprogram may interact with the library subprogram, often responsible for generating random variates. Next, the system state is reviewed and statistical counters compile system performance. Information about the occurrence of future events is collected and added to an event list. This cycle is followed by a review of the condition that terminates

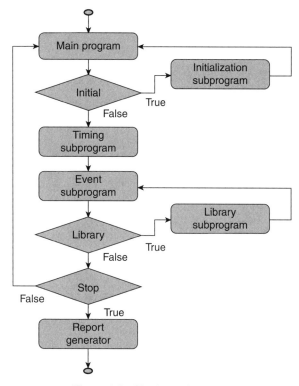

**Figure 3.2** Simulation flowchart.

the simulation. The catalyst for termination may be any number of conditions including the meeting of a threshold number of entities serviced or passage of time. The iterative, closed process repeats until the termination condition is satisfied, whence the report generator is activated. Thus, using data stored by the statistical counters, estimates of measures of performance are calculated.

## DES EXAMPLE

### Problem Description

DES models involve queuing systems in which entities arrive, are processed, and then leave. A single-server queuing system illustrates a simple, yet representative, model in which the general principles of DES can be demonstrated. In this section, we render the logic and mechanics of this basic model understandable.

The queuing system we consider is a one-teller banking facility where customers arrive, are processed by the single teller, and then leave. Figure 3.3

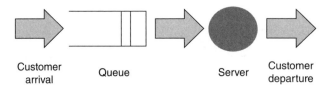

**Figure 3.3** A single-server queuing system.

illustrates this scheme. If a customer arrives and finds the teller idle, then servicing at the teller starts immediately; otherwise, the customer waits in queue according to the FCFS discipline. The interarrival times are random and IID. The service times associated with the arriving customers are also IID random variables and are independent of the interarrival times. The simulation will analyze 20 min of business time at the bank immediately following the opening at 9:00 a.m.

Upon simulation initialization at time zero, the queue is empty (no customers are present) and the teller is idle. Although the arrival of the first customer could occur after time zero, in this scenario, a single customer arrives and enters the bank upon opening. During the 20-min simulation period, customers will arrive, be either serviced or queued and serviced, and depart. While we will be collecting a range of measures of performance for analysis, the following is a selection of four common measures of performance that require explication.

*Average delay in queue* refers to the time in queue that customers wait for service excluding the time spent being served. The formula provides the average delay in a queuing system in which $N$ customers leave the queue during the 20-min replication and $D$ is the waiting time in queue of the $i$th customer:

$$\text{Average waiting time} = \frac{\sum_{i=1}^{N} D}{N}.$$

*Time-average number of customers waiting in the queue* refers to the weighted average of the possible queue lengths without including anyone being served by the teller. This is weighted by the proportion of time during the replication that the queue was at such lengths. It indicates the average queue length. If $Q(t)$ is the number of customers in the queue at any moment $t$, then the time-average queue length is represented by the area under the curve $Q(t)$. This is divided by the length of the replication $t = 20$ min. Formally stated,

$$\text{Time-average number of customers waiting in the queue} = \frac{\int_0^t Q(t)\,dt}{t}.$$

*Maximum flow time and average flow time through the system* refers to customers that have finished being processed by the teller and have left the system. Also known as cycle time, this is the time that passes between a customer's arrival and his or her departure. Simply, it is the addition of the customer's waiting time in queue and service time at the teller.

*Resource utilization* refers to the amount of time that the teller is busy when simulating the operation. Thus, at time $t$, the utilization function can be defined as

$$B(t) = \begin{cases} 1 & \text{busy} \\ 0 & \text{idle.} \end{cases}$$

The area under $B(t)$ divided by the length of the run $t$ represents the resource utilization. Formally stated,

$$\text{Utilization} = \frac{\int_0^t B(t)\,dt}{t}.$$

## HAND SIMULATION—SPREADSHEET IMPLEMENTATION

In this section, we execute the above described simulation through a series of logically sequenced tables. Recall, first, the parameters of the system we are modeling: (1) one-teller and single-queue banking facility, (2) FCFS discipline, (3) interarrival and service times are random and IID, and (4) simulation time represents 20 min of business at the bank. Recall, second, that our modeling potentially relies on foundational information about the real-world system. The choice of foundational information reflects the object of interest, or goal, of our modeling and simulation efforts: an understanding of the treatment of bank customers by the system. In this case, we have collected representative data from the processing of customers by the system. These data express the arrival time, interarrival time, and service time, and are presented in Table 3.1. We are interested in the system's performance within a very specific, defined

**Table 3.1 Arrival time, interarrival time, and service time**

| Customer Number | Arrival Time | Interarrival | Service Times |
|---|---|---|---|
| 1 | 0.00 | 2.81 | 0.58 |
| 2 | 2.81 | 1.19 | 1.69 |
| 3 | 4.00 | 2.35 | 4.01 |
| 4 | 6.35 | 1.59 | 2.13 |
| 5 | 7.94 | 2.71 | 3.81 |
| 6 | 8.87 | 0.93 | 6.08 |

time window. The simulation begins at time moment zero when the first customer, or entity, is scheduled to arrive; the collection of simulated output data begins precisely at this moment. As you can see, customer 1 is expected to enter the bank at time zero and require 0.58 units of service time (measured in minutes) from the single teller. The distance in time between the arrival of the first customer and the arrival of the second customer is captured in the interarrival time statistic. In this case, the entry of entity 1 is followed 2.81 min later by the entry of entity 2 which, in turn, is followed 1.19 min later by the entry of entity 3, and so forth. As you can see, there is a fair amount of variation in the service times among the six modeled customers.

Next, we present 13 additional tables; each table represents a discrete step, or event, in the simulation process; each table is accompanied by a narrative explaining change in the figures from one table to the next. Remember that each event within the simulation is characterized by the updating, or expansion, of the statistical accumulators; the change in figures from one table to the next reflects snapshots of the evolving flow of customers through the system. Each "snapshot" is triggered by an event: initialization, arrival, departure, or end.

The resultant spreadsheets from the simulation are shown in Tables 3.2 through 3.14. Each table is subdivided into five areas including a description of the current event, state variables, entity attributes, statistical accumulators, and the next-event list. Below is a brief description of each of these parts:

(1) *Current Event.* This section provides the number of the event that has just occurred, the timing and type of event, and the entity that triggered the event.
(2) *State Variables.* This section addresses the number of entities in queue at the moment of the event (represented by $Q(t)$) and the status of the servers at the moment of the event (represented by $B(t)$).
(3) *Entity Attributes.* This section lists the main attributes of arrival time at queue, arrival time at server, and departure time.
(4) *Statistical Accumulators.* As the simulation advances, the statistical accumulators track the main measures of performance up to the moment of the most recent event. The following nine stats are accumulated: (1) number processed by queue (this is a running total of all entities who have passed through the queue, independent of whether or not they waited in queue, and began service); (2) maximum number of entities queued at any single moment (i.e., max $Q(t)$); (3) sum of all delay times for those entities that have been processed through the queue; (4) maximum delay time of those entities that have been processed through the queue; (5) the area under the $Q(t)$ curve denoted by $\int Q(t)$ (with the execution of an event, this is the total amount of time spent in queue thus far by both those entities that have advanced to service and those entities that have not yet been advanced out of the queue); (6) number served (this is a running total of those entities that

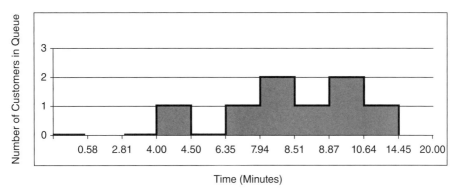

**Figure 3.4** Time-average delay in queue of customers waiting in queue.

have completed service); (7) area under $B(t)$ curve denoted by $\int B(t)$ (with the execution of an event, this is the total amount of time spent in service thus far by both those entities that have departed the system and those entities that have not yet finished being served); (8) sum of flow time of those entities that have departed the system; and (9) maximum flow time from those entities that have departed the system.

(5) *Next-Event List.* Also called the event calendar, this provides a list of imminent events with the top record being the most proximate event.

Finally, presented below are the trends for two state variables, $Q(t)$ and $B(t)$, respectively, representing the number of customers in queue and the state of teller utilization over a series of moments. As illustrated in Figure 3.4, the simulation takes place within a 20-min real-world time window (horizontal axis) and tracks the number of customers in queue (vertical axis). Since several customers may enter the bank as early as moment zero (the beginning of the simulation), potentially a queue may form immediately. However, in the current scenario, it is not until 4.00 min following the start of the simulation that an entering customer finds the teller occupied with a previous customer and, hence, is positioned in queue. In addition, there are just two time intervals, beginning with 7.94 and 8.87, when we witness two customers in queue. As illustrated in Figure 3.5, over the period of the 20-min simulation (horizontal axis), the single teller is in continuous service with the exception of the time interval beginning at moment 0.58.

The general approach is to accumulate data and document the state of both entities and servers as the simulation progresses. The progress of a simulation is bookended by the initialization of the simulation (time 0.00) and the termination of the simulation (stop event at time 20.00). Between these is a series of events beginning with entry into the system by the first entity at time zero, and continuing with subsequent arrivals or departures of entities. With the advent of each event, the state variables and the accumulated stats need to be updated. Thus, Table 3.2 offers the template for the sequential collection and

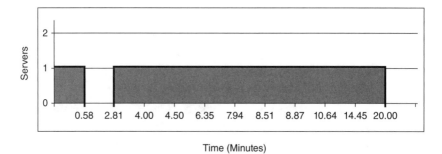

**Figure 3.5** Utilization.

presentation of information associated with each event. In the upper-right corner, we find the event number and event type (neither the event number nor the event type is yet specified); each event is given a number corresponding to its chronological order (event number) and described as either arrival or departure (event type). The state variables, by default initially set at zero, show that there are no customers in the queue (i.e., $Q(t) = 0$) and the teller is idle ($B(t) = 0$). Near the bottom is what is generally referred to as the event calendar or next-event list. The next-event list section is itself a form of a queue that starts at the top with the next, most proximate pending event to be executed in the simulation. By logical extension, an entity no longer appears on the next-event list with the execution of the event, whether it be an entity's arrival or departure. With the passing of each event, the entity attributes section is updated to reflect those entities that have already either entered or exited the system (notice that the section contains six rows corresponding to the six entities we proffered in Table 3.1 above). The statistical accumulators are designed to accumulate stats with the progression of each event; all are set at zero reflecting the absence of collected data thus far. With this brief exposition now complete, we are ready to turn attention toward the hand calculation of the simulation.

*Event 1: Arrival of Entity 1. Time = 0.00.* Table 3.3 indicates the occurrence of the first event, the entry of entity 1 into the system. This is expressed by the event number and event type showing "1" and "arrival," respectively. The state variables, representing the current status of the queue and the teller at the moment of arrival of entity 1, have been advanced to reflect the entry of the first customer into the bank. Upon entry, customer 1 finds no other customers in the system and the teller waiting to provide service, so the customer passes through the queue and goes directly to the teller for service. Thus, the queue remains empty (i.e., $Q(t) = 0$), the teller's status changes from idle to busy (i.e., $B(t) = 1$), and the number processed by queue changes to 1. With the exceptions of the time-persistent stats integral $Q(t)$ and integral $B(t)$, the stats within the statistical accumulators are updated when an entity is either finished waiting in queue or finished being served; completion of waiting or

**Table 3.2** Hand simulation spreadsheet template and initialization

| **Current Event** | | | **Event Number** = | |
|---|---|---|---|---|
| Time = 0.00 | Entity = | | Event Type = | |

**State Variables**
$Q(t) = 0$  $B(t) = 0$

**Entity Attributes**

| Entity | Arrival Time at Queue | Arrival Time at Server | | Departure Time |
|---|---|---|---|---|
| 1 | | | | |
| 2 | | | | |
| 3 | | | | |
| 4 | | | | |
| 5 | | | | |
| 6 | | | | |

**Statistical Accumulators**

| # Served = | 0 | # Processed in queue = | 0 | Sum of delays = | 0 |
|---|---|---|---|---|---|
| Max delay = | 0 | Sum of flow times = | 0 | Max flow time = | 0 |
| Integral $Q(t)$ = | 0 | Max $Q(t)$ = | 0 | Integral $B(t)$ = | 0 |

**Next-Event List (Event Calendar)**

| Number | Time | Entity | Event Type |
|---|---|---|---|
| 1 | | | |
| 2 | | | |
| 3 | | | |

**Table 3.3  Event 1**

| Current Event | | | Event Number = | **1** |
|---|---|---|---|---|
| Time = | 0.00 | Entity = 1 | Event Type = | Arrival |

**State Variables**
$Q(t) = 0$  $B(t) = 1$

**Entity Attributes**

| Entity | Arrival Time at Queue | Arrival Time at Server | Departure Time |
|---|---|---|---|
| 1 | 0.00 | 0.00 | |
| 2 | | | |
| 3 | | | |
| 4 | | | |
| 5 | | | |
| 6 | | | |

**Statistical Accumulators**

| # Served = | 0 | # Processed in queue = | 0 | Sum of delays = | 0 |
|---|---|---|---|---|---|
| Max delay = | 0 | Sum of flow times = | 0 | Max flow time = | 0 |
| Integral $Q(t)$ = | 0 | Max $Q(t)$ = | 0 | Integral $B(t)$ = | 0 |

**Next-Event List (Event Calendar)**

| Number | Time | Entity | Event Type |
|---|---|---|---|
| 1 | 0.58 | 1 | Departure |
| 2 | 2.81 | 2 | Arrival |
| 3 | 20.00 | – | Stop |

being served is indicated by the advancement of the entity to the next stage (either entering into service after waiting or departing the system after being served). That is, for example, the # served will remain at zero until the occurrence of the event that signifies the completion of the interaction between the teller and the customer, the sum of flow times will remain at zero until a customer completes his or her banking transaction and departs the system, and the sum of delays will remain at zero until a customer is advanced beyond the queue. Notice that, according to the next-event list, the next scheduled event at time 0.58 will be the departure of entity 1 from the system and, following that, entity 2, representing the second customer, is poised to arrive at 2.81 min.

*Event 2: Departure of Entity 1. Time = 0.58.* This event indicates the completion of service for customer 1 and departure from the system (Table 3.4). The event clock records the departure event at time 0.58 since entity 1 entered the system at time 0.00, proceeded directly to the teller, and received 0.58 min of service. There are no other customers in the bank at this moment and, thus, the teller is idled and the state of the queue remains at zero. The statistical accumulators have been updated to reflect the departure: number served = 1, sum of flow times, reflecting the time from entry to departure, as well as max flow time, is set to 0.58, and the integral $\int B(t)$, reflecting the time-persistent measure is also set at 0.58 since $1 \times (0.58 - 0) = 0.58$. Given that the single customer did not enter a queue and did not experience delay, the max delay, sum of delays, integral $\int Q(t)$ will be $0 \times (0.58 - 0) = 0$, and max $Q(t)$ all remain at zero. The next event, shown within the next-event list, will be the arrival of entity 2 at time 2.81 (Table 3.5).

*Event 3: Arrival of Entity 2. Time = 2.81.* This event denotes the arrival of entity 2 to the system. Similar to the experience of customer 1, customer 2 arrives to find the bank empty of other customers and the teller in an idle state and, thus, customer 2 immediately passes through the queue and begins the service interaction expected to take 1.69 min. We can look ahead and expect the future departure of entity 2 at time 4.50 because we know both the arrival time and the service time $(2.81 + 1.69 = 4.50)$. Notice that stats within the statistical accumulators remain unchanged; the absence of other customers in the bank at the time of arrival of customer 2 translates into no queuing or service activities in need of updating. The next-event list has been advanced to show the proximate arrival of entity 3.

*Event 4: Arrival of Entity 3. Time = 4.00.* This table presents event 4, the arrival of customer 3 at time 4.00 (Table 3.6). We know from the above that the teller will be occupied with customer 2 until time 4.50. Therefore, the newly arrived customer 3 will have to wait, in queue, for the availability of the teller. Notice that the max $Q(t)$ has now been updated to "1," representing the fact that a single customer is now queuing and since no customer time has been spent waiting in queue up to this moment, the integral $\int Q(t) = 0 \times (4.00 - 2.81) = 0$. Although customer 2 has not completed his or her transaction with the teller, up to this moment some time indeed has been spent with the teller; the integral $\int B(t)$ is advanced to $0.58 + 1 \times (4.00 - 2.81) = 1.77$ min capturing at this moment the fact that the teller, while still in the

**Table 3.4** Event 2

| Current Event | | | Event Number = | **2** |
|---|---|---|---|---|
| Time = 0.58 | Entity = 1 | | Event Type = | Departure |

**State Variables**
$Q(t) = 0$    $B(t) = 0$

**Entity Attributes**

| Entity | Arrival Time at Queue | | Arrival Time at Server | Departure Time |
|---|---|---|---|---|
| 1 | 0.00 | | 0.00 | 0.58 |
| 2 | | | | |
| 3 | | | | |
| 4 | | | | |
| 5 | | | | |
| 6 | | | | |

**Statistical Accumulators**

| # Served = | 1 | # Processed in queue = | 1 | Sum of delays = | 0 |
|---|---|---|---|---|---|
| Max delay = | 0 | Sum of flow times = | 0.58 | Max flow time = | 0.58 |
| Integral $Q(t)$ = | 0 | Max $Q(t)$ = | 0 | Integral $B(t)$ = | 0.58 |

**Next-Event List**

| Number | Time | Entity | Event Type |
|---|---|---|---|
| 1 | 2.81 | 2 | Arrival |
| 2 | 4.00 | 3 | Arrival |
| 3 | 20.00 | – | Stop |
| 4 | | | |

**Table 3.5 Event 3**

| Current Event | | | | Event Number = | 3 |
|---|---|---|---|---|---|
| Time = | 2.81 | Entity = | 2 | Event Type = | Arrival |

**State Variables**

$Q(t) = 0 \quad\quad B(t) = 1$

**Entity Attributes**

| Entity | Arrival Time at Queue | | Arrival Time at Server | | Departure Time |
|---|---|---|---|---|---|
| 1 | 0.00 | | 0.00 | | 0.58 |
| 2 | 2.81 | | 2.81 | | |
| 3 | | | | | |
| 4 | | | | | |
| 5 | | | | | |
| 6 | | | | | |

**Statistical Accumulators**

| # Served = | 1 | # Processed in queue = | 2 | Sum of delays = | 0 |
|---|---|---|---|---|---|
| Max delay = | 0 | Sum of flow times = | 0.58 | Max flow time = | 0.58 |
| Integral $Q(t)$ = | 0 | Max $Q(t)$ = | 0 | Integral $B(t)$ = | 0.58 |

**Next-Event List**

| Number | Time | Entity | Event Type |
|---|---|---|---|
| 1 | 4.00 | 3 | Arrival |
| 2 | 4.50 | 2 | Departure |
| 3 | 20.00 | — | Stop |
| 4 | | | |

**Table 3.6** Event 4

**Current Event**　　　　　　　　　　　　　　　　　　　　**Event Number =** 4
Time = 4.00　　　　　Entity = 3　　　　　　　　　　　　Event Type = Arrival

**State Variables** *Entities in queue*　　*storing time*
$Q(t) =$ 　　$B(t) = 1$

**Entity Attributes**

| Entity | Arrival Time at Queue | Arrival Time at Server | Departure Time |
|---|---|---|---|
| 1 | 0.00 | 0.00 | |
| 2 | 2.81 | 2.81 | 0.58 |
| 3 | 4.00 | | |
| 4 | | | |
| 5 | | | |
| 6 | | | |

**Statistical Accumulators**

| | | | |
|---|---|---|---|
| # Served = | 1 | # Processed in queue = | 2 |
| Max delay = | 0 | Sum of flow times = / | 0.58 |
| Integral $Q(t) =$ | 0 | Max $Q(t) =$ | 1 |
| | | Sum of delays = | 0 |
| | | Max flow time = | 0.58 |
| | | Integral $B(t) =$ | 1.77 |

**Next-Event List**

| Number | Time | Entity | Event type |
|---|---|---|---|
| 1 | 4.50 | 2 | Departure |
| 2 | 6.35 | 4 | Arrival |
| 3 | 20.00 | – | Stop |
| 4 | | | |

process of servicing entity 2, has nonetheless devoted some service time to customer 2. We expect, looking at the next-event list, that the departure of entity 2 will be the next event.

*Event 5: Departure of Entity 2. Time = 4.50.* At this moment, customer 2 has completed his or her transaction with the teller and departed the system and customer 3 has been advanced from the queue to the teller window (Table 3.7). Both the # served and the # processed by queue have increased by 1. It stands to reason that the statistical accumulators referencing both wait and service times will also show change. The sum of delays and max delay have now been changed from zero to 0.50 (4.50 − 4.00 = 0.50) just as the integral $Q(t)$ has been changed from zero to 0.50 ($\int Q(t)$ will be $1 \times (4.50 - 4.00) = 0.50$). Previously, with the departure of customer 1, the max flow time was 0.58, but has now been updated to reflect the lengthier flow time of customer 2, which is 4.50 − 2.81 = 1.69 min. Notice also that the sum of flow times has been updated to 2.27 (0.58 + 1.69 = 2.27). The integral $\int B(t)$ will have the same value, but as a product of $1.77 + 1 \times (4.50 - 4.00) = 2.27$. The next-event list informs us that entity 4 is expected to arrive at 6.35.

*Event 6: Arrival of Entity 4. Time = 6.35.* This illustrates event 6, the arrival of the fourth customer at 6.35 (Table 3.8). Customer 4 arrives to find no other queued customers and the teller occupied with customer 3 and, thus, customer 4 enters into the queue; the state variables $Q(t)$ and $B(t)$ both register "1." Since no others were in queue at the moment of this event, the integral $Q(t)$ remains at 0.50 by $\int Q(t) = 0.50 + 0 \times (6.35 - 4.50) = 0.50$. Customer 3 is in service at the time of this event (i.e., has not completed his or her transaction with the teller) so the integral $\int B(t)$ changes from 2.27 to 4.12 min by $\int B(t) = 2.27 + 1 \times (6.35 - 4.50) = 4.12$. We can see that the next-event list specifies the arrival of entity 5 at 7.94.

*Event 7: Arrival of Entity 5. Time = 7.94.* This event presents the arrival of customer 5, who finds the teller busy with customer 3 and customer 4 waiting in queue (Table 3.9). Since entity 5 joins the queue, the state variable $Q(t)$ changes from "1" to "2." The continued service of entity 3 means that the integral $\int B(t)$ is increased from 4.12 to $4.12 + 1 \times (7.94 - 6.35) = 5.71$, and the continued delay of entity 3 means that the integral $\int Q(t)$ is increased from 0.50 to $0.50 + 1 \times (7.94 - 6.35) = 2.09$. The next event is expected to be the departure of entity 3 at time 8.51.

*Event 8: Departure of Entity 3. Time = 8.51.* Table 3.10 provides the updates associated with event 8, the departure of customer 3. With the departure of entity 3, we follow the FCFS queue discipline and advance customer 4 to the teller service station while customer 5 remains in queue. As occasioned by any departure, the stats symbolizing queue, service, and flow times are updated. The length of flow time for entity 2 has now been trumped by the length of flow time for entity 3 and, thus, the max flow time adjusts to 4.51 (i.e., 1.69 vs. 4.51) and the sum of flow times jumps to 6.78 (2.27 + 4.51 = 6.78). With the advancement of entity 4 from queue to service, the max delay, experienced by customer 4, is now registered at 2.16 min (8.51 − 6.35 = 2.16) rather than the earlier delay of 0.50 min experienced by customer 3. Thus, the sum of delays increases to 2.66

**Table 3.7** Event 5

| Current Event | | | Event Number = | 5 |
|---|---|---|---|---|
| Time = | 4.50 | Entity = 2 | Event Type = | Departure |

**State Variables**

$Q(t) = 0 \qquad B(t) = 1$

**Entity Attributes**

| Entity | Arrival Time at Queue | | Arrival Time at Server | | Departure Time |
|---|---|---|---|---|---|
| 1 | 0.00 | | 0.00 | | 2.81 |
| 2 | 2.81 | | 2.81 | | 4.50 |
| 3 | 4.00 | | 4.50 | | |
| 4 | | | | | |
| 5 | | | | | |
| 6 | | | | | |

**Statistical Accumulators**

| # Served = | 2 | # Processed in queue = | 3 | Sum of delays = | 0.50 |
|---|---|---|---|---|---|
| Max delay = | 0.50 | Sum of flow times = | 2.27 | Max flow time = | 1.69 |
| Integral $Q(t)$ = | 0.50 | Max $Q(t)$ = | 1 | Integral $B(t)$ = | 2.27 |

**Next-Event List**

| Number | Time | Entity | Event Type |
|---|---|---|---|
| 1 | 6.35 | 4 | Arrival |
| 2 | 7.94 | 5 | Arrival |
| 3 | 20.00 | — | Stop |
| 4 | | | |

**Table 3.8** Event 6

**Current Event**                                          **Event Number =** 6
Time = 6.35    Entity = 4                                  Event Type = Arrival

**State Variables**
$Q(t) = 1$     $B(t) = 1$

**Entity Attributes**

| Entity | Arrival Time at Queue | Arrival Time at Server | Departure Time |
|---|---|---|---|
| 1 | 0.00 | 0.00 | 2.81 |
| 2 | 2.81 | 2.81 | 4.50 |
| 3 | 4.00 | 4.50 | |
| 4 | 6.35 | | |
| 5 | | | |
| 6 | | | |

**Statistical Accumulators**

| # Served = | 2 | # Processed in queue = | 3 | Sum of delays = | 0.50 |
| Max delay = | 0.50 | Sum of flow times = | 2.27 | Max flow time = | 1.69 |
| Integral $Q(t)$ = | 0.50 | Max $Q(t)$ = | 1 | Integral $B(t)$ = | 4.12 |

**Next-Event List**

| Number | Time | Entity | Event Type |
|---|---|---|---|
| 1 | 7.94 | 5 | Arrival |
| 2 | 8.51 | 3 | Departure |
| 3 | 20.00 | — | Stop |
| 4 | | | |

**Table 3.9** Event 7

**Current Event**                                 **Event Number =** 7

| Time = | 7.94 | Entity = | 5 | Event Type = | Arrival |
|---|---|---|---|---|---|

**State Variables**

$Q(t) = 2$      $B(t) = 1$

**Entity Attributes**

| Entity | Arrival Time at Queue | | Arrival Time at Server | | Departure Time |
|---|---|---|---|---|---|
| 1 | 0.00 | | 0.00 | | 2.81 |
| 2 | 2.81 | | 2.81 | | 4.50 |
| 3 | 4.00 | | 4.50 | | |
| 4 | 6.35 | | | | |
| 5 | 7.94 | | | | |
| 6 | | | | | |

**Statistical Accumulators**

| # Served = | 2 | # Processed in queue = | 3 | Sum of delays = | 0.50 |
|---|---|---|---|---|---|
| Max delay = | 0.50 | Sum of flow times = | 2.27 | Max flow time = | 1.69 |
| Integral $Q(t)$ = | 2.09 | Max $Q(t)$ = | 2 | Integral $B(t)$ = | 5.71 |

**Next-Event List**

| Number | Time | Entity | Event Type |
|---|---|---|---|
| 1 | 8.51 | 3 | Departure |
| 2 | 8.87 | 6 | Arrival |
| 3 | 20.00 | — | Stop |
| 4 | | | |

**Table 3.10  Event 8**

| Current Event | | | | Event Number = | 8 |
|---|---|---|---|---|---|
| Time = | 8.51 | Entity = | 3 | Event Type = | Departure |

**State Variables**

$Q(t) = $ 1  $B(t) = $ 1

**Entity Attributes**

| Entity | Arrival Time at Queue | | Arrival Time at Server | | Departure Time |
|---|---|---|---|---|---|
| 1 | 0.00 | | 0.00 | | 2.81 |
| 2 | 2.81 | | 2.81 | | 4.50 |
| 3 | 4.00 | | 4.50 | | 8.51 |
| 4 | 6.35 | | 8.51 | | |
| 5 | 7.94 | | | | |
| 6 | | | | | |

**Statistical Accumulators**

| # Served = | 3 | # Processed in queue = | 4 | Sum of delays = | 2.66 |
|---|---|---|---|---|---|
| Max delay = | 2.16 | Sum of flow times = | 6.78 | Max flow time = | 4.51 |
| Integral $Q(t)$ = | 3.23 | Max $Q(t)$ = | 2 | Integral $B(t)$ = | 6.28 |

**Next-Event List**

| Number | Time | Entity | Event Type |
|---|---|---|---|
| 1 | 8.87 | 6 | Arrival |
| 2 | 10.64 | 4 | Departure |
| 3 | 20.00 | — | Stop |
| 4 | | | |

(0.50 + 2.16 = 2.66) and the integral $\int Q(t)$ changes to 2.09 + 2 × (8.51 − 7.94) = 3.23, while the integral $\int B(t)$ changes to 5.71 + 1 × (8.51 − 7.94) = 6.28. The next-event list informs us that entity 6 is expected to arrive at 8.87.

*Event 9: Arrival of Entity 6. Time = 8.87.* This table (Table 3.11) illustrates the arrival of the sixth customer, who immediately stands in queue behind customer 5. As would be expected, without a departure, the only stats within the statistical accumulators to change are the time-resistant averages for waiting and service. Thus, the integral $\int Q(t)$ changes to 3.59 and the integral $\int B(t)$ changes to 6.64. Among the state variables, the teller remains busy (i.e., $B(t) = 1$) and the queue increases by one (i.e., $Q(t) = 2$). The next expected event, according to the next-event list, is the departure of entity 4 at 10.64.

*Event 10: Departure of Entity 4. Time = 10.64.* This event shows the departure of customer 4, the advancement of customer 5 to the teller, and the continued waiting for service by customer 6 (Table 3.12). The max delay, experienced by entity 5, is recorded at 10.64 − 7.94 = 2.70. The sum of delays will be (2.66 + 2.70) = 5.36. The integral $\int Q(t)$ is increased to 3.59 + 2 × (10.64 − 8.87) = 7.13, while integral $\int B(t) = 6.64 + 1 \times (10.64 - 8.87) = 8.41$. The max flow time remains the same, but the sum of flow times elevates to 6.78 + 4.29 = 11.07. The departure of entity 5 is the next most proximate event.

*Event 11: Departure of Entity 5. Time = 14.45.* Table 3.13 shows the updating of figures for event 11 denoted by the departure of customer 5. With this departure, the queue empties as customer 6 is advanced to the teller and the state variable $Q(t)$ is reduced to zero. The max delay is increased to 14.45 − 8.87 = 5.58 as entity 6, relative to the others to date, has now experienced the longest delay. Thus, the sum of the delays will be (5.36 + 5.58) = 10.94. The integral $\int Q(t)$ is advanced to 7.13 + 1 × (14.45 − 10.64) = 10.94. While the max flow time remains unchanged, the integral $\int B(t)$ increases to 8.41 + 1 × (14.45 − 10.64) = 12.22 and the sum of flow times increases to 11.07 + 6.51 = 17.58. The next event is the termination of the simulation.

*Event 12: Stop. Time = 20.* Customer 6 is being serviced by the teller (i.e., $B(t) = 1$) and no other customers are in the bank (Table 3.14). Customer 6 began interacting with the teller at 14.45, and the expected service time is 6.08 min. This knowledge informs us that entity 6 will exit the system at (14.45 + 6.08) = 20.53. Since the simulation is initialized to terminate at 20.00, the departure event will not take place within the simulation window. However, the time-persistent statistics are updated through the end of the simulation at 20.00 min. Thus, since no customer are in queue, $\int Q(t)$ will remain the same (10.94), while $\int B(t)$ is increased to 12.22 + 1 × (20.00 − 14.45) = 17.77.

*System Performance Summary.* The final values of the output performance measures include:

(1) number of customer served = 5
(2) maximum delay experienced = 5.58
(3) maximum flow time = 6.51
(4) maximum number of customers in queue = 2

**Table 3.11** Event 9

| Current Event | | | | Event Number = | 9 |
|---|---|---|---|---|---|
| Time = | 8.87 | Entity = | 6 | Event Type = | Arrival |

**State Variables**

$Q(t) = $ 2  $B(t) = $ 1

**Entity Attributes**

| Entity | Arrival Time at Queue | Arrival Time at Server | Departure Time |
|---|---|---|---|
| 1 | 0.00 | 0.00 | 2.81 |
| 2 | 2.81 | 2.81 | 4.50 |
| 3 | 4.00 | 4.50 | 8.51 |
| 4 | 6.35 | 8.51 | |
| 5 | 7.94 | | |
| 6 | 8.87 | | |

**Statistical Accumulators**

| # Served = | 3 | # Processed in queue = | 4 | Sum of delays = | 2.66 |
|---|---|---|---|---|---|
| Max delay = | 2.16 | Sum of flow times = | 6.78 | Max flow time = | 4.51 |
| Integral $Q(t) = $ | 3.59 | Max $Q(t) = $ | 2 | Integral $B(t) = $ | 6.64 |

**Next-Event List**

| Number | Time | Entity | Event Type |
|---|---|---|---|
| 1 | 10.64 | 4 | Departure |
| 2 | 14.45 | 5 | Departure |
| 3 | 20.00 | – | Stop |
| 4 | | | |

**Table 3.12  Event 10**

**Current Event**

| Time = | 10.64 | Entity = | 4 | **Event Number =** | **10** |
|---|---|---|---|---|---|
| | | | | Event Type = | Departure |

**State Variables**

$Q(t) = $ 1    $B(t) = $ 1

**Entity Attributes**

| Entity | Arrival Time at Queue | | Arrival Time at Server | | Departure Time |
|---|---|---|---|---|---|
| 1 | 0.00 | | 0.00 | | 2.81 |
| 2 | 2.81 | | 2.81 | | 4.50 |
| 3 | 4.00 | | 4.50 | | 8.51 |
| 4 | 6.35 | | 8.51 | | 10.64 |
| 5 | 7.94 | | 10.64 | | |
| 6 | 8.87 | | | | |

**Statistical Accumulators**

| # Served = | 4 | # Processed in queue = | 5 | Sum of delays = | 5.36 |
|---|---|---|---|---|---|
| Max delay = | 2.70 | Sum of flow times = | 11.07 | Max flow time = | 4.51 |
| Integral $Q(t)$ = | 7.13 | Max $Q(t)$ = | 2 | Integral $B(t)$ = | 8.41 |

**Next-Event List**

| Number | Time | Entity | Event Type |
|---|---|---|---|
| 1 | 14.45 | 5 | Departure |
| 2 | 20.00 | – | Stop |
| 3 | 20.53 | 6 | Departure |
| 4 | | | |

Table 3.13  Event 11

## Current Event
Time = 14.45   Entity = 5   Event Number = 11
            Event Type = Departure

## State Variables
$Q(t) = 0$   $B(t) = 1$

## Entity Attributes

| Entity | Arrival Time at Queue | Arrival Time at Server | Departure Time |
|---|---|---|---|
| 1 | 0.00 | 0.00 | 2.81 |
| 2 | 2.81 | 2.81 | 4.50 |
| 3 | 4.00 | 4.50 | 8.51 |
| 4 | 6.35 | 8.51 | 10.64 |
| 5 | 7.94 | 10.64 | 14.45 |
| 6 | 8.87 | 14.45 | |

## Statistical Accumulators

| | | |
|---|---|---|
| # Served = 5 | # Processed in queue = 6 | Sum of delays = 10.94 |
| Max delay = 5.58 | Sum of flow times = 17.58 | Max flow time = 6.51 |
| Integral $Q(t)$ = 10.94 | Max $Q(t)$ = 2 | Integral $B(t)$ = 12.22 |

## Next-Event List

| Number | Time | Entity | Event Type |
|---|---|---|---|
| 1 | 20.00 | – | Stop |
| 2 | 20.53 | 6 | Departure |
| 3 | | | |
| 4 | | | |

**Table 3.14  Event 12**

**Current Event**
Time = 20.00                 Entity = —               **Event Number = 12**
                                                       Event Type = Stop

**State Variables**
$Q(t) = 0$                   $B(t) = 1$

**Entity Attributes**

| Entity | Arrival Time at Queue | Arrival Time at Server | Departure Time |
|---|---|---|---|
| 1 | 0.00 | 0.00 | 2.81 |
| 2 | 2.81 | 2.81 | 4.50 |
| 3 | 4.00 | 4.50 | 8.51 |
| 4 | 6.35 | 8.51 | 10.64 |
| 5 | 7.94 | 10.64 | 14.45 |
| 6 | 8.87 | 14.45 | |

**Statistical Accumulators**

| | | | |
|---|---|---|---|
| # Served = | 5 | # Processed in queue = | 6 |
| Max delay = | 5.58 | Sum of flow times = | 17.58 |
| Integral $Q(t)$ = | 10.94 | Max $Q(t)$ = | 2 |
| | | Sum of delays = | 10.94 |
| | | Max flow time = | 6.51 |
| | | Integral $B(t)$ = | 17.77 |

**Next-Event List**

| Number | Time | Entity | Event type |
|---|---|---|---|
| 1 | 20.53 | 6 | Departure |
| 2 | | | |
| 3 | | | |
| 4 | | | |

(5) the average delay in queue is sum of delays/# processed in queue, or 10.94/6 = 1.82 min/customer
(6) the average flow time through the system is sum of flow times/# served, or 17.58/5 = 3.52 min/customer
(7) the time-average number of customers waiting in the queue is $\int B(t)/t$ = 10.94/20.00 = 0.55 customers
(8) the resource utilization is $\int B(t)/t$ = 17.77/20.00 = 0.89.

## ARENA SIMULATION

*Arena simulation software* employs the SIMAN simulation language and is used to construct experimental models that mimic the behavior of a real-world system. The approach of the software is both visual and intuitive; users can choose from a range of shapes representing distinct stages in the flow of entities through a system (e.g., create, process, and dispose), place these objects on an empty field, specify parameters relating to the arrival and service times for entities, and simulate the system. Arena also generates and compiles statistical data that describe the performance of the system. In addition, it can support a range of technology from Microsoft VBA to general-purpose languages such as C++. As a result, users have the flexibility of building a diverse set of simulation models. What follows is a brief introduction to the Arena software [3].

### Arena Components—Terminology

***Entities, Attributes, Variables, Resources, and Queues*** In the Arena simulation environment, the term entity may refer to an object, such as a mattress under assembly, a person, such as a customer engaged in the process of banking, or an intangible. In many cases, to assist modeling operations, it is necessary to create entities that represent intangibles. For example, in a flow shop, an entity that represents a cycle time may be released periodically to move semifinished products from one workstation to another. Thus, in the most basic model, entities are created, enter the system, and, following service, depart the system, while in other models, an entity may remain in the system, as is the case with a cycle-time entity, until the simulation terminates. Intuitively, entities condition performance measures.

There are many ways to classify entities. For example, customers may be classified according to the concept gender, while a factory's mattress product may be classified according to the concept mattress type. Associated with each concept is a set of attributes. An attribute is a common characteristic shared by one or more entities. Gender can be assigned the two attributes, male and female, while the concept mattress type may be assigned the attributes king, queen, and single. Using the real-world system as a referent, the modeler is responsible for identifying the relevant concepts and associated attributes meant to characterize the entities.

A variable is meant to model features of the real-world system. The process of exiting a bank, for example, may be modeled by a variable labeled, "leaving the bank time." This variable may represent the time interval between the moment the customer completes the transaction with the teller and the moment he or she exits the bank. In addition, variables can also correspond to a system condition that is modified when running the simulation. Again, looking to the bank as an example, for security purposes, a bank may choose to limit the number of customers entering the building at any particular time. A modeler can create a variable called "security limit" that establishes a maximal number of combined customers, waiting and in service, allowed within the bank. This variable can increase incrementally to reflect those customers currently in a queue and being served; once the ceiling number is reached, the system bars entry into the bank of any additional entities. As the number of entities in queue draws down and reaches a more tolerable condition, the "security limit" variable can be reset to again allow entry into the bank of customers seeking service.

Resources represent servers and are usually constrained in number. Entities compete for, or are routed to, resources. Resources can be machinery, workstations, space, or people that provide the service. Once an entity is served, the resource is released and available to be accessed by another entity. The teller is a resource that provides banking services to customers and the quilting machine is a resource that stitches fabric covering on mattresses. A system can have multiple resources. For example, the bank can have several tellers in addition to a credit specialist, while a mattress factory may have both a quilting station and a label-tagging station.

There are many system environments where either a person is necessitated to "mark time" or a product is placed in a "holding pattern" prior to interface with a resource. Queues are the proverbial line or, in the case of a product, the staging area where the entity awaits interaction. Within any system, a modeler may construct several queues and assign to each a unique property and name. A modeled bank may have two teller queues, a single credit queue, and an overcapacity queue. Since there are only four chairs in the waiting area outside the credit manager's office, the credit queue may be assigned the unique finite capacity of four, the larger lobby area can manage a finite capacity of eight customers in each teller queue, and the overcapacity queue, located just outside the bank's front entrance, is designated an infinite queue.

**Arena Statistics and Hierarchical Structure** Arena produces two types of essential statistics: time-dependent and observational. Time-dependent statistics report system status as a function of time, such as the number of entities in a queue at a discrete time. Observational statistics, called tallies, are initialized at zero and report statistics that accumulate with each new event to describe entity activity to the most proximate moment. Observational statistics include, among others, the number of customers served, the combined waiting time in queue, the number of customers that have passed through the queue, and the longest time spent in the system among all departed customers.

The modeling flexibility of Arena is rooted in its hierarchical structure, a structure found in high-level simulators. At the base of the structure is programming code represented through a simulation language such as the general-purpose language C++. This code is used to construct modules such as the block/element templates and advanced-process/transfer templates. Since the templates are interchangeable, there is flexibility in the construction of simulation models that vary according to the desired level of fidelity and the goal. The following presents the hierarchical structure of Arena in descending order:

(1) user-developed templates
(2) application–solution templates
(3) basic process templates
(4) advanced-process/transfer templates
(5) block/element templates
(6) code (e.g., VB, C, and C++)

The versatility of Arena allows the modeler to combine a low-level modeling module, such as a block/element template, with a high-level template, such as an application–solution template. In the situation where a modeler requires specialized algorithms, he or she can code such a piece by employing a procedural language, thus creating a module that may be saved under either a new or common template.

## User Interface

***Menu Items, Toolbars, Model Flowchart, and Spreadsheet Window*** As with other MS Windows applications, Arena includes the basic Microsoft concepts and operations such as disks, files, and folders as well as employs the mouse and the key word. The sequence of accessing the different files and levels within Arena typically includes Menu > Choose > Submenu > Tab labels. One can cut-and-paste either by employing the regular menu options or by "alt + shift."

Once the Arena program is opened, Figure 3.6 illustrates the typical screen view, called the Arena Window. With Arena launched, one can see File, View, Tools, and Help menus at the top. If a blank model is already opened, then there are other menus. The most popular mechanism to create a new model involves File > New. To open an existing saved model, one clicks File > Open, which will display a dialog box in which one navigates through folders and subfolders until the desired file has been accessed. In the event that one is required to have more than one model at a time, Arena displays a separate window for each model. To save an existing model, click File > Save, or to save a new model, click File > Save As. To end an Arena session, one simply clicks "x" at the upper-right corner of the Arena window or File > Exit.

The Arena Window is subdivided into several parts. The Model Window is divided between the flowchart view and the spreadsheet view. The flowchart

**90** DISCRETE-EVENT SIMULATION

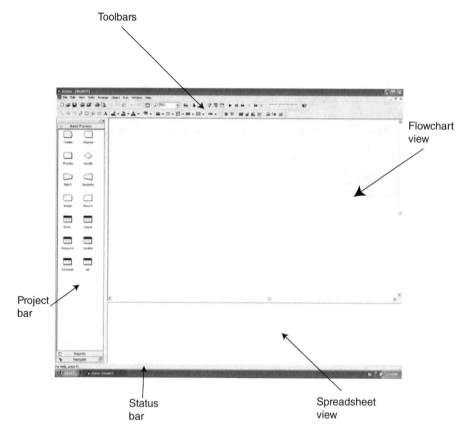

**Figure 3.6** Arena screen overview.

view involves the process flowchart, animation, and other drawing elements. Here, many parameters can be accessed, viewed, and modified. The spreadsheet view provides access to most of the model parameters and can be edited. When working with larger models, it is desirable to access at once a category of parameters instead of opening element-by-element in the flowchart view.

The project bar is located to the left of the Arena Window. Attached to this bar are panels that contain a variety of simulation modeling objects. For example, the panel called "Basic Process" displays some essential objects, or modules, that are common to most simulations models, such as Create and Assign. Notice the distinct shape of each object. In general, the lowest horizontal button contains the panel "Reports." In this panel, simulation outputs can be accessed. In addition, the panel called "Navigate" allows one to move through different existing submodels that compose a larger model. This project bar can be floated anywhere or can be docked to the right edge. One can hide/unhide the project bar, if required, by clicking on the small "x" button or by clicking on View > Project Bar. The number of panels depends upon the features licensed at time of software purchase. Some panels, such as the Basic Process, are common. Others, however, such as the Call Center panel, are more specialized. One can attach templates that contain different modeling elements by way of File > Template Panel > Attach. In the case that one requires the removal of a panel, File > Template Panel > Detach may be employed.

The status bar is usually located at the bottom of the Arena Window. It displays a variety of information related to the current condition of the simulation model. For example, if the model is not being run, it shows the "x" and "y" coordinates when the pointer is positioned over the flowchart view. In the case that the model is being replicated, the status bar exhibits the simulation clock, the replication number, and the number of replications left to end the run. This bar can be hidden or unhidden by selecting View > Status Bar. To show the status bar again, one checks the View > Status Bar options.

**Arena Modules** Modules describe the process to be simulated and are the primary building blocks used in the modeling process. They are categorized as flowchart and data modules. A general description of each type of module follows.

An Arena flowchart within the Model Window will contain objects, visually represented by various shapes, through which entities flow. Within a simple system, typically, entities enter the system, are serviced, and exit the system. Within a flowchart, this sequential process is represented by placing within the flowchart view the modules "Create" representing the point of entity entry into the system, "Process" representing the point of treatment or service of the entity within the system, and "Dispose" representing the entity's point of exit from the system. Each of these modules may be placed into the flowchart view by dragging the chosen module from the panel and dropping it into the flowchart view. Generally, once dropped into the View, objects are automatically tethered to one another with a connecting line.

In addition to the Create, Process, and Dispose modules, there are many more flowchart modules on other panels; they can have the shape of the operation they perform, but some are simple rectangles without any flowchart association. In some cases, a flowchart module within a panel can pose a distinct color that differentiates it from the other modules; some may have more elaborated graphics that distinctly describe their purpose or operation. To parameterize, edit, or access the properties of a module that has been placed in the model within the flowchart view, one double-clicks revealing a dialog box where such operations can be performed. Alternatively, one can see and access the parameters of a module by clicking only once the flowchart module. This shows the spreadsheet view and the entries related to that module. Then, one can edit such entries.

In addition to the flowchart modules, the other central building block is the data module. A data module is used to describe the features of a range of process elements. These process elements can involve entities, resources, and queues. Data modules can define other variables, values, expressions, and conditions that are related to the modeled system. In the project bar, data modules are symbolized as spreadsheets. For example, in the basic process panel, two such spreadsheet icons are Resource and Variables. Unlike flowchart modules, a data module cannot be placed into the flowchart view. However, data modules can be edited by simply clicking on the spreadsheet icon; the data module will then become visible in the spreadsheet view; once in View, existing rows may be modified or new rows added.

As noted by Kelton et al., data and flowchart modules are connected to one another by the names of each shared structure such as resources, variables, and entity types. When one is defining or editing a data or flowchart module, Arena presents such related names in a drop-down list menu [3]. This convenient list minimizes the potential for assigning duplicate names.

### Simulation of Simple Queuing System Using Arena

***Problem Description: Analyzing an Airline Check-in Process*** In this portion of the chapter, the reader will be introduced to a typical flow scenario found in an airline check-in system. The flow of this check-in process will be modeled and performance measures analyzed using Arena. We present the following foundational information relating to the customers and servers:

(1) The simulation will take place over a period of 16 real-world hours.
(2) Travelers arrive at the main entrance door of the airline terminal according to a Weibull distribution with *Beta* = 4.5 and *Alpha* = 10.
(3) At the check-in, travelers wait in a single queue.
(4) The service time follows a triangular distribution with the values 1.5 for minimum, 3.5 for most likely, and 7 for maximum; note that the triangular distribution is often employed when a more precise distribution is unknown, and there are, however, informed or educated estimates for the minimum, maximum, and most likely values.

We now turn to modeling and simulating the process by presenting the step-by-step mechanics found in Arena. There are three basic steps: first, define the types of entity that will flow through the system; second, using the point of view of the entity, define basic information such as the point of entry into the system, the entity transformation at each step, and the resources that will be used to complete the service; third, run the simulation and report the performance measures.

Entities represent the customers that flow through the check-in system. These entities are created using the "Create" module relying on the timing information provided. The Create module is located in the Basic Process panel. This module represents the origin for entities that will flow through the model. As described above, to place this module, one drags the Create module into the Model Window as illustrated in Figure 3.7.

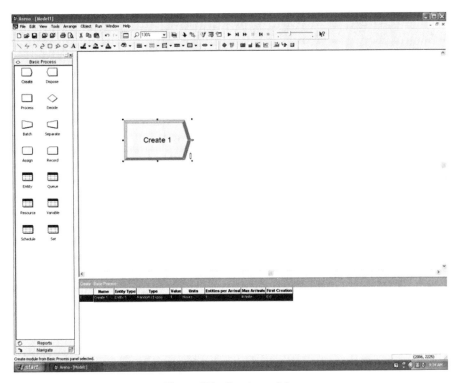

**Figure 3.7** Create module.

Next, model the check-in service process by placing a Process module. To do this, first ensure that the Create module has already been placed within the flowchart view and has been selected. When dragging the Process module, a connector will link automatically both the selected Create and the newly dragged Process modules. In the case that this connection does not occur, one can either select Object > Connect or select the Connect button from the toolbar and click between the exit point of the Create module to the entry point of the Process module. Figure 3.8 presents the connected modules.

The flow of the entity through the system finishes with a Dispose module. This module is necessary to remove entities after processing. To place this flowchart module into the model, one first selects the Process module in the flowchart view such that the Dispose module will be connected. Drag a Dispose module from the Basic Process panel and place it next to the Process module. Again, if it is not connected, follow the above steps to connect objects.

Now that the basic flowchart for the airline check-in process has been constructed, one can define the data associated with the modules, including the name of the module and the information that will be used in the simulation.

**94**  DISCRETE-EVENT SIMULATION

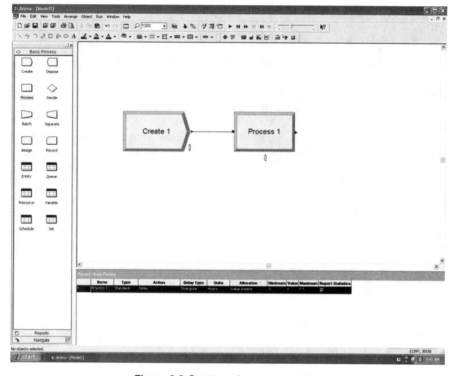

**Figure 3.8** Create and process module.

*Customer Arrival for Check-In (Create Module)*  First, one can assign a meaningful name to the Create module, such as "Customers Arrival" to symbolize the process of customers entering the system. In this module, it is necessary to define how people arrive. Since this has been defined as a random activity with a Weibull distribution (Beta = 4.5 and Alpha = 10), one characterizes accordingly.

To open the property dialog box of the Create module, double-click the Create module, then in the Name field, type "Customer Arrival." Next, for the Entity Type, type in "Customer." In the field type, select "Expression." In the field "Expression," select WEIB (Beta, Alpha) and substitute Beta and Alpha with 4.5 and 10, respectively. Change the units to Minutes. Then click "OK" to close the dialog box.

*Check-In Process (Process Module)*  After entities that represent the customer have been created, the customers will undertake a process to be identified, bags and suitcases checked in, and boarding passes issued. Previously, it was defined that the check-in process could be described using a triangular distribution. Since the check-in process per customer takes some time, the Process module will hold the customer entity for a delay while involving a resource to execute the operation. Thus, during a replication, every time a customer enters for service, a sample is drawn from the probability distribution that represents service time. A resource called "Airline Agent" will perform the check-in operation.

The mechanics of this process is straightforward. First, double-click to open the property dialog box of the Process module. Then, type "Check-in Process" in the Name field. To create a resource to perform this operation, select "Seize Delay Release" from the action list. Thus, after a customer entity has waited for the availability of the resource, the customer entity seizes the resource and a delay for the process time is experienced. Finally, the entity releases the resource to perform another operation. Click "Add" to add a resource to perform this check-in operation. Type "Airline Agent" in the field called Resource Name within the dialog box and click OK to close the window.

To parameterize the experienced delay, first verify that the distribution is triangular in Delay Type, select Minutes in Units, and type "1.5" for minimum, "3.5" value (most likely), and "7" for maximum. Finally, close the dialog box by selecting OK.

*Customer Leave (Dispose Module)* Once customers are served, the customer leaves the check-in system. The process is ended with a Dispose module. The Dispose module counts the number of customers that leave the system. To rename the Dispose module, double-click to open the property dialog box, and, then, type "Customer Leave" and click OK to close the window.

*Airline Agent (Resource Module)* In addition to the flowchart modules used to model the above process, we can also define additional parameters using data modules. In this case, one can define the cost rate for the airline agent such that the output includes this cost as part of the report. The Airport Agent cost is constant at $20 per hour. To accomplish this, one clicks the Resource module from the Basic Process panel, which displays the Resource spreadsheet. The resource, Airline Agent, which was defined (added) when setting the Process module, now appears in the first row. Assign $20 per hour when the resource is either busy or idle by clicking and typing 20 on the Busy/Hour and Idle/Hour cells.

**Executing the Simulation** To simulate the model, some general project information as well as the duration of the replications must be defined. As mentioned above, the investigation requires a simulation window of 16 h. To accomplish this, open the Project Parameters window by Run > Setup. Click on the Project Parameters tab and type "Check-in Process Analysis" in the Project Title field. The Statistics Collection box is kept checked and, additionally, the Costing box is checked. Next, click on the "Replication Parameters" tab and type "16" in the Replication Length field. Keep Hours in the Time Units field. Finally, to close the dialog box, click OK.

To start the simulation, click Run > Go. The software will verify the integrity of the model and start the simulation. When the simulation starts, we may see the entities' pictures moving through the flowchart; counters adjust values while the simulation creates, process, and removes entities. Additionally, one may adjust the animation scale factor either by clicking Run > Speed > Animation Speed Factor or by moving back or forward the run speed button. The simulation can be paused or stopped at any time by clicking on the Pause or End buttons. In addition, one may walk through the simulation one event at a time; at each step, an entity moves in the flowchart. One can bypass the animation feature and obtain faster reports by clicking the fast-forward button. Once a run ends, the simulation software asks the user if he or she would like to see the report.

***Measures of Performance*** Once it is verified by Arena that the user would like to see the report, it displays the Category Overview Report. This report provides a summary of the results across all replications in terms of common measures of performance. A map that categorizes the types of existing information in the report appears at the leftmost hand. The project name is listed on the top. Entries for each type of data, as well as detailed reports for each replication, are listed in the report map. Figure 3.9 illustrates the Category Overview Report.

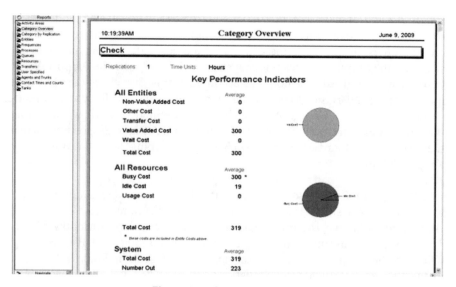

**Figure 3.9** Category overview.

## CONCLUSION

This chapter has introduced the reader to DES as an approach to answering efficiency-related questions, especially those inquiries involving the flow of entities through queues and servers. We began with a conceptual overview that defined queues and queuing and illustrated the relevance of these to customer service and operations. In this overview, a common general approach—explicated throughout this chapter—is put forth. The first part of this approach is the development of a logical representation, or model, of the real-world queuing environment with special emphasis on the importance of quality foundational information. The second part of this approach is a process where the model is executed, or simulated, to allow it to play out over time revealing the dynamic behavior of the system. The simulation is first initialized, followed next by the introduction of entities into the system, either probabilistically or deterministically, and then data are accumulated. Next, the simulation is terminated and the resultant performance measures evaluated. If the performance measures generated by this run do not fall within established parameters, then it may be necessary to make adjustments to the model and replicate the simulation process.

The next part of this chapter offers a more detailed look at simulation methodology. A critical distinction is made between a discrete system, where the state variables are updated instantaneously at specific times, and a continuous system, where the state variables change continuously with the progression of time. Some common terms, or main components, associated with DES are then introduced. In addition, a detailed discussion of the two most common time-advance mechanisms is provided; the advantages of the next-event time-advance mechanism, relative to the fixed-increment mechanism, are explicated. This section concludes with an illustration and description of a typical simulation flowchart.

A DES example is proffered in the next portion of the chapter. Here we render understandable how to simulate a basic model by introducing the example of a one-teller banking facility where customers arrive, are processed by the single teller, and then leave. The concepts such as FCFS queue discipline and IID variates are presented in the context of the example system. Also presented are the mathematical formulas for four measures of performance that are of particular interest in queuing systems.

Next, we offer practical pedagogy through the hand simulation of the banking system. This section is designed to be both informative and instructive, addressing some of the most common questions and pitfalls associated with an understanding of DES. We first present a simulation template that contains most of the common statistical accumulators useful to assessing a system's performance. The five basic sections within the template are explained. The template is followed by a series of logically sequenced tables and a detailed narrative that carefully walks the reader through the simulation process, event-by-event. Changes in state variables, attributes, accumulators, and the event calendar are tracked and explained both conceptually and operationally.

Building upon the knowledge gained through the practical execution of the banking system model, we then offer the reader a more sophisticated—and powerful—tool to model and simulate a queuing system. We introduce the reader to Arena simulation software capable of constructing experimental models. We familiarize the reader with the building blocks necessary to construct and execute models, referred to as Arena components. We then carefully illustrate the mechanics of modeling and simulating using Arena by analyzing the performance of an airline check-in system. Through the use of modules, we begin by defining and creating entities and modeling the check-in service process. We then run the simulation and generate reports.

We have introduced, and made reference to, a few of the many applications of modeling and simulation including transportation, operations management, medical, engineering, and social sciences. The approach offered here is but one of several approaches that may be employed to gain better insight into the complexities inherent in a queuing system. Clearly, DES, when properly parameterized and girded theoretically, can yield information useful to decision making.

## REFERENCES

[1] Law AM, Kelton WD. *Simulation, Modeling, and Analysis*. 4th ed. New York: McGraw-Hill; 2006.

[2] Banks J, Carson JS, Nelson BL. *Discrete-Event System Simulation*. 2nd ed. Upper Saddle River, NJ: Prentice Hall; 1996.

[3] Kelton WD, Sadowski RP, Sturrock DT. *Simulation with Arena*. 4th ed. New York: McGraw-Hill; 2004.

# 4

# MODELING CONTINUOUS SYSTEMS

Wesley N. Colley

The very title of this chapter may seem oxymoronic. Digital computers cannot, in general, represent continuous (real) numbers, so how can we possibly use computers to simulate continuous systems? Mathematicians define degree of continuity by the ability to take a derivative, but when we look at Leibniz's formula (Eq. (4.1)) for computing the derivative,

$$\frac{dx}{dt} = \lim_{\Delta t \to 0} \frac{x(t + \Delta t) - x(t)}{\Delta t}, \tag{4.1}$$

we discover something very troubling indeed—a first-class numerical "no-no." We are asked to subtract two very similar numbers in the numerator and divide that result by a tiny number. This experiment is custom-made to exploit the rounding errors of digital computation.

Imagine we are simply taking the derivative of $t^2$ at $t = 1$. The limit demands that we use ever smaller values of $\Delta t$ to get better and better estimates of the derivative. Table 4.1 shows what we might find if we were to carry out this exercise in Microsoft Excel. As we begin to decrease $\Delta t$, the error decreases nicely, until $\Delta t$ is around $10^{-8}$, but then the error gets worse again as we decrease $\Delta t$ to a level where the error introduced by round-off limit of the

---

*Modeling and Simulation Fundamentals: Theoretical Underpinnings and Practical Domains*,
Edited by John A. Sokolowski and Catherine M. Banks
Copyright © 2010 John Wiley & Sons, Inc.

**Table 4.1 Estimation of $dx/dt$ for $x = t^2$ in Microsoft Excel for different values of $\Delta t$**

| $\Delta t$ | $t$ | $t + \Delta t$ | $t{\wedge}2$ | $(t + \Delta t){\wedge}2$ | Approximate $dx/dt$ | Error |
|---|---|---|---|---|---|---|
| 1 | 1 | 2 | 1 | 4 | 3 | 0.5 |
| 0.01 | 1 | 1.01 | 1 | 1.0201 | 2.01 | 0.005 |
| 0.0001 | 1 | 1.0001 | 1 | 1.0002 | 2.0001 | 5E-05 |
| 0.000001 | 1 | 1.000001 | 1 | 1.000002 | 2.000001 | 5E-07 |
| 1E-08 | 1 | 1 | 1 | 1 | 1.999999988 | −6.1E-09 |
| 1E-10 | 1 | 1 | 1 | 1 | 2.000000165 | 8.27E-08 |
| 1E-12 | 1 | 1 | 1 | 1 | 2.000177801 | 8.89E-05 |
| 1E-14 | 1 | 1 | 1 | 1 | 1.998401444 | −0.0008 |
| 1E-16 | 1 | 1 | 1 | 1 | 0 | −1 |

machine dominates over the mathematical error itself. So, if one wants to estimate the derivative natively numerically on a typical PC, the best precision possible is about a few parts per billion. For some areas of numerical computation, this round-off limit can be a fundamental limitation to the stability of the code and the quality of its output. (Curiously, in some cases, the errors introduced can actually *add* stability, e.g., Jain et al. [1].)

*Round-off* is but one example of the broader problem of discretization. Just as the computer discretizes the real numbers themselves, the computational techniques for handling continuous systems inherently involve discretization into chunks much larger than the machine round-off limit. As such, managing the errors associated with discretization quickly becomes the primary concern when simulating continuous systems.

At their worst, discretization errors can lead to extreme instability in coding—finite differencing codes for partial differential equations may yield acceptable results at one cell size, but complete nonsense at just a slightly different cell size. In this chapter, we focus on a class of problems that is rather tame by contrast—integration of ordinary differential equations. Here, the only derivatives in the problem are taken with respect to the single independent variable (in our case, time), and those derivatives are known analytically. Use of the analytic derivative removes the necessity of estimating derivatives numerically and nicely isolates the discretization solely to the time step.

## SYSTEM CLASS

We now consider some classes of continuous systems, according to the mathematical methods, and therefore the numerical techniques, associated with modeling their members. Table 4.2 presents a few common examples. Very often, simulation or evaluation of continuous physical systems boils down to integration of a first- or second-order differential equation (ordinary or partial). In fact, a very large fraction of continuous systems are governed

Table 4.2 Mathematical classes of some continuous systems

| Name | Some Uses | Formula |
|---|---|---|
| First-order ordinary | Predator–prey models (others) | $\dot{x} = xf(x,y)$ <br> $\dot{y} = yg(x,y)$ |
| Second-order ordinary | Orbits <br> Oscillators <br> Ballistics <br> (many, many more) | $\mathbf{F} = m\mathbf{a}$ |
| | RLC circuits | $L\ddot{q} + R\dot{q} + \dfrac{1}{C}q = V(t)$ |
| Second-order partial (not covered herein) | Water waves <br> Sound waves <br> Electromagnetic radiation <br> Heat transfer <br> Chemical diffusion <br> Schrödinger's equation | $\nabla^2 f = \dfrac{1}{D}\dfrac{\partial f}{\partial t} - \dfrac{1}{v^2}\dfrac{\partial^2 f}{\partial t^2}$ <br> $i\hbar \dfrac{\partial \psi}{\partial t} = -\dfrac{\hbar^2}{2m}\nabla^2 \psi + V\psi$ <br> etc. |

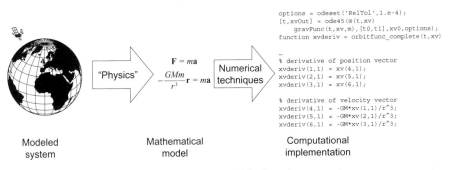

Figure 4.1 Basic strategy for M&S of continuous systems.

by second-order differential equations, which is due to something of a coincidence, that several different essential equations in physics are of second order. For the purposes of this chapter, we will focus on the first three items in the table, which all involve ordinary differential equations of first or second order.

## MODELING AND SIMULATION (M&S) STRATEGY

The basic M&S strategy for modeling continuous systems is dictated by two key steps: understanding the *physics of the system* and identifying the appropriate numerical implementation for solving the resulting equations (see Fig. 4.1). We use the term "physics" very broadly, to mean translation of a real-world system into a set of quantitative mathematical equations, a process that is the very essence of physics, in particular. Of course, this is not to say that we may not use equations from biology, chemistry, economics, sociology, or other sciences, as applicable to the system at hand.

## MODELING APPROACH

The "physics" part of the problem is really the modeling phase. There is no way around developing a great enough understanding of the science of the problem to pose it as a set of mathematical equations. This is a fundamental and repeating challenge to the modeler over a career—on project after project, the modeler is presented with new problems for which he or she must provide credible models with perhaps only cursory (or even no) training in the relevant scientific fields. This is why successful modelers often call themselves "generalists." Remember, the modeler is not trying to make breakthrough discoveries in biochemistry or economics, but is trying to use the relevant knowledge base from such fields to provide accurate models in a particular context. Fortunately for modelers, reliable sources of such knowledge are more accessible than ever.

### State Variables

The usual assumption in continuous modeling, and for that matter, much of science, is that a system can be defined, at any particular instant, by its state. This state is, in turn, defined by a suite of state variables. We regard that the state variables of the real-world system evolve in time continuously (hence the name of the chapter), and we further regard that almost always, the state variables themselves are continuous, meaning real or complex rather than integer valued. Common examples of state variables are position, velocity, mass, rate, angle, fraction, current, and voltage. All of these quantities are real (or possibly complex) and evolve continuously with time.

Perhaps most important in selection of state variables is the notion of *completeness*. The selected state variables must contain sufficient information to elicit all the relevant phenomena associated with the system with adequate accuracy and precision, as set forth during requirements definition. Identifying all of the correct (i.e., necessary and sufficient) state variables is obviously critical.

Just as a simple example, consider the problem of how long it takes an object on earth's surface to drop 10 ft through the air (under "normal" conditions) after starting at rest. If the object is an anvil, the only two state variables necessary to obtain an excellent result are position and speed. If, however, the object is a big-leaf maple seed ("whirligig" type seed), which generates lift by spinning, we must know about its rotation rate to estimate its lift. Leaving out rotation would give completely invalid results.

On the other hand, one can certainly be guilty of overkill when striving for completeness. A particular orbit propagator encountered by the author had the purpose of maintaining tracks of satellites in geosynchronous orbit. The code correctly provided for the radiation pressure from the sun, which has a fairly small, but nonnegligible effect on the orbit; position of the sun and orientation of the satellite were necessary state variables. Borderline overkill was that the code provided for this pressure to be modified by solar eclipses (extremely rare at the position of a given satellite and very short in duration). Definite overkill was that the code provided for the (very small) oblateness and position angle of the moon during that eclipse. The code's author should have estimated the contributions to error for each of these effects before plopping them into the code. He would have recognized that the position angle of the moon was simply unnecessary to keep as a state variable.

## State Variable Equations

A task closely interrelated with identification of the state variables is identification of the equations that govern them. These equations dictate how the state of the system changes with respect to the independent variable (for the purposes of this chapter, time). These equations may be akin to any of those seen in Table 4.1, or may be completely different, depending on the model selected. There is no fundamental limit to the form the state variable equations may take, but fortunately, as we have previously discussed, most real-world systems obey first- or second-order differential equations one way or the other. Quite often, auxiliary variables must be computed during the computation of the state variable equations, and these variables, with their associated equations, must be treated with equal care to the state variables and state variable equations themselves.

## Output Equations

Finally, we must consider what the relevant outputs are: what quantities does the consumer of our simulation want or need? These issues are usually addressed during the requirements definition phase of development.

In the simple examples given in this chapter, the output variables we report are state variables themselves, and sometimes that is the case with other models. However, much of the time, the desired outputs must be computed from the state variables using a new set of equations, called output equations. Output equations often require the same level of expertise and rigor as the state variable equations and, as such, should always be treated with appropriate care.

As an example, let us consider planetary orbits. The state variables at hand are simply the (three-dimensional) position and velocity of the planet relative to the sun, collectively called the state vector. Often, however, consumers of the orbital data prefer the "orbital elements" as output. To move from state vector to the orbital elements requires a fair bit of algebra, including the computation of energy and angular momentum as intermediate values. Just as a flavor, if one wants the semimajor axis $a$ (which describes the size of an orbit), we solve for it by computing the energy $E$:

$$\begin{aligned} r &= \sqrt{x^2 + y^2 + z^2} \\ v &= \sqrt{v_x^2 + v_y^2 + v_z^2} \\ E &= -\frac{GMm}{r} + \frac{1}{2}mv^2 = -\frac{GMm}{2a}, \end{aligned} \quad (4.2)$$

where $G$ is Newton's constant, $M$ is the mass of the sun, and $m$ is the mass of the planet.

## MODEL EXAMPLES

### Predator–Prey Models

*Predator–prey models* are intended to model two species whose populations are very closely dependent on each other, much in the way we might expect "real" predator–prey systems to work in the wild [2, 3]. Perfectly isolated two-species predator–prey situations are fairly rare in the wild, since so many species usually interact in a given biome; however, there are many cases where the natural situation is well approximated by the simple two-species model (e.g., Canadian lynx and Snowshoe rabbit [4]). Furthermore, there are a great many systems in which the fundamental dynamic is the same, even when the system has nothing to do with two species of animals in the wild.

At its simplest, the predator–prey scenario usually produces cycles in the population numbers of both species:

(1) Prey population increases.
(2) Predators feast on plentiful prey.
(3) Predator population increases.
(4) Copious predators overhunt the prey population.
(5) Prey population decreases.
(6) Predators starve.
(7) Predator population goes down.
(8) Prey is less vulnerable to dwindled predator population.
(9) Prey species thrives.
   (Repeat from step 1.)

Usually the predator–prey model is specified mathematically by the following two equations,

$$\dot{x} = xf(x, y)$$
$$\dot{y} = yg(x, y), \qquad (4.3)$$

where $x$ is the population of prey and $y$ is the population of predator. Note that in each equation, the change in population is proportional to the current population, such that $f(x,y)$ and $g(x,y)$ define the fractional, or per capita, growth rates in the populations (equivalently, $f$ is the derivative of $\ln x$).

The cagey reader may be asking why, in a chapter where we are so sensitive to discretization errors, are we taking a fundamentally discrete system of, say, lynxes and rabbits, and representing their populations as continuous real numbers, only to come back and grouse about the discrete algorithms we will use to compute those real numbers. The answer is twofold: (1) the predator–prey problem has applications in truly continuous areas such as chemical abundances and (2) even in the case of lynxes and rabbits, the assumption is that the populations are large enough that the continuous approximation is valid; this permits the use of calculus, which in turn permits much greater understanding and means of computation.

Getting back to the math, it is assumed that $df/dy < 0$, which says that as the predator population, $y$, increases, the growth rate of the prey population of $x$ declines (steps 3–5). However, we assume that $dg/dx > 0$, which says that as the prey population, $x$, increases, the predator growth rate increases (steps 1–3). These conditions ensure the basic cyclic conditions we described above; together with Equation (4.3), these are often called Kolmogorov's predator–prey model [5].

Perhaps the most direct way to ensure the Kolmogorov conditions is simply to design a model in which $f$ is linear in $y$ with a negative slope, and $g$ is linear in $x$ with a positive slope:

$$f = b - py$$
$$g = rx - d. \qquad (4.4)$$

This particular version of the model is called the *Lotka–Volterra model*, after two scientists who very much independently arrived at the form in the 1920s—Lotka modeled fish stocks in the Adriatic Sea and Volterra modeled chemical abundances under certain types of reactions [2]. The behaviors can be quite varied depending on the selection of the constants, as Figure 4.2 shows.

The Lotka–Volterra model (as well as some other particular models) permit construction of an integral of the "motion," or conserved quantity. This conserved quantity is analogous to more familiar conserved quantities in physics, such as energy and momentum. Using Equations (4.3) and (4.4), we have

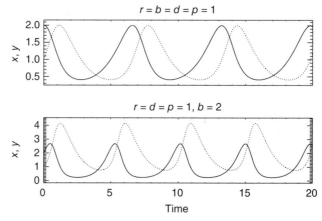

**Figure 4.2** Predator–prey behavior for the Lotka–Volterra model, with starting conditions x (prey, solid) = 2 and y (predator, dotted) = 1. Note the differing values for the governing constants at the top and bottom.

$$\dot{x} = x(b - py)$$

Use
$$\frac{\dot{y}}{y} = g(x, y) \Rightarrow y = \frac{\dot{y}}{rx - d}$$

$$\dot{x} = x\left(b - p\frac{\dot{y}}{rx - d}\right)$$

$$\frac{\dot{x}}{x}(rx - d) = b(rx - d) - p\dot{y}$$

$$r\dot{x} - d\frac{\dot{x}}{x} = b(rx - d) - p\dot{y} \qquad (4.5)$$

Use
$$rx - d = g(x, y) = \frac{\dot{y}}{y}$$

$$r\dot{x} - d\frac{\dot{x}}{x} = b\frac{\dot{y}}{y} - p\dot{y}$$

$$r\dot{x} - d\frac{\dot{x}}{x} - b\frac{\dot{y}}{y} + p\dot{y} = 0$$

$$C = b\ln y - py - rx + d\ln x.$$

This is a conservation law that holds for all $x(t)$ and $y(t)$ that obey Equations (4.3) and (4.4). The particular forms of $x(t)$ and $y(t)$ depend on the choices of $b, p, r$, and $d$, and the initial conditions at time zero. Figure 4.3 shows "orbits" in the $x$–$y$ plane for different values of $C$.

There is a host of other particular predator–prey models that are also very well studied (e.g., Hoppensteadt [2]), many with conservation laws similar to Equation (4.5).

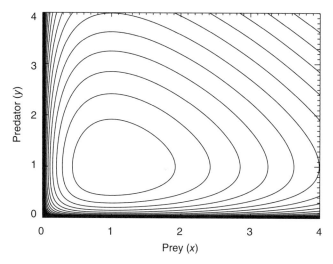

**Figure 4.3** Lotka–Volterra predator–prey model: "orbits" for different values of C, with $r = b = d = p = 1$.

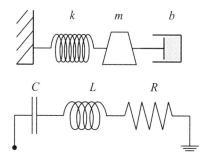

**Figure 4.4** Two simple oscillator systems: the mass-spring-dashpot system (top) and the RLC circuit (bottom).

## Oscillators

The solution of second-order differential equations very often leads to oscillating solutions. The second-order nature of Newton's second law, $\mathbf{F} = m\mathbf{a}$, therefore, yields oscillating solutions for a large number of mechanical systems. We shall see that for somewhat related reasons, analog electronic circuits can also have such solutions.

Let us consider the canonical oscillator, the mass-spring-dashpot system (Fig. 4.4, top) (e.g., see Tipler [6]). The spring provides a force that is negative and proportional to the displacement, $x$, of the mass; the constant of proportionality is simply called the spring constant $k$. There is also a "dashpot," which acts as a damper. In a shock absorber (a real-life version of a mass-spring-dashpot system), the dashpot is very often a piston that moves through a gas chamber. The dashpot presents a force that is proportional to the negative of velocity, with a constant of proportionality, $b$. So let us consider our forces, to derive the equation of motion for the system:

$$F = ma = m\ddot{x}$$
$$-kx - b\dot{x} = m\ddot{x} \qquad (4.6)$$
$$m\ddot{x} + b\dot{x} + kx = 0.$$

As it turns out, the system has exact analytic solutions,

$$x = Ae^{zt}, \qquad (4.7)$$

where $z$ is a complex number. Substituting the solution, we find

$$x = Ae^z; \quad \dot{x} = zAe^z; \quad \ddot{x} = z^2 Ae^z$$
$$A(mz^2 + bz + k)e^{zt} = 0 \qquad (4.8)$$
$$mz^2 + bz + k = 0.$$

Using the quadratic formula:

$$z = \frac{-b \pm \sqrt{b^2 - 4km}}{2m}. \qquad (4.9)$$

The sign of the discriminant gives three different solution types:

$$\begin{aligned} b^2 - 4km &< 0: \quad \text{underdamped (oscillating)} \\ b^2 - 4km &= 0: \quad \text{critically damped} \\ b^2 - 4km &> 0: \quad \text{overdamped.} \end{aligned} \qquad (4.10)$$

**Underdamped Solution** The underdamped solution will feature some level of oscillatory behavior, because the exponential of an imaginary is a sum of trigonometric functions ($e^{i\theta} = \cos\theta + i\sin\theta$). The solution has the form

$$x = A\cos(\omega t + \phi)e^{-bt/2m}$$
$$\omega^2 = \frac{k}{m} - \frac{b^2}{4m^2}. \qquad (4.11)$$

Note that in the case of no dashpot ($b = 0$), we have the simple harmonic oscillator, $x = A\cos(\omega t + \phi)$ with $\omega = \sqrt{k/m}$ The phase angle $\phi$ and the amplitude $A$ are determined by the initial conditions. In the simplest case, where the mass is simply displaced by an amount $A$ and then released at time zero, $\phi = 0$. The dashed line in Figure 4.5 shows the motion of such an oscillator.

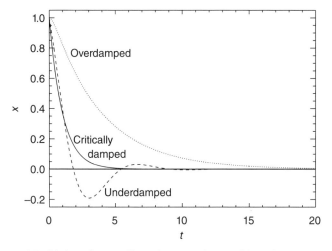

**Figure 4.5** Motion of an oscillator, for a starting condition of $x_0 = 1$, $v_0 = 0$.

***Critically Damped Solution*** The critically damped solution guarantees that the displacement will return exponentially to zero, but does allow for some growth from any initial velocity:

$$x = (At + B)e^{-bt/2m}. \tag{4.12}$$

The smooth and rapid return to the zero displacement value is a sought-after feature in shock-absorbing systems, and so most such systems are tuned to reside in the critical damping regime. The solid line in Figure 4.5 shows the behavior.

***Overdamped Solution*** The overdamped solution is also guaranteed to decline to zero, but, in general, does so more slowly than the critically damped solution. This seems counterintuitive at first, but what is happening is that the overbearing dashpot is actually slowing down the return to zero displacement.

The two available solutions to the quadratic, again allow for different initial conditions. The dotted line in Figure 4.5 shows the behavior of and overdamped system:

$$\begin{aligned} x &= Ae^{\lambda_+ t} + Be^{\lambda_- t}; \\ \lambda_\pm &= \frac{-b \pm \sqrt{b^2 - 4km}}{2m}. \end{aligned} \tag{4.13}$$

In all three of the oscillator cases, the constants in the solutions are determined by setting the initial position and velocity equal to the desired values at time zero, and solving for the relevant constants.

**RLC Circuit** In something of a coincidence, a very common analog electronic circuit demonstrates the very much the same behavior as the mass-spring-dashpot system. The RLC circuit (see Fig. 4.4) consists of a resistor (resistance $R$), an inductor (inductance $L$), and a capacitor (capacitance $C$). Here, we are not measuring displacement, $x$, of the spring, but the charge, $q$, on the capacitor. As it turns out, according to Ohm's law, a resistor "resists" motion of charge, or current, in direct proportion to the current, just as the dashpot resists motion in direct proportion to the velocity. The inductor opposes the second derivative of charge (or first derivative of current), just as the mass opposes acceleration. The capacitor stores the energy of accumulated charge, just as the spring stores energy of the displacement. The capacitor works a bit differently than the spring, in that the stiffer the spring (the higher the value of $k$) becomes, the greater force it applies, but the greater the capacitance, $C$, the less voltage it applies—as the name implies, a greater capacitance means a greater ability to accommodate more charge and thus provide less opposing voltage. (For further discussion on RLC circuits, see Purcell [7].) Figure 4.4 is drawn so that the analogy of each of the three components to the mechanical system is clear. We are left with an equation for the charge that is mathematically identical to Equation (4.1):

$$m\ddot{x} + b\dot{x} + kx = 0 \quad \text{(mechanical system)};$$
$$L\ddot{q} + R\dot{q} + \frac{1}{C}q = 0 \quad \text{(RLC circuit)}. \tag{4.14}$$

Of particular interest is the underdamped case, where we inherit a fundamental frequency to the system, $\omega$, given below:

$$q = A\cos(\omega t + \phi)e^{-Rt/2L}$$
$$\omega^2 = \frac{1}{LC} - \frac{R^2}{4L^2}. \tag{4.15}$$

The behavior of these circuits when stimulated with an external oscillating voltage, near this fundamental frequency is extremely useful in electronics, most notably for construction of frequency filters.

## SIMULATING CONTINUOUS SYSTEMS

### General Solution Approach

In our discussions thus far, we have had a single independent variable: time. The assumption that time is a universal independent variable in the real world is a loaded one that came crashing down with the theory of relativity [8]; however, for our purposes, in which we are making classical macroscopic assumptions, the assumption that time is a fully independent time variable is quite justified.

Because we have couched our problems as ordinary differential equations in time, simulation is really a matter of integration of the differential equations. One often thinks of "integration" as simply taking the integral of a function ("the integral of $2t$ is $t^2$"). More broadly, however, *integration* includes solution of differential equations, of which the integral is just a special case in which the function depends only on the single independent variable.

The reader may recall from a calculus II course that analytic integration of functions is, in general, much more difficult than differentiation of functions, and, in fact, is often just not doable. Unfortunately, the situation only gets worse when we move to differential equations. These are very rarely analytically integrable. One reason we chose the oscillator model is that it is a rare example of an analytically integrable system.

## Numerical Solution Techniques

Since most systems yield equations that are not integrable, we have little choice but to turn to numerical integration, but as we discussed in the introduction to the chapter, translation of continuous systems onto fundamentally discrete computers is a task fraught with peril. As such, the rest of this section addresses how to carry out discretization in time, and how to minimize the damage.

## Euler's Method

The simplest technique for solving an ordinary differential equation is the use of *Euler's method*. Here, we simply refer back to the formula for the derivative in Equation (4.1) in the form of an approximation (where the limit has not truly gone to zero), then invert to estimate the value of $x$ at the next time step, $x(t + \Delta t)$ based on the value of $x(t)$ and $dx/dt$ evaluated at the current values of $x$, and $t$:

$$\frac{dx}{dt} \approx \frac{x(t+\Delta t)-x(t)}{\Delta t}$$
$$x(t+\Delta t) \approx x(t) + \left.\frac{dx}{dt}\right|_{x,t} \Delta t. \tag{4.16}$$

For ordinary differential equations, we have the derivative available to us from some function $f(x,t)$, which leaves us with this final form of the Eulerian method:

$$x(t+\Delta t) \approx x(t) + f(x,t)\Delta t. \tag{4.17}$$

One may recognize this as a truly linear formula, in that one could write Equation (4.17), with some changes in notation, as simply the equation of a

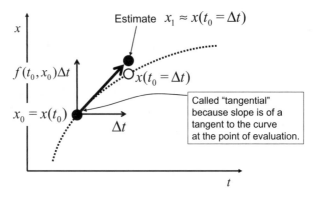

**Figure 4.6** Eulerian method for integration of an ODE.

line as $y = b + mx$, where the slope $m$ is $dx/dt = f(x,t)$. Because $dx/dt$ is defined to be the slope of the tangent to the curve, Euler's method is also often called the "tangential method." The ostensibly linear form also makes clear that errors will be of order $O(\Delta t^2)$. This means that for a decrease by a factor of two in $\Delta t$, we should expect an estimate four times more accurate at a given time step. Unfortunately for the same total time interval, halving the time step doubles the number of time steps, which multiplies the error back by one factor of $\Delta t$, leaving the error over a given interval at $O(\Delta t)$. If you want errors 10 times smaller, you just have to carry out 10 times as many iterations.

The fundamental limitation to the Eulerian method is revealed by its other name, the tangential method—the derivative is always approximated as the tangent to the curve. This means that if there is any curvature to the solution function, the tangent method will always end up "shooting wide" of the target, which is clearly visible in Figure 4.6. In practice, this effect is also quite easy to see. Figure 4.7 shows the integration of the simple first-order ordinary differential equation (ODE) $dx/dt = 0.3x$, where we have integrated with time steps of one unit for 16 time units. At each time step, the Eulerian method undershoots its estimate of the derivative, because the tangential slope at each point fails to accommodate the curvature of the true solution.

What makes matters worse is that the errors compound. By under- or overshooting the derivative estimate, one incorrectly computes the next value of $x$ (bad enough), but that also means that the next evaluation of the derivative is carried out at the wrong value of $x$. In general, these errors carry through and compound during the course of the integration.

### Runge–Kutta Methods

Fortunately, there are alternatives to the Euler method that are straightforward to implement. A large suite of such techniques are called *Runge–Kutta methods*. With these methods, one evaluates the derivative at multiple

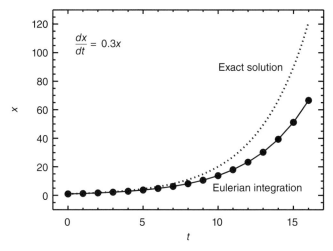

**Figure 4.7** Eulerian integration of a simple ODE.

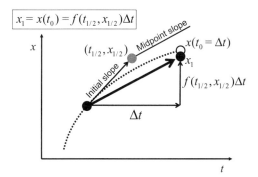

**Figure 4.8** Midpoint integration method; compare with Figure 4.6.

substeps per iteration to evaluate and accommodate curvature and higher-order effects, before advancing to the next full time step. We base the following discussion on Runge–Kutta methods largely on the excellent account in Press et al. [9].

The first of these methods is the midpoint method, which Figure 4.8 illustrates. This method uses two estimates of the derivative during a time step. First, the derivative is evaluated at the current time step (just as with Euler integration), but then we multiply only by $\Delta t/2$ to form an estimate $x_{1/2}$. We now evaluate the derivative again at this intermediate point $(t_{1/2}, x_{1/2})$, then simply use $f(t_{1/2}, x_{1/2})$ as our estimate of the derivative for the full time step.

Notice that the time step is twice as long in this figure. This is because we have called the derivatives function twice, whereas the Euler technique only calls the derivatives function once.

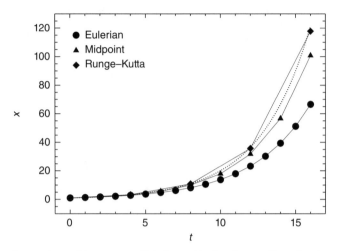

**Figure 4.9** Comparison of integrators: Eulerian, midpoint, Runge–Kutta fourth order for the ODE, $dx/dt = 0.3x$.

We summarize this in the equations below:

$$t_{1/2} = t_0 + \frac{\Delta t}{2};$$
$$x_{1/2} = x_0 + f(t_0, x_0)\frac{\Delta t}{2}; \quad (4.18)$$
$$x_1 = x_0 + f(t_{1/2}, x_{1/2})\Delta t.$$

Figure 4.9 shows the significant improvement made by using the midpoint method. Note that we have taken only eight steps of 2 time units, rather than 16 steps of 1 time unit. This is to hold constant the number of calls to the derivative function when comparing to the Euler integration.

It is useful here to make a notational change, in which we use $k_i$ to represent an estimate of the change in $x$ over the full time step. The Euler integration equation becomes

$$k_1 = f(t_0, x_0)\Delta t$$
$$x_1 = x_0 + k_1, \quad (4.19)$$

while the midpoint integration becomes

$$k_1 = f(t_0, x_0)\Delta t$$
$$k_2 = f\left(t_0 + \tfrac{1}{2}\Delta t, x_0 + \tfrac{1}{2}k_1\right)\Delta t \quad (4.20)$$
$$x_1 = x_0 + k_2.$$

The Euler formula and the midpoint formula are actually the first- and second-order Runge–Kutta integration formulas.

By far the most commonly used Runge–Kutta formula is the fourth-order formula,

$$k_1 = f(t_0, x_0) \Delta t$$
$$k_2 = f(t_0 + \tfrac{1}{2}\Delta t, x_0 + \tfrac{1}{2}k_1) \Delta t \quad (4.21)$$
$$k_3 = f(t_0 + \tfrac{1}{2}\Delta t, x_0 + \tfrac{1}{2}k_2) \Delta t$$
$$k_4 = f(t_0 + \Delta t, x_0 + k_3) \Delta t$$
$$x_1 = x_0 + (k_1 + 2k_2 + 2k_3 + k_4)/6.$$

This fourth-order formula, as you may expect, introduces an error at each time step of order $O(\Delta t^5)$. Figure 4.9 illustrates the performance of the fourth-order Runge–Kutta formulas versus the first- and second-order formulas. Note again that we have equalized the computational load for each of the curves, by holding constant the same number of calls to the derivative function in each case.

Figure 4.10 shows that as we increase the number of iterations (decrease $\Delta t$), the errors decrease for each method according to their expected order. For a factor of 10 decrease in time step, the Eulerian error decreases by a factor of 10, the midpoint error by a factor of $10^2 = 100$, and the fourth-order Runge–Kutta error by a factor of $10^4 = 10{,}000$.

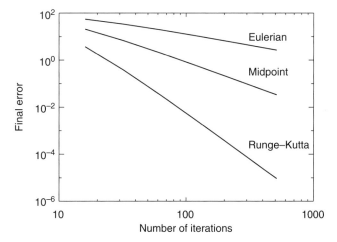

**Figure 4.10** Comparison of integrator errors versus stepsize: Eulerian, midpoint, Runge–Kutta fourth order. Note that the stepsize decreases to the right as the number of iterations (calls to the derivative function) increases to the right. Note also the log-log scale.

It is clear enough that the higher-order formulas outperform the lower-order formulas in this case; however, one must not assume that is always true. A method of order $n$ assumes that the function is continuously differentiable to order $n$ (a notion we alluded to in the chapter introduction). If the derivatives function has a continuous second derivative, but not a continuous third derivative, the Runge–Kutta (fourth order) precision will only improve at the same rate as the midpoint method (second order). Most analytic functions, such as exp, sine, and ln, are infinitely continuously differentiable, but many functions used in modeling the real world are not. Nonetheless, the Runge–Kutta fourth-order formula very, very rarely leads to *worse* results than the midpoint or Euler formula, and it therefore remains recommendable, even if the performance is not quite as good as possible.

Finally, Figure 4.10 suggests that with Runge–Kutta methods, one could fairly quickly approach machine precision with just a few more factors of 10 in iteration number. In cases where the compute time is very small to start with, perhaps that is a tenable plan. When compute time impacts the utility of a model, however, careful trades must be made in desired precision versus compute time, and charts like Figure 4.10 become an important instrument for such studies.

**Adaptive Time Step**

Runge–Kutta techniques are certainly capable of integrating ODEs within expected error margins; however, we have thus far considered only systems that are quite stable in their derivatives over time. That is to say, the magnitude of the derivative does not change dramatically over the course of the run. In many systems, this assumption cannot be made. A problematic example may be one involving sporadic impulse forces, such as a shock absorber in a car that runs over occasional large potholes at high speed. Less obvious are cases where the derivative varies very smoothly but still varies a great deal in magnitude.

Figure 4.11 shows the case of a simple two-body gravitational orbit around the sun. During the orbit, no sudden changes occur in the derivative, but the performance of the fourth-order Runge–Kutta integration is disastrous in the vicinity of the sun. Instead of staying in the proper orbit, the planet begins to loop almost randomly and eventually flies completely off into space. The reason is that far from the sun, the gravitational field is weak, but near the sun, the field is very strong (due to Newton's inverse square law). In this example, where we have a highly elliptical (eccentric) orbit, the planet is 20 times nearer the sun at closest pass than farthest past, creating a ratio of force of a factor of 400. The result in reality is Kepler's second law, which very loosely states that objects on eccentric orbits whiz around the sun at closest approach, but linger almost idly at farthest approach. In fact, at the top of Figure 4.11, one can see how with equal time steps, the planet is barely moving at farthest approach (the initial condition), but finally starts to race around the orbit much more rapidly near the sun. And this is the problem; the time

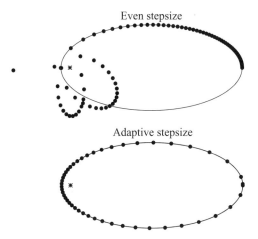

**Figure 4.11** Runge–Kutta integration over one period of a high eccentricity orbit, with even time step (top) and adaptive time step (bottom). Each point is a time step taken by the integrator. Overplotted as a solid curve is the orbit predicted by Kepler's laws. The sun is represented by the snowflake symbol.

step that was appropriate far from the sun, is not appropriate near the sun. This is a classic case in which adaptive stepsize is recommended.

With adaptive stepsize, one uses a remarkable Runge–Kutta formulation, first found by Fehlberg, and generalized by Cash and Karp, in which six evaluations of the derivative yield *both* fourth- and fifth-order precise estimates of the integration over a time step, using different coefficients on the six $k$ values [10]. So we now have one estimate whose error is of order $O(\Delta t^5)$ and a second estimate whose error is of order $O(\Delta t^6)$. Their difference, $\Delta$, is an estimate of the error made for this time step and has error of order $O(\Delta t^5)$. So, we compute our integration for stepsize $\Delta t_{\text{try}}$ and find an error, $\Delta_{\text{try}}$. The $\Delta$ is of order $O(\Delta t^5)$, so if we had some desired error limit, $\Delta_0$, the desired stepsize, would be found this way:

$$\Delta t_0 = \Delta t_{\text{try}} \left| \frac{\Delta_0}{\Delta_{\text{try}}} \right|^{1/5}. \tag{4.22}$$

If the $\Delta_{\text{try}}$ is too large, $\Delta t_0$ is now the recommended time step to try, so that we stay within the desired error budget. Just as important, if $\Delta_{\text{try}}$ is too small (by some factor), $\Delta t_0$ tells us how much we can afford to increase the time step without violating our error criteria—this means that the integrator does not spend extra time slogging through very dull portions of the simulation.

Now we can reconsider Figure 4.11. As we have seen, in the top figure, with even time step, the integrator laboriously lingered far from the sun where little was going on, only to race much too fast past the sun and make fatal errors. This is, essentially, the exact opposite of what we want. At the bottom, we have used adaptive stepsize. One can readily see that the integrator quickly recognizes that there is little need to spend effort far from the sun, and so it

**118** MODELING CONTINUOUS SYSTEMS

races the planet around the orbit quickly until things get more interesting near the sun. Then, the integrator hits the brakes and chooses much smaller time steps to move very carefully past the sun. Once around the sun, the integrator again speeds things up and zips through the dull part of the orbit. Not only is this a much more judicious use of computer resources, but one can see the very dramatic improvement in precision.

The lesson is to remember that the discretization need not remain fixed throughout the course of the simulation. It is much more important to maintain performance and error budget than constant discretization.

## SIMULATION IMPLEMENTATION

Implementation of these integration techniques we have mentioned does not require a particularly special computing environment. Java, C++, FORTRAN, and other general-purpose computing languages very often have code libraries available that allow rapid development with these techniques (of which Press et al. [9] is an example in C++).

There are also mathematical programming environments, such as MATLAB (produced by The MathWorks, Inc.) and IDL (produced by ITT Visual Information Systems), which are custom-made for just this type of work. We now provide discussion of implementation in MATLAB and briefly discuss how these methods can even be implemented in Microsoft Excel.

### First-Order Simulation by Integration—Predator–Prey

***MATLAB Eulerian Simulation*** Implementing the Eulerian integration is straightforward in MATLAB (or virtually any other coding environment). We now discuss the implementation of the predator–prey model in MATLAB, using Eulerian integration.

Referring to Figure 4.6 and the surrounding text, we can quickly delineate what needs to be done:

1. Evaluate the derivative $\dot{x}(t, x)$ at time $t$ and position $x$.
2. Add $\dot{x}(t, x)\Delta t$ to $x$.
3. Add $\Delta t$ to $t$.
4. Repeat from step 1.

Below is the MATLAB code for Eulerian integration of the Lotka–Volterra predator–prey model. The code is a straightforward implementation of the Eulerian method, with one additional nuance. We have two variables $x$ and $y$ that are updated each iteration. To ensure proper behavior, we need to store a temporary version of x (xTmp) for the computation of the y derivative—otherwise, we would be updating the velocity with the subsequent iteration's prey population, which is not valid. For plotting, we simply grow the arrays tStore, and xStore and yStore with each iteration:

```
% Eulerian integration of predator-prey system

% start and end times
t0 = 0.;    % sec
t1 = 20.;   % sec
dt = 0.05;  % sec

% constants (Lotka-Volterra)
b = 1.;
p = 1.;
r = 1.;
d = 1.;

% initial condition
x0 = 1.0;   % meters
y0 = 2.0;   % meters/sec

% perform the integration
t = t0;
x = x0;
y = y0;
while (t <= t1)
    if (t == t0)
        tStore = [t];
        xStore = [x];
        yStore = [y];
    else
        tStore = [tStore,t];
        xStore = [xStore,x];
        yStore = [yStore,y];
    end
    xTmp = x;
    x = x + x*(b-p*y)*dt;
    y = y + y*(r*xTmp-d)*dt;
    t = t + dt;
end

% plot the results
plot(tStore,xStore,'k');
hold on;
plot(tStore,yStore,'k:');
hold off;
```

Figure 4.12 shows the results of the integration by this code. The overall form is quite similar to the top plot of Figure 4.2, although there appears to be an overall growth in amplitude as time progresses. This is because the linear Eulerian method is overshooting at the maxima and minima, as we have discussed.

**120** MODELING CONTINUOUS SYSTEMS

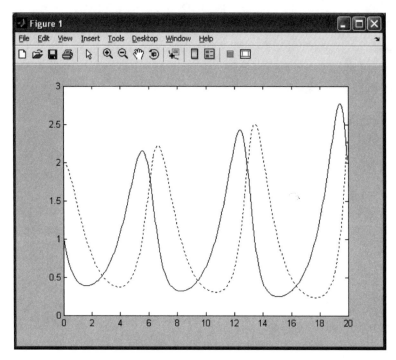

**Figure 4.12** Screenshot of Eulerian integration of a Lotka–Volterra predator–prey model in MATLAB. The prey population (*x*) is solid, and the predator population (*y*) is dotted.

***MATLAB Runge–Kutta Simulation*** Having seen the errors in the Eulerian integration, the next logical step is to implement a Runge–Kutta integrator. Fortunately, in MATLAB, aside from some ungainly syntax, this task could hardly be more straightforward. All of the supporting code remains identical to that in our Eulerian integrator; we simply replace the while loop that carries out the iterations with a single call to the prepackaged Runge–Kutta integrator (in this case a fourth- and fifth-order adaptive stepsize integrator called ode45):

```
% Runge-Kutta integration of predator-prey system

% start and end times
t0 = 0.; % sec
t1 = 20.; % sec

% constants (Lotka-Volterra)
b = 1.;
p = 1.;
r = 1.;
d = 1.;
```

```
% initial condition
x0 = 1.0; % meters
y0 = 2.0; % meters/sec
xy0 = [x0;y0];

% perform the integration
options = odeset('RelTol',1.e-4);
[t,xyOut] = ...
   ode45(@(t,xy)ppFunc(t,xy,b,p,r,d),[t0,t1],xy0,options);

% plot the results
plot(t,xyOut(:,1),'k');
hold on;
plot(t,xyOut(:,2),'k:');
hold off;
```

Note that the first argument to `ode45` is the callback to a user-provided function:

```
function xyderiv = ppFunc(t,xy,b,p,r,d)

xyderiv = [xy(1)*(b-p*xy(2));xy(2)*(r*xy(1)-d)];
```

The cumbersome syntax for `ode45` in the main code arises from the fact that the integrator expects to be passed only the time value and the vector of variables to be integrated (in our case, $x$ and $y$, stored as a column vector `xy`), but our callback derivatives function expects to hear about the Lotka–Volterra parameters `b`, `p`, `r`, and `d` as well. This syntax is simply the way MATLAB handles such situations. Cumbersome though it may be, this style has some distinct advantages over many others, which demand use of global variables or common blocks, or the like, to achieve the same result.

The second argument of `ode45` is a row vector of the start and end times for the integration; and the third argument contains the column vector of the initial values of the variables to be integrated. Finally, we include an optional argument, `options`, which was set in the previous line, using the `odeset` interface. In our case, we have set the relative tolerance for error to a part in 10,000 for either $x$ or $y$. This tolerance governs the stepsize adjustments made by the integrator. Figure 4.13 gives the results for this run. Note that the increasing amplitude problems seen in the Eulerian integration have gone away completely.

## Second-Order Simulation by Integration—Oscillators

Because Newton's second law involves acceleration, the second derivative of position, the better fraction of time-evolving physical systems require second-order treatment. In our discussion, so far, however, we have not discussed methods for treating second-order systems numerically. Fortunately, implementation is straightforward. We simply maintain track of both the position and the velocity as independent variables. The derivatives of position and

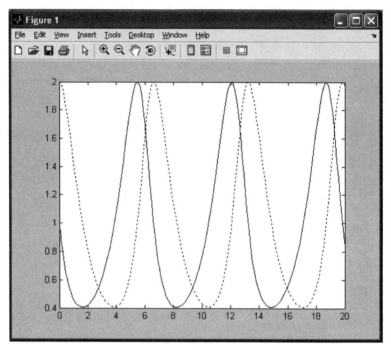

**Figure 4.13** Screenshot of adaptive stepsize fourth- and fifth-order Runge–Kutta integration of Lotka–Volterra predator–prey model in MATLAB. The prey population ($x$) is solid, and the predator population ($y$) is dotted.

velocity are velocity and acceleration. This trick may seem groundless, since very often position and velocity are not independent at all, but remember, we are working under the premise that the system has been linearized for very small changes in time, and in that limit, the two are independent. A single Eulerian step for a mass under and applied force looks like this,

$$\begin{aligned} x_1 &= x_0 + v_0 \Delta t \\ v_1 &= v_0 + [F(x_0, t_0)/m] \Delta t \\ t_1 &= t_0 + \Delta t, \end{aligned} \quad (4.23)$$

where $F(x,t)$ is the force applied and $m$ is the mass.

**MATLAB Eulerian Simulation** We provide now a code listing for Eulerian integration of the mass-spring-dashpot system. The variables t, x, and v store current time, position, and speed. We have parameterized the behavior of the system according to how we set $b$ (b) in terms of the critical value of $b$ at which the system would be critically damped. When $b < b_{\text{crit}}$, the system is underdamped; when $b > b_{\text{crit}}$, it is overdamped; and when $b = b_{\text{crit}}$, it is critically damped:

```
% start and end times
t0 = 0.; % sec
t1 = 20.; % sec
dt = 0.1; % sec

% spring systems constants
m = 1.; % kg
k = 1.; % Newtons/meter

bCrit = sqrt(4*k*m);
b = bCrit*0.5;
% Eulerian integration of oscillator system

% initial conditions
x0 = 1.0; % meters
v0 = 0.0; % meters/sec

% perform the integration
t = t0; % initial time
x = x0; % initial position
v = v0; % initial speed
while (t <= t1)
   if (t == t0)
       tStore = [t];
       xStore = [x];
   else
       tStore = [tStore,t];
       xStore = [xStore,x];
   end
   xTmp = x;
   x = x + v*dt;
   v = v - (k*xTmp + b*v)/m*dt;
   t = t + dt;
end

xExact = springExact(tStore,m,k,b,x0,v0);

% plot the results
subplot(2,1,1);
plot(tStore,tStore*0,'k');
hold on;
plot(tStore,xExact,'r');
plot(tStore,xStore,'o');
xlabel('time (sec)');
ylabel('position (m)');
hold off;

subplot(2,1,2);
plot(tStore,tStore*0,'r');
```

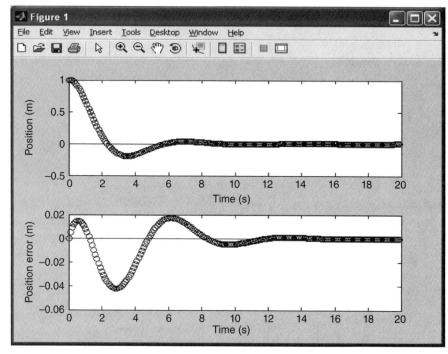

**Figure 4.14** Screenshot of Eulerian integration of the mass-spring-dashpot system in MATLAB, where $b = b_{crit}/2$.

```
hold on;
plot(tStore,xStore-xExact,'o');
xlabel('time (sec)');
ylabel('position err (m)');
hold off;
```

Figure 4.14 shows the results of this run. There are two plots: the position as a function of time and the error of the integration when compared with the exact solution. The errors we might expect for each time step should be of order $O(\Delta t^2)$. Since $\Delta t = 0.1$ s, and all other quantities in the system are of order unity, the errors peaking in magnitude at about 0.04 m appear quite reasonable. Checking the performance against exact solutions (when available), as we have done here, is a critical step in the verification of a code. Because this is such an important step, we also include now our code for computing the exact solution for a given system, following Equations (4.11–4.13). For each case (over-, under-, and critically damped), the code adjusts the various coefficients to ensure that the initial conditions are met:

```
function xExact = springExact(t,m,k,b,x0,v0)

discrim = b*b - 4*k*m;
```

```
if (discrim > 0.)
    lamPlus = (-b+sqrt(discrim))/(2*m);
    lamMinus = (-b-sqrt(discrim))/(2*m);
    Mat = [1.,1.;lamPlus,lamMinus];
    soln = inv(Mat)*[x0;v0];
    A = soln(1);
    B = soln(2);
    xExact = A*exp(lamPlus*t) + B*exp(lamMinus*t);
else
    if (discrim < 0.)
        omega = sqrt(-discrim)/(2*m);
        lambda = -b/(2*m);
        phi = atan((lambda-(v0/x0))/omega);
        if (abs(phi) < pi*0.25)
            A = x0/cos(phi);
        else
            A = v0/(lambda*cos(phi)-omega*sin(phi));
        end
        xExact = A.*cos(omega.*t+phi).*exp(lambda.*t);
    else
        lambda = -b/(2*m);
        B = x0;
        A = (v0-lambda*B);
        xExact = (A.*t+B).*exp(lambda.*t);
    end
end
```

***MATLAB Runge–Kutta Simulation*** The Runge-Kutta implementation of the oscillator system follows directly:

```
% Runge-Kutta integration of oscillator system

% initial condition
x0 = 1.0; % meters
v0 = 0.0; % meters/sec
xv0 = [x0;v0];

% perform the integration
options = odeset('RelTol',1.e-4);
[t,xvOut] = ...
    ode45(@(t,xv)springFunc(t,xv,m,k,b),[t0,t1],xv0,options);
...
```

The derivatives function is also straightforward, following Equation (4.6):

```
function xvderiv = springFunc(t,xv,m,k,b)

xvderiv = [xv(2);-(b*xv(2)+k*xv(1))/m];
```

**126** MODELING CONTINUOUS SYSTEMS

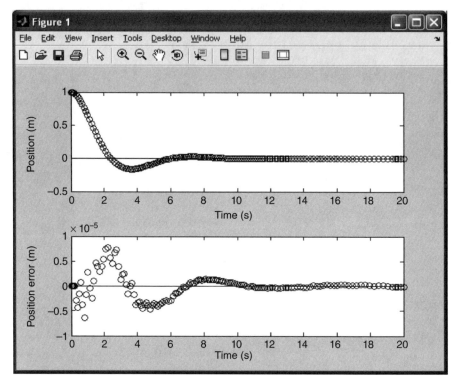

**Figure 4.15** Screenshot of adaptive stepsize Runge–Kutta fourth/fifth-order integration of the mass-spring-dashpot system in MATLAB, where $b = b_{crit}/2$.

Figure 4.15 shows the results of the adaptive stepsize Runge–Kutta run. The errors shown in the bottom should remain better than a part in 10,000 or so, and we see that, in fact, our fractional errors do remain below that threshold as desired.

I recommend that readers try out the code and play with lots of values for bCrit.

***Spreadsheet Simulation*** The MATLAB environment is an excellent one for implementing simulations of continuous systems, but not every system has MATLAB installed. We therefore pause to give a quick discussion of carrying out a simulation in a common spreadsheet environment.

The Eulerian integration technique is particularly straightforward to implement. Table 4.3 shows the actual spreadsheet.

After the parameter headings, column A stores the time variable, incrementing by $\Delta t$ set in cell B8. Column B is the position of the mass. The first entry, B15, is simply set equal to $x_0$ (B10), but thereafter has the form seen in B16 = B15 + D15 * $B$8, which says that we will add to the previous

**Table 4.3 Microsoft Excel spreadsheet for integrating the oscillator system**

|    | A       | B       | C     | D        | E      | F | G                  |
|----|---------|---------|-------|----------|--------|---|--------------------|
| 1  | m =     | 1       | kg    |          |        |   |                    |
| 2  | k =     | 1       | N/m   |          |        |   |                    |
| 3  | bCrit = | 2       |       |          |        |   |                    |
| 4  | b =     | 1       |       |          |        |   |                    |
| 5  |         |         |       |          |        |   |                    |
| 6  | t0 =    | 0       | s     |          |        |   |                    |
| 7  | t1 =    | 20      | s     |          |        |   |                    |
| 8  | dt =    | 0.1     | s     |          |        |   |                    |
| 9  |         |         |       |          |        |   |                    |
| 10 | x0 =    | 1       | m     |          |        |   |                    |
| 11 | v0 =    | 0       | m/s   |          |        |   |                    |
| 12 |         |         |       |          |        |   |                    |
| 13 |         |         |       |          |        |   |                    |
| 14 | Time    | x       | v     | x-dot    | v-dot  |   | Position err (mm)  |
| 15 | 0       | 1       | 0     | 0        | −1     |   | 0                  |
| 16 | 0.1     | 1       | −0.1  | −0.1     | −0.9   |   | 4.833415278        |
| 17 | 0.2     | 0.99    | −0.19 | −0.19    | −0.8   |   | 8.669244507        |
| 18 | 0.3     | 0.971   | −0.27 | −0.27    | −0.701 |   | 11.51923907        |
| 19 | 0.4     | 0.944   | −0.3401 | −0.3401 | −0.6039 |  | 13.41299331       |
| 20 | 0.5     | 0.90999 | −0.40049 | −0.40049 | −0.5095 | | 14.39547346      |
| 21 | 0.6     | 0.869941 | −0.45144 | −0.45144 | −0.4185 | | 14.52458889     |
| 22 | 0.7     | 0.824797 | −0.49329 | −0.49329 | −0.33151 | | 13.8688343     |
| 23 | 0.8     | 0.775468 | −0.52644 | −0.52644 | −0.24903 | | 12.5050276     |
| 24 | 0.9     | 0.722824 | −0.55134 | −0.55134 | −0.17148 | | 10.51616442    |
| 25 | 1       | 0.66769  | −0.56849 | −0.56849 | −0.0992 | | 7.989406608      |

row's value the value of $\dot{x}_{prev}\Delta t$ = D15 * $B$8. Similarly, C15 is set to $v_0$, but C16 is =C15 + E15 * $B$8, the previous velocity plus the time derivative of velocity (acceleration) times $\Delta t$. Column D features our "trick" for computing the second derivative; D15 contains simply =C15, which says that $\dot{x} = v$. It is in column E where the actual physics happens, because, again, $\dot{v} = a$. So, we have in E15 "= −($B$2 * B15 + $B$4 * C15)/$B$1," which is $-(kx + bv)/m$, as prescribed by Equation (4.6). Figure 4.16 shows the results of the Excel implementation for the case where $b = b_{crit}/2$ (compare to Fig. 4.14).

Figure 4.16 shows the results from the Eulerian integration of mass-spring-dashpot system in Microsoft Excel, with $b = b_{crit}/2$. Note that the scale of the y-axis in the error plot is in millimeters.

One can even implement a Runge–Kutta integration in Excel. The spreadsheet is similar to the Eulerian spreadsheet; however, instead of having two columns for and , one has eight columns for the $k_1$, $k_2$, $k_3$, and $k_4$ terms for both x and v. Column B is changed from $x_{prev} + \dot{x}_{prev}\Delta t$ to the familiar Runge–Kutta formula $x_{prev} + (k_1 + 2k_2 + 2k_3 + k_4)/6$, or =B15 + (D15 + 2 * F15 + 2 * H15 + J15)/6 in B16, with a similar implementation in column C for the

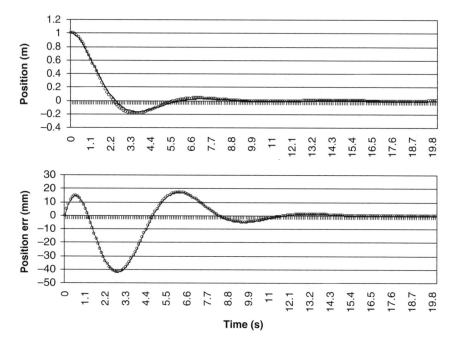

**Figure 4.16** Results from the Eulerian integration of the mass-spring-dashpot system in Microsoft Excel, with $b = b_{crit}/2$. Note that the scale of the y-axis in the error plot is in millimeters.

velocity. It is worth giving as an example the code inside the $k_2$ column for $x$ (cell F15), just to show how one proceeds through the $k$'s. Cell F15 contains =$B$8 * (C15 + 0.5 * E15). As prescribed by the Runge–Kutta fourth-order method, $k_{2,x} = \Delta t \cdot \dot{x}(x + k_{1,x}/2, t + \Delta t/2)$, which for our system is simply $k_{2,x} = \Delta t \cdot \dot{x}(x + k_{1,x}/2) = \Delta t \cdot (v + k_{1,v}/2)$. The entry in G15 for $k_{2,v}$ is somewhat more complicated, though the principal is the same: =$B$8*(−($B$2 * ($B15 + 0.5 * $D15) + $B$4 * ($C15 + 0.5 * $E15))/$B$1).

The results of this integration can be seen in Figure 4.17. The errors reported are tiny indeed. We note here, however, that adaptive stepsize (as seen in Fig. 4.15) would be a significantly greater challenge to implement in a spreadsheet.

Figure 4.17 shows the results from the fourth-order Runge–Kutta integration of mass-spring-dashpot system in Microsoft Excel, with $b = b_{crit}/2$. Note that the scale of the y-axis in the error plot is in millimeters.

## CONCLUSION

Simulating continuous systems on fundamentally discrete digital computers seems, on its face, to be problematic from the start. Fortunately, when care is used in the discretization process, errors can be budgeted carefully and effec-

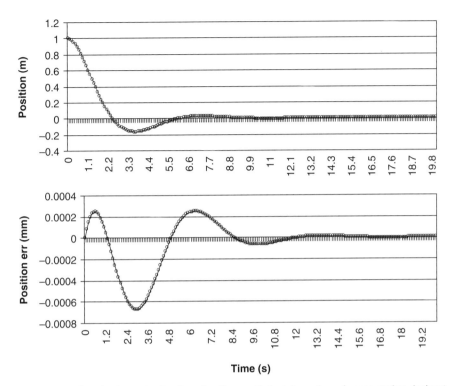

**Figure 4.17** Results from the fourth-order Runge–Kutta integration of mass-spring-dashpot system in Microsoft Excel, with $b = b_{crit}/2$. Note that the scale of the y-axis in the error plot is in millimeters.

tively. We have presented the relatively simple case of integrating ordinary differential equations, where we know the analytic derivatives of our solution functions, and the only derivatives are taken with respect to the independent variable. We have found that results with predictable precision can be calculated with straightforward implementations in a host of different programming environments, even spreadsheets. As the reader ventures into more complex continuous systems, continued diligence must be exercised in monitoring errors and numerical instabilities.

## REFERENCES

[1] Jain B, Seljak U, White S. Ray-tracing simulations of weak lensing by large-scale structure. *Astrophysical Journal*, 530:547–577; 2000.

[2] Hoppensteadt F. Predator-prey model. *Scholarpedia*, 1(10):1563; 2006. Available at http://www.scholarpedia.org/article/Predator-prey_model. Accessed June 1, 2009.

[3] Sadava D, Heller HC, Orians GH, Purves WK, Hillis DM. *Life: The Science of Biology*. New York: W.H. Freeman & Company; 2008.

[4] Allan D. Trophic Links: Predation and Parasitism (lecture in Global Change 1, taught at the University of Michigan, Fall 2008). Available at http://www.globalchange.umich.edu/globalchange1/current/lectures/predation/predation.html. Accessed June 1, 2009.

[5] Freedman HI. *Deterministic Mathematical Models in Population Ecology*. New York: Marcel Dekker; 1980.

[6] Tipler PA. *Physics*. New York: Worth; 1982, pp. 311–339.

[7] Purcell EM. *Electricity and Magnetism, Berkeley Physics Course*. Vol. 2. New York: McGraw-Hill; 1985, pp. 298–319.

[8] Einstein A. Zur Elektrodynamik bewegter Körper. *Annalen der Physik*, 17:891; 1905.

[9] Press WH, Teukolsky SA, Vetterling WT, Flannery BP. *Numerical Recipes in C++*. Cambridge: Cambridge University Press; 1988, pp. 717–727.

[10] Cash JR, Karp AH. A variable order Runge-Kutta method for initial value problems with rapidly varying right-hand sides. *ACM Transactions on Mathematical Software*, 16:201–222; 1990.

# 5

# MONTE CARLO SIMULATION

John A. Sokolowski

When one hears the name *Monte Carlo*, one often thinks of the gambling locale in the country of Monaco. It is the home of the famous *Le Grand Casino* as well as many other gambling resorts and Formula One Racing. This chapter, however, is not about gambling or racing. It is, however, about a concept that underlies gambling, that is, probability, hence, its association and designation with the well-known gambling region. The scientific study of *probability* concerns itself with the occurrence of random events and the characterization of those random happenings. Gambling casinos rely on probability to ensure, over the long run, that they are profitable. For this to happen, the odds or chance of the casino winning has to be in its favor. This is where probability comes into play because *the theory of probability provides a mathematical way to set the rules for each one of its games to make sure the odds are in its favor.* As a simulation technique, Monte Carlo simulation relies very heavily on probability.

Monte Carlo simulation, also known as the Monte Carlo method, originated in the 1940s at Los Alamos National Laboratory. Physicists Stanislaw Ulman, Enrico Fermi, John von Neumann, and Nicholas Metropolis had to perform repeated simulations of their atomic physics models to understand how these models would behave given the large number of uncertain input variable values. As random samples of the input variables were chosen for

*Modeling and Simulation Fundamentals: Theoretical Underpinnings and Practical Domains*, Edited by John A. Sokolowski and Catherine M. Banks
Copyright © 2010 John Wiley & Sons, Inc.

each simulation run, a statistical description of the model output emerged that provided evidence as to how the real-world system would behave. It is this concept of *repeated random samples of model input variables over many simulation runs* that defines *Monte Carlo simulation*. Essentially, we are creating an artificial world (model) that is meant to closely resemble the real world in all relevant aspects.

Monte Carlo simulation is often superior to a deterministic simulation of a system when that system has input variables that are random. Deterministic simulations are referred to as *what-if* simulations. In these simulations, a single value is chosen for each input random variable (a particular what-if scenario) based on a best guess by the modeler. The simulation is then run and the output is observed. This output is a single value or a single set of values based on the chosen input. But because the input variables are random variables, they can take on any number of values defined by their probability distributions. So to have a sense of how the system would respond over the complete range of input values, more than one set of inputs must be evaluated. Monte Carlo simulation randomly samples values from each input variable distribution and uses that sample to calculate the model's output. This process is repeated many times until the modeler obtains a sense of how the output varies given the random input values. One should readily see that when the simulation contains input random variables, Monte Carlo simulation will yield a result that is likely to be more representative of the true behavior of the system. The next section formally defines Monte Carlo simulation and provides examples of its use.

## THE MONTE CARLO METHOD

When setting up a Monte Carlo simulation or employing the Monte Carlo Method, one follows a four-step process. These four steps are:

*Step 1* Define a distribution of possible inputs for each input random variable.

*Step 2* Generate inputs randomly from those distributions.

*Step 3* Perform a deterministic computation using that set of inputs.

*Step 4* Aggregate the results of the individual computations into the final result.

While these steps may seem overly simplistic, they are necessary to capture the essence of how Monte Carlo simulations are set up and run.

This four-step method requires having the necessary components in place to achieve the final result. These components may include:

(1) *probability distribution functions* (pdfs) for each random variable
(2) a *random number generator*

(3) a *sampling rule*—a prescription for sampling from the pdfs
(4) *scoring*—a method for combining the results of each run into the final result
(5) *error estimation*—an estimate of the statistical error of the simulation output as a function of the number of simulation runs and other parameters.

Step 1 requires the modeler to match a statistical distribution to each input random variable. If this distribution is known or sufficient data exist to derive it, then this step is straightforward. However, if the behavior of an input variable is not well understood, then the modeler might have to estimate this distribution based on empirical observation or subject matter expertise.* The modeler may also use a uniform distribution if he or she is lacking any specific knowledge of the variable's characteristics. When additional information is gathered to define the variable, then the uniform distribution can be replaced.

Step 2 requires randomly sampling each input variable's distribution many times to develop a vector of inputs for each variable. Suppose we have two input random variables $X$ and $Z$. After sampling $n$ times, we have $X = (x_1, x_2, \ldots, x_n)$ and $Z = (z_1, z_2, \ldots, z_n)$. Elements from these vectors are then sequentially chosen as inputs to the function defining the model. The question of how large $n$ should be is an important one because the number of samples determines the power of the output test statistic. As the number of samples increases, the standard deviation of the test statistic decreases. In other words, there is less variance in the output with larger sample sizes. However, the increase in power is not linear with the number of samples. The incremental improvement of power decreases by a factor of about $1/\sqrt{n}$, so there is a point when more sampling provides little improvement. Determining the number of trials needed for a desired accuracy is addressed below.

Step 3 is straightforward. It involves sequentially choosing elements from the randomly generated input vectors and computing the value of the output variable or variables until all $n$ outputs are generated for each output variable.

Step 4 involves aggregating all these outputs. Suppose we have one output variable $Y$. Then we would have as a result of step 4 an output vector $Y = (y_1, y_2, \ldots, y_n)$. We can then perform a variety of statistical tests on $Y$ to analyze this output. These tests will be described later in the chapter.

The following is a simple example of how this method works.

---

*When modeling systems, especially those in the social sciences, subject matter experts may be the only source of data available to characterize the behavior of a variable. This is true when no scientific data or data collection is available. Subject matter expertise may also be called upon as a method to validate the output of the simulation. See Chapter 10 for a further discussion on validation techniques.

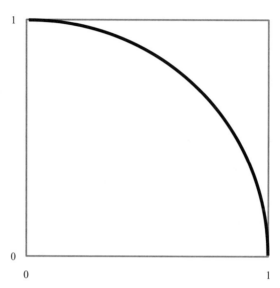

**Figure 5.1** Unit circle arc for calculation of $\pi$.

## Example 1: Determining the Value of $\pi$

Recall that the value of $\pi$ is the ratio of a circle's circumference to its diameter. To calculate this value, we can set up a Monte Carlo simulation that employs a geometric representation of the circle.

1. To start, draw a unit circle arc, that is, an arc of radius one circumscribed by a square as shown in Figure 5.1.
2. Then, randomly choose an $x$ and $y$ coordinate inside the square, and place a dot at that location.
3. Repeat step 2 at a given number of times. See Figure 5.2.
4. Count the total number of dots inside the square and the number of dots inside the quarter circle. With a large number of dots generated, these values will approximate the area of the circle and the area of the square. From mathematics, this result can be represented as

$$\frac{\text{\# of dots inside circle}}{\text{\# of dots inside square}} = \frac{\frac{1}{4}\pi r^2}{r^2} = \frac{1}{4}\pi.$$

Step 1 of our example represents step 1 of the above method, that is, determining the domain of possible inputs. Steps 2 and 3 correspond to method step 2, and step 4 encompasses steps 3 and 4 of the method.

Our example relied on several components mentioned above. A random number generator was necessary to select the coordinates for each dot. The coordinates were selected from a uniform distribution that provided the prob-

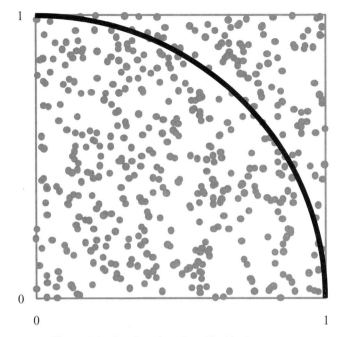

**Figure 5.2** Random dots placed inside the square.

ability density function. A sampling rule existed that used the random numbers to select values from the uniform distribution. The scoring method was given by the formula in step 4 above. Finally, error estimation can be performed by comparing the computed value of $\pi$ to an authoritative source for its value.

This simulation can be set up using a spreadsheet and the built in functions of *rand()* that generates uniform random numbers between 0 and 1 and the *countif(range, criteria)* function that can count the number of random numbers that meet the specified criteria. The author generated 500 uniform random numbers between zero and one for the $x$ coordinate of each point and the same for the $y$ coordinate. These numbers were paired up and plotted. Precisely 340 of the 500 points fell inside the circle giving a simulated value for $\pi$ of 2.76. This method produced an error of 12.1 percent. Using a larger set of generated dots can help reduce the error to an acceptable range realizing that it requires a trade-off for extra computation.

From example 1, you can see the necessary components that are central to Monte Carlo simulations. These components are one or more input random variables, one or more output variables, and a function that computes the outputs from the inputs. This configuration is shown in Figure 5.3.

In this figure, notice that there are three input random variables $x_1, x_2$, and $x_3$, all with different distributions. There are two output variables, $y_1$ and $y_2$, that have resulting distributions created by the repeated sampling of the input

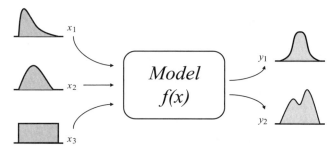

**Figure 5.3** Basic Monte Carlo model.

**Table 5.1** Product earnings by year

|  | Year | | | | |
|---|---|---|---|---|---|
|  | 1 | 2 | 3 | 4 | 5 |
| Unit price | 50 | 52 | 55 | 57 | 65 |
| Unit sales | 2000 | 2200 | 2700 | 2500 | 2800 |
| Variable costs | 50,000 | 55,000 | 56,000 | 57,000 | 58,000 |
| Fixed costs | 10,000 | 12,000 | 15,000 | 16,000 | 17,000 |
| Earnings | 40,000 | 47,400 | 77,500 | 69,500 | 107,000 |

and feeding those samples into the function $f(x)$. The next example builds on this model to illustrate how a *what-if* scenario outcome can differ from one produced via a Monte Carlo approach.

### Example 2: Computing Product Earnings

Let us suppose we want to predict a product's earning in future years given sales data accumulated over the last 5 years. A product's earning is a function of *unit price, unit sales, variable costs*, and *fixed costs*. Specifically, earning = (unit price) × (unit costs) − (variable costs + fixed costs). We will assume that variables used to calculate earnings are all independent of one another. The last 5 years of data for these variables are shown in Table 5.1.

From these data, one can develop a *probability distribution* to represent each of the input variables. An appropriate distribution representation would be a *triangular distribution*, which is typically used when only a small amount of data is available to characterize the input variables. These distributions may be constructed from Table 5.2.

Constructing a triangular distribution requires three values: a minimum, a maximum, and a most likely or mode. We can represent these values by $a, b$, and $c$, respectively. Then, the *probability density function* for this distribution is defined as follows:

**Table 5.2 Triangular distribution data**

|  | Min | Most Likely | Max |
|---|---|---|---|
| Unit price | 50 | 55 | 70 |
| Unit sales | 2000 | 2440 | 3000 |
| Variable costs | 50,000 | 55,200 | 65,000 |
| Fixed costs | 10,000 | 14,000 | 20,000 |
| Earnings | 40,000 | 65,000 | 125,000 |

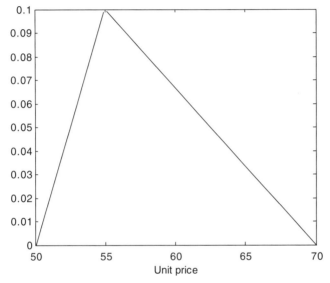

**Figure 5.4** Unit price triangular distribution.

$$f(x|a,b,c) = \begin{cases} \dfrac{2(x-a)}{(b-a)(c-a)} & \text{for } a \leq x \leq c \\ \dfrac{2(b-x)}{(b-a)(b-c)} & \text{for } c \leq x \leq b \\ 0 & \text{otherwise.} \end{cases} \quad (5.1)$$

Figure 5.4 illustrates the triangular distribution for *unit price* as computed from Equation (5.1). The other variables have similar triangular representations. The *max values* were chosen as best guess estimates of the highest values these parameters will reach. This is often done using subject matter experts intuitively familiar with how these variables are likely to behave. The *min values* were the minimum numbers found in Table 5.1 for each variable. The most likely values were calculated by averaging the 5 years of data for each factor.

**Table 5.3 Earnings summary statistics**

| Parameter | Value |
|---|---:|
| Mean | 73,206 |
| Median | 71,215 |
| Standard deviation | 16,523 |
| Variance | 273,009,529 |
| Min | 21,155 |
| Max | 137,930 |

Using these triangular distributions, 10,000 samples were generated and used to compute predicted future earnings. The following MATLAB® code was used to generate the samples.

**MATLAB Program**

```
h = sqrt(rand(1,10000));
unit_price = (70-50)*h.*rand(1,10000)+55-(55-50)*h;
variable_costs = (65000-50000)*h.*rand(1,10000)+55200-(55200-50000)*h;
fixed_costs = (20000-10000)*h.*rand(1,10000)+14000-(14000-10000)*h;
unit_sales = (3000-2000)*h.*rand(1,10000)+2440-(2440-2000)*h;
for i = 1:10000
earnings(i) = unit_price(i)*unit_sales(i)-(variable_costs(i)+fixed_costs(i));
end
```

The summary statistics for the output variable *earnings* are shown in Table 5.3.

A plot of the computed earnings is shown in Figure 5.5. This plot represents the probability density function for the output distribution.

From the Monte Carlo simulation results, one can see a difference between the *most likely earnings value* (65,000) from Table 5.2 and the *mean earnings value* (73,206) of Table 5.3. In other words, the simple *what-if* analysis of a deterministic computation of earnings differs from the Monte Carlo computation that takes into account many combinations of input variable values that could occur in predicting future earnings. Instead of a single point analysis, the modeler has the results of 10,000 points on which to base his or her estimate of future earnings. These simulation runs take into account 10,000 different combinations of input variables, which provide a much broader picture of the possible values that *earnings* could take on given the possible variability in the real-world data. One can also compare the minimum and maximum expected earnings from both the single-point estimate and the Monte Carlo estimate to get an understanding of the possible extreme values that may result.

How good of an estimate of the true population mean is the Monte Carlo *computed mean earnings value*? One way to assess this is to compute a

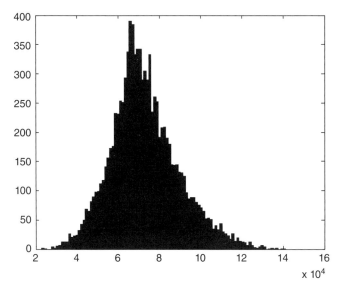

**Figure 5.5** Computed earnings distribution plot.

confidence interval for the population mean based on sample data. This computation is based on the important statistical concept of the *central limit theorem*. This theorem is expressed as follows.

**Theorem 1 (Central Limit Theorem)** Suppose $Y_1, \ldots, Y_n$ are independent and identically distributed (IID) samples and $E[Y_i^2] < \infty$. Then

$$\frac{\hat{\theta}_n - \theta}{\sigma/\sqrt{n}} \Rightarrow N(0,1) \text{ as } n \to \infty, \tag{5.2}$$

where $\hat{\theta}_n = \sum_{i=1}^n Y_i/n$, $\theta := E[Y_i]$ and $\sigma^2 := Var(Y_i)$.

Note that nothing is assumed about the distribution of the $Y_i$'s other than their variance is less than infinity. So from Equation (5.2), if $n$ is sufficiently large, then we can compute a confidence interval for $\theta$ based on a standard normal distribution. The confidence interval is computed as follows.

Let $z_{1-\alpha/2}$ be the $(1-\alpha/2)$ percentile point of the $N(0, 1)$ distribution such that

$$P(-z_{1-\alpha/2} \leq Z \leq z_{1-\alpha/2}) = 1 - \alpha,$$

where $Z \sim N(0,1)$. Now take the simulated IID samples $Y_i$ and construct a $100(1-\alpha)\%$ confidence interval for $\theta = E[Y]$. Essentially, we are constructing a lower and upper bound $L(Y)$ and $U(Y)$ such that

$$P(L(Y) \le \theta \le U(Y)) = 1-\alpha.$$

The central limit theorem tells us that $\sqrt{n}(\hat{\theta}_n - \theta)/n$ is approximately a standard normal distribution for large $n$, so we have

$$P\left(-z_{1-\alpha/2} \le \frac{\sqrt{n}(\hat{\theta}_n - \theta)}{\sigma} \le z_{1-\alpha/2}\right) \approx 1-\alpha$$

$$\Rightarrow P\left(-z_{1-\alpha/2}\frac{\sigma}{\sqrt{n}} \le \hat{\theta}_n - \theta \le z_{1-\alpha/2}\frac{\sigma}{\sqrt{n}}\right) \approx 1-a$$

$$\Rightarrow P\left(\hat{\theta}_n - z_{1-\alpha/2}\frac{\sigma}{\sqrt{n}} \le \theta \le \hat{\theta}_n + z_{1-\alpha/2}\frac{\sigma}{\sqrt{n}}\right) \approx 1-\alpha.$$

So the approximate $100(1-\alpha)\%$ confidence interval for $\theta$ is given by Equation (5.3):

$$[L(Y), U(Y)] = \left[\hat{\theta}_n - z_{1-\alpha/2}\frac{\sigma}{\sqrt{n}}, \hat{\theta}_n + z_{1-\alpha/2}\frac{\sigma}{\sqrt{n}}\right]. \tag{5.3}$$

One other issue must be addressed before computing our confidence interval, that is, $\sigma^2$ is usually not known. However, it can be estimated by the following formula:

$$\hat{\sigma}_n^2 = \frac{\sum_{i=1}^{n}(Y_i - \hat{\theta}_n)^2}{n-1}.$$

So replacing $\sigma$ with $\hat{\sigma}$, we arrive at

$$[L(Y), U(Y)] = \left[\hat{\theta}_n - z_{1-\alpha/2}\frac{\hat{\sigma}_n}{\sqrt{n}}, \hat{\theta}_n + z_{1-\alpha/2}\frac{\hat{\sigma}_n}{\sqrt{n}}\right] \tag{5.4}$$

as the final equation for computing the *confidence interval*.

Equation (5.4) can now be applied to the results of Monte Carlo simulation in example 2. For $\alpha = 0.05$, the $z_{1-\alpha/2}$ value for the standard normal distribution is 1.96. Table 5.3 provides the standard deviation for the 10,000 sample points so the confidence interval for the mean earnings is [72,999, 73,446]. One should interpret this interval as *we are 95 percent confident that the interval contains the actual population mean*. The smaller this interval is, the more confidence we have in the estimate of the actual population mean.

Even though we have high confidence that the population mean falls in the above interval, that does not necessarily indicate that is what *earnings* will be. Figure 5.5 shows how widely *earnings* could vary given a specific set of sales and price conditions. It is because of the Monte Carlo method that we are able to represent and to visualize the possible outcomes.

The width of the confidence interval is a function of the number of sample points chosen. If one wants to achieve a certain level of confidence, then one must be able to determine the number of samples necessary to achieve that accuracy. The error between the actual mean and the computed mean can be represented by an absolute error $E_a = |\hat{\theta}_n - \theta|$. Thus, we want to choose a value for $n$ such that $P(E_a \leq \varepsilon) = 1 - \alpha$, where $\varepsilon$ is the actual error. Recall from above,

$$P\left(\hat{\theta}_n - z_{1-\alpha/2}\frac{\sigma}{\sqrt{n}} \leq \theta \leq \hat{\theta}_n + z_{1-\alpha/2}\frac{\sigma}{\sqrt{n}}\right) \approx 1 - \alpha.$$

This implies that

$$P\left(|\hat{\theta}_n - \theta| \leq z_{1-\alpha/2}\frac{\sigma}{\sqrt{n}}\right) \approx 1 - \alpha.$$

So in terms of $E_a$, we have

$$P\left(E_a \leq +z_{1-\alpha/2}\frac{\sigma}{\sqrt{n}}\right) \approx 1 - \alpha.$$

If we want $P(E_a \leq \varepsilon) \approx 1 - \alpha$, then we must choose $n$ such that

$$n = \frac{\sigma^2 z_{1-\alpha/2}^2}{\varepsilon^2}. \tag{5.5}$$

Just as with the computation of the confidence interval, $\sigma^2$ is usually not known. One way to solve this problem is to estimate it by doing a *pilot simulation*. Here, the modeler conducts a small number of runs and uses the results of those runs to estimate $\sigma^2$. The estimate is then used to compute an $\hat{n}$. This number of runs is then performed, the output variable's statistics are gathered, and a confidence interval is computed. If the modeler follows this two-stage procedure, it is likely that $\hat{n}$ runs will produce the desired level of accuracy. For this method to work, the initial number of runs to estimate $\sigma^2$ must be sufficiently large ($\geq 50$). The following pseudocode describes this procedure:

**Two-Stage Procedure for Estimating the Number of Simulation Runs**

```
/*Do pilot simulation first*/
for i=1 to p
generate X^i
end for
set  θ̂ = Σ h(X^i)/p
set  σ̂² = Σ (h(X^i) - θ̂)² /(p - 1)
```

```
set  n = σ̂²z²_{1-α/2} / ε²
/*Now do main simulation*/
for i=1 to n
generate Xⁱ
end for
set  θ̂_n = Σ h(Xⁱ)/n
set  σ̂²_n = Σ (h(Xⁱ) - θ̂_n)² / (n - 1)
set  100(1 - α)%Ci = [θ̂_n - z_{1-α/2} σ̂_n/√n , θ̂_n + z_{1-α/2} σ̂_n/√n]
```

To illustrate the procedure, we will repeat the Monte Carlo simulation of *earnings* from example 2. Suppose we want to control the absolute error so that

$$P(E_a \leq 1000) = 1 - \alpha.$$

Note that this is equivalent to saying that we want the confidence interval to have a width of less than or equal to $2 \times 1000 = \$2000$. For the pilot simulation, we choose $p = 100$ and $\alpha = 0.05$. Using the two-stage procedure above produces a $\hat{\theta} = 70{,}586$ and $\hat{\sigma}^2 = 1.8897e8$. Applying Equation (5.5) gives us an $\hat{n} \approx 726$. Using this number for our second stage, we obtain the following: $\hat{\theta}_n = 73{,}424$, $\hat{\sigma}_n^2 = 2.6731e8$, and a confidence interval of [72,235, 74,613], which is about $2400 wide. Note that the first Monte Carlo simulation using 10,000 samples produces a confidence interval width of $447.

From the example above, one can see that this two-stage procedure provides a method for determining a close approximation for the number of runs needed to achieve a certain absolute error value.

## SENSITIVITY ANALYSIS

An important analytic concept based on Monte Carlo simulation is that of *sensitivity analysis*. For our purposes, we will define sensitivity analysis as the study of how uncertainty in a model's output can be assigned to the various sources of input uncertainty. As one can see from the discussion above, input and output uncertainties are at the heart of the Monte Carlo simulation. Gauging which input random variables have the most influence on the output random variables is an important fact to know when trying to analyze a model's behavior. This section will introduce concepts for performing sensitivity analysis based on Monte Carlo simulation and how sensitivity analysis can be used to adjust the Monte Carlo simulation.

Sensitivity analysis is important for several reasons. It can help uncover model errors and identify important bounds on input variables. This analysis can also help identify research priorities and simplify models. Thus, sensitivity analysis plays a significant role as a tool to assess model validity.

The most common method for conducting sensitivity analysis is based on derivatives. For example, given $\partial Y_j/\partial X_i$ where $Y_j$ is a output random variable and $X_i$ is a input random variable, one can see that this partial derivative can be interpreted as the change in $Y_j$ with respect to $X_i$, which is consistent with our definition of sensitivity analysis. *Derivative-based approaches* are very efficient from a computational standpoint; however, it does have one serious flaw. Derivative-based approaches are only valid at the point that they are computed. This is acceptable for linear systems but would be of little value for systems exhibiting nonlinear behavior. There are, however, other methods that can be applied for all systems.

One simple method involves a visual assessment of an input variable's effect on an output variable. This method employs a *scatter plot* where each input variable in the Monte Carlo simulation is individually plotted against the output variable and the resulting pattern is analyzed. The more structured the output pattern, the more sensitive is the output variable to that input variable.

Referring back to example 2, we had four random input variables that contributed to computing the *earnings* random output variable. If we plot each of the 10,000 randomly generated inputs against the corresponding output using a scatter plot, we get the results shown in Figure 5.6.

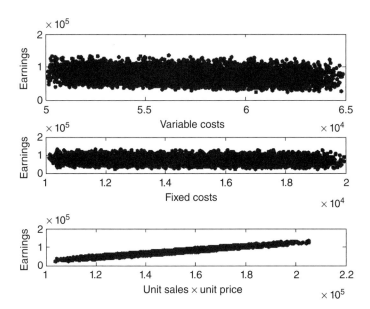

**Figure 5.6** Scatter plots of earnings versus each input variable.

**Table 5.4 Earnings summary statistics for simplified model**

| Parameter | Value |
|---|---|
| Mean | 75,551 |
| Median | 73,102 |
| Standard deviation | 16,219 |
| Variance | 263,040,000 |
| Min | 33,223 |
| Max | 135,790 |

The scatter plots show that the variable *earnings* is more sensitive to the product of *unit_price* and *unit_sales* than the *fixed and variable costs* because of its structured pattern. From the results of this analysis, one could set the *fixed_costs* and *variable_costs* inputs to their mean values and rerun the Monte Carlo simulation using only *unit_sales* and *unit_price* as random input variables with little loss in accuracy. The results of this Monte Carlo simulation are shown in Table 5.4.

The resultant $100(1 - \alpha)\%$ confidence interval for the mean is [75,223, 75,869]. The mean in Table 5.4 is within $2400 of the mean of the full Monte Carlo model from Table 5.3 and is still a better predictor than just using the single-point *what-if* analysis, which produced an *earnings* prediction of $65,000. Thus, sensitivity analysis allowed us to reduce the complexity of our model with only about 3 percent change in results.

While scatter plots provide a good visual means for identifying the relative sensitivity among input variables, other computational methods are available that improve on the derivative-based approach mentioned above. One such approach is the *sigma-normalized derivatives*. This is defined as follows:

$$S_{X_i}^{\sigma} = \frac{\sigma_{X_i} \partial Y}{\sigma_Y \partial X_i}. \tag{5.6}$$

The derivative is normalized by the input–output standard deviations. The larger the result of this computation, the more sensitive the output is to this input. The sensitivity measure of Equation (5.6) is widely recognized and is recommended for sensitivity analysis by a guideline of the Intergovernmental Panel for Climate Change [1]. When the results of Equation (5.6) are squared and summed across all input variables, the following equation holds:

$$\sum_{i=1}^{r} (S_{X_i}^{\sigma})^2 = 1.$$

This will be illustrated by again revisiting example 2, our *earnings* computation. Table 5.5 provides the sigma-normalized derivatives for the *earnings* model.

Table 5.5 Sigma-normalized derivatives

| Variable | $\left(S_{X_i}^\sigma\right)^2$ |
|---|---|
| variable_costs | 0.03 |
| fixed_costs | 0.01 |
| unit_sales × unit_price | 0.96 |

As one can see, the product of *unit_sales* and *unit_price* is the most sensitive parameter and bears out the results of the scatter plot analysis.

There are other sensitivity analysis techniques based on the Monte Carlo simulation. The reader is referred to Santelli et al. for a discussion of the most prominent techniques as well as a comparison of the two techniques presented above [2].

## CONCLUSION

This chapter explored the Monte Carlo simulation method for characterizing a model's behavior in the face of one or more input random variables. This method provides a more representative way to understand the behavior of such models compared with a fixed set of parameters under *what-if* analysis. Additionally, the concept of a confidence interval was introduced as a measure of the accuracy of the Monte Carlo simulation in relation to the actual population mean of the system under study. A technique for estimating the sample size required to achieve a specified accuracy was also described. The chapter concluded with a discussion of sensitivity analysis based on Monte Carlo techniques and introduced two methods for assessing the contribution of each input random variable to the model's output.

## REFERENCES

[1] IPCC. Good Practice Guidance and Uncertainty Management in National Greenhouse Gas Inventories. 2000. Available at http://www.ipcc-nggip.iges.or.jp/public/gp/gpgaum.htm. Accessed May 2, 2009.

[2] Santelli A, Ratto M, Andres T, Campolongo F, Cariboni J, Gatelli D, Saisana M, Tarantola S. *Global Sensitivity Analysis: The Primer*. West Sussex: John Wiley & Sons; 2008.

# 6

# SYSTEMS MODELING: ANALYSIS AND OPERATIONS RESEARCH

Frederic D. McKenzie

Model engineering is the process of determining the appropriate model type and implementation methodology to use for representing a system of interest and then designing and implementing the model. This chapter introduces the concept of system model types and discusses in detail various types of models and particular methodologies that create models of the types discussed. In those sections, we will draw upon terminology from Fishwick and relate those terms to other generalized nomenclature [1]. Formal and semiformal tools to aid analysis are presented in subsequent sections as well as analytic methods and procedures. We complete the chapter with a discussion of operations research (OR) methods and a familiar extended example. Throughout the chapter, examples have a common thread that is used to emphasize the fact that a system may be modeled in many different ways depending upon what questions are being asked.

## SYSTEM MODEL TYPES

Models are representations and, therefore, their depictions and specifications can take many forms. Probably the most convenient way to represent a system is by using a textual description. System requirements try to capture all the

*Modeling and Simulation Fundamentals: Theoretical Underpinnings and Practical Domains*,
Edited by John A. Sokolowski and Catherine M. Banks
Copyright © 2010 John Wiley & Sons, Inc.

needed capabilities of a model but do not easily convey the overall functionality of the system as these specifications are usually at a very low level and one can rarely see the forest through the trees. So, a better way to conceive and perceive the overall functionality of a model is by using a graphic depiction. Conceptual models are generally informal and typically graphic depictions of systems that quickly and easily convey the overall functionality of a system. This type of model is created and used early in the design process whether the design is of a system to be built or of a model to be abstracted from an already existing system. We build models to learn or to communicate something about a system, and, depending on what we want to learn or to communicate, we can use different types of models.

Conceptual models are often used only for communicating the overall functionality of a system. When we want to analyze a system to learn something more about it, we use different types of models that may be (1) based on the state of the system as it evolves over time; (2) focused on the stochastic nature of the model; (3) representative of the dynamic, physics-based processes of the system; (4) described according to the systems' multidomain or multielement makeup; or (5) composed of a hybrid of more than one of these modeling flavors. As terminology may differ among the varied backgrounds of simulation professionals, we will choose to describe these modeling flavors using the terminology from Fishwick's perspective on model types [1]. Fishwick uses the term declarative to connote state-event focused models and the term functional to describe a system using directionally connected components with mathematical relationships of equations and the variables that relate them. Fishwick's constraint models are similar to functional models except that the directed nature of the connected components is less important than the physics-based relationships among system components and the balancing of those physical properties. Additionally, systems that are modeled using a multielement approach is said to be spatially modeled. Such spatial models either focus on the space the system occupies by dividing that space into many units with procedures to update elements in that unit or spatial models may focus on the elements within the space occupied by the system and the elements follow their own rules while occupying the system space. Of course, a major philosophy of Fishwick is that a single modeling technique may not fulfill the needs of every question to be learned from modeling the system especially if the system is complex. Therefore, a system model is likely to be a hybrid, composed of different types of models, or, in Fishwick's terms, a multimodel. In the following sections, we will describe these various types of models and provide examples to enhance the explanations.

## MODELING METHODOLOGIES AND TOOLS

In this section, the types of models outlined above are defined and discussed. One or two methodologies that create the various model types are discussed in considerable detail with unifying examples. We start with high-level, low-detail conceptual models and then consider several other more detailed model types that answer particular questions we want answered about the example systems presented.

**Conceptual Models**

A system concept is a generalized idea of one or a group of interacting components and its desired functionality articulated by textual or graphic means. This expression of the concept is the conceptual model. The degree of generalization distinguishes a conceptual model from other types of models. *Conceptual models* are typically very informal in terms of detail and accuracy. The focus is on quickly communicating the main qualities and capabilities of the target system. Therefore, in one or two paragraphs or figures, a conceptual model should convey what is the system, what does the system do, and what if anything is unique about the target system.

The simplest form of conceptual model is a picture of the system. If the system does not exist yet, a picture of a similar or analogous system to the target system may be used. A picture is indeed an abstraction of the actual subject but still may contain too many details so as to detract from the main qualities intended to be conveyed. Perhaps the most ubiquitous format for conceptual models is the simple *sketch* that is a rough drawing of a system leaving out unimportant details. A sketch conforms to no particular rules, but, generally, elements of a sketch will represent objects or actions in the real world. A sketch may combine text along with symbols as necessary. This first draft conceptual model representation may be further refined into a diagram, plan, graph, and map, which are essentially more organized conceptual models.

The following paragraph is a textual representation of a conceptual model:

> The system will be placed on a two-lane highway with sufficient space to construct double bypass lanes for 500 m. The bypass lanes will contain a two-lane toll area that will provide cash and credit toll payment and change on a 24 h basis. The lanes of the main highway will contain our new Easy Pass electronic toll collection capability. Drivers signed up for the Easy Pass system will remain on the main highway lanes and maintain the posted speed limit. Their Easy Pass accounts will automatically be credited as the cars pass by using the constructed highway sensors. The bypass toll lanes will have sufficient space for non-Easy Pass users to queue while paying the toll. We believe that Easy Pass will be a significant advantage to local drivers and our subscription accounts will quickly rise due to the obvious convenience. An early advertising campaign will ensure significant enrollment before the system comes on-line.

The text-based model above is detailed enough to convey the necessary information. What is the system? The system is an automatic toll collector. What does it do? It senses the cars from drivers that have enrolled in the system and collects the toll from their accounts. What is unique? It provides

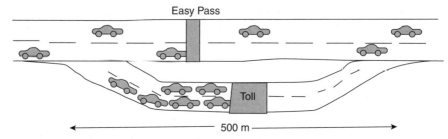

**Figure 6.1** Easy Pass (EP) toll collection system sketch.

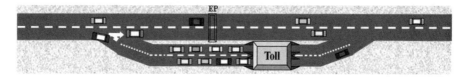

**Figure 6.2** Easy Pass (EP) toll collection system visualization.

the convenience to local drivers of not having to stop to pay the toll. Therefore, this is a good conceptual model. However, it may have taken too much time to convey the information particularly for busy executives who may be making the decision to fund the project.

Consider Figure 6.1, which illustrates a sketch of our target Easy Pass system. Almost immediately, the overall concept of the system can be understood. The sketch conveys a system that has some method of allowing some cars to pass a toll area seemingly without slowing down while other cars are required to stop and queue to pay the toll. The sketch includes the words "toll" and "Easy Pass" to convey some of the uniqueness of the system and a 500-m scale to provide added detail. However, there are still a few details not garnered from the sketch that are present in the textual description, such as the statements that both cash and credit payments will be accepted, that no slowing down will be needed, and that there will be an advertising campaign for early enrollment.

The combination of the sketch and an accompanying textual description will capture the advantages and mitigate the disadvantages of both conceptual model formats. Of course, our sketch should probably be refined a bit more before being presented to the decision-making executives. So, Figure 6.2 shows a visual model or visualization of our Easy Pass system. Certainly, more polished but not necessarily conveying any more information than our sketch. In fact, the diagram leaves off the 500-m scale—something the executives probably will not be interested in but engineers intending to build the system will definitely want to know.

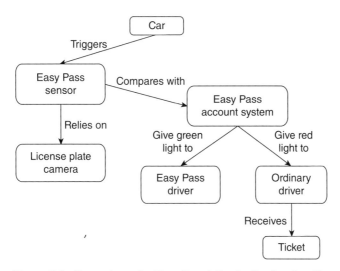

**Figure 6.3** Concept map for Easy Pass toll collection functionality.

So far, we have captured much of the structure of the system and implied some of the functionality. If we want to be more explicit in the various aspects of the functionality and interactions among system elements, we can allow for other constructs that are also conceptual models. One such construct that will capture overall functionality is called a *concept map*. Concept maps are very simple and informal directed graphs, where the nodes of the graphs capture the concepts or elements of the system and directed arcs (arrows) imply interactions between connected elements. Labels on the directed arcs define the nature of the interaction, and the element nodes are shown as labeled rectangles or circles. Figure 6.3 is a concept map that captures the toll collection functionality of the Easy Pass system. The flow in this case starts with the node that does not have an input arc namely the "car" node. Notice how all of the nodes map easily to real objects in the system and the arcs map nicely to operations on the objects. This follows the object-oriented nature of conceptual modeling. In fact, it is easy to see how a textual description can be derived from the concept map and vice versa by turning objects into nouns and operations into verbs. From Figure 6.3, we can obtain the following: (1) a car triggers the Easy Pass sensor that relies on a license plate camera, (2) the Easy Pass sensor compares some information with the Easy Pass account system, (3) the Easy Pass account system gives a green light to Easy Pass drivers and a red light to ordinary drivers, and (4) ordinary drivers receive a traffic ticket for not paying the toll.

This diagram provides an overall or high-level view of one element of the system's functionality. There are still many details that engineers would need to know in order to implement the system. How quickly does the system need to respond to give the drivers green or red lights? What happens if the license plate is unreadable? How much memory is needed in the account system? For these and other questions, more formal representation methods may be needed to remove ambiguity especially if automated analytic methods are to be used. Formal conceptual models include conceptual graphs and concept graphs that are essentially the same [2]. These models utilize additional rules and notations to construct unambiguous logic graphs that are typically used by the artificial intelligence community to support reasoning methodologies. The reader is left to explore independently these powerful techniques of representing conceptual models.

**Declarative Models**

Declarative models follow closely from conceptual models in that the system and elements of the system are described at a fairly high level of detail as opposed to a physics-based model of the dynamics of the system. Essentially, these models declare the status or state of each element of the system as they interact over time. Mappings are also provided to show how inputs to the system affect changes in the state of these elements. Inputs to the system are events that are considered instantaneous in contrast to states that have duration. Note that the beginning of a new state in itself can be considered an event, which leads one to the realization that events can be considered instantaneous states.

Real-world systems can be very complex with many elements, so keeping track of the changing states of these elements through discrete points in time may be a daunting task. Therefore, an initial challenge for declarative models is to parsimoniously choose appropriate elements, states, and events to appropriately model the aspects of the system of interest to study. This grouping of discrete states associated with the modeled system is known as its state space. We will illustrate these concepts by examining a few methodologies to implement declarative models—finite-state automata (FSA), Markov chains, and queuing simulations.

***FSA*** FSA are also widely known as finite-state machines. This modeling methodology represents states as circles and transitions between these states as arrows. Events that can trigger transitions to new states can be as simple as the end of the current state or as complex as a combination of conditions and behaviors of objects within the system. *Conditions* are requirements that must be satisfied before a transition can be made to another state. Outputs associated with FSA can be based on either the current state (*Moore machine*) or both the current state and a triggered event (*Mealy machine*). Figure 6.4 is an FSA model of our Easy Pass toll collection functionality using notation from Harel [3]. Harel is the name of an FSA notation that has been incorporated into a useful tool for modeling systems called the *Unified Modeling Language* (UML). UML is a standardized, semiformal, object-oriented, graphic notation that is used widely for modeling the design, implementation,

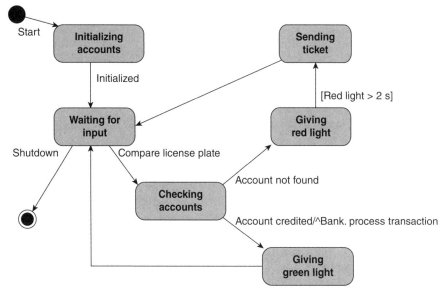

**Figure 6.4** FSA for Easy Pass account system object behavior.

and deployment of primarily software-focused systems. We will revisit UML later.

The figure describes the functionality or behavior of the account system object from the Easy Pass system model. The small black circle at the top left-hand corner of the figure is the start state, while the black circle with another concentric circle around it (near bottom left) symbolizes the end state. The other rounded rectangle shapes are also states. These states identify how the account system object changes over time. Labels on the transition arrows in most cases identify events that will transition the object from one state to another. A notable exception is the "[red light > 2 s]" label, which is a condition that has to be satisfied before the transition can take place. Another notable exception is the transition between the checking accounts state and the giving green light state. The transition occurs based on the "account credited" event, but then there is also an action that is performed. The slash symbol identifies that there is an activity to be performed by the account system object, while the "^" symbol indicates that the activity is the sending or invoking of an event on the bank object. The event invoked on the bank object is called processTransaction. Note that there are many parallels with the object-oriented design such as events being analogous to procedure or method calls on instantiated classes. A class is the name given to the component that defines the general structure and functionality (methods) for elements or entities that are of a common type.

Finally, the arrows without labels indicate that the transition will occur upon completion of the current state. Note that outputs in this model occur both on the transitions, as evidenced by the action of sending an event, and also within some states as implied by the states that show green and red lights. The green and red light indicators would be activated during the time within the respective states. Therefore, this model uses both Moore and Mealy finite-state machine constructs.

The model can be used to communicate the behavior of the various elements of the Easy Pass system as well as aid in analyzing the correctness of the behavior by tracing through the variety of scenarios and behaviors that can occur and by ensuring that all possible uses and faults are accounted for. But, what if we are unsure of various aspects of the system that can only be represented by a stochastic process? In such a case, we can make use of a different declarative modeling methodology such as Markov chains.

**Markov Chains** Professor Andrey Markov was a Russian mathematician whose work focused on stochastic processes. From him we get the *Markov property* that says that transitioning to a future state depends only on the current state and not on any of the previously visited states. A *Markov chain* is a stochastic process that exhibits the Markov property. Thus, a Markov model is a system model that can be described by state changing stochastic processes using a Markov chain.

Returning to our Easy Pass system, suppose that the license plate camera had only an 80 percent success rate and the combined probability of unknowing drivers and unscrupulous drivers is estimated to be 30 percent during the first 6 months, which will be the grace period for giving tickets. The executives want to know how much ticket revenue is lost during the grace period. We will use the FSA model in Figure 6.5 to help with this analysis. This time the

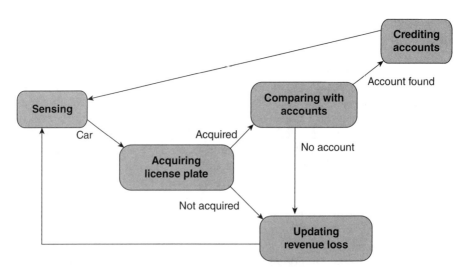

**Figure 6.5** FSA for Easy Pass revenue loss.

## MODELING METHODOLOGIES AND TOOLS 155

Table 6.1 Sequential state labels

| | |
|---|---|
| S1 | Sensing |
| S2 | Acquiring license plate |
| S3 | Comparing with accounts |
| S4 | Updating revenue loss |
| S5 | Crediting accounts |

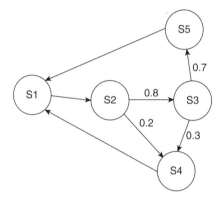

Figure 6.6  Markov model for Easy Pass revenue loss.

model does not show the behavior of a particular object but rather the overall system functionality that we are interested in. To simplify the model, we will substitute a sequential labeling convention for the names of the states and then utilize a Markov chain to create a Markov model.

The sequential labeling we will use is S1 for the first state, S2 for state 2, and so on. The mappings between sequential state labels and the state names are shown in Table 6.1.

Using this mapping and including the given probabilities, we have the following Markov model shown in Figure 6.6.

In Figure 6.6, arrows without labels are assumed to have a transition probability of 1.0. If a car is detected, the system will transition to S2, which is the acquiring of the license plate. At this point, there is a random chance of properly acquiring the plate number of 80 percent. Otherwise, the number of failures is incremented in S4. In general, each car will either transition through states {S1, S2, S4} with probability $1.0 \times 0.2 = 0.2$, {S1, S2, S3, S4} with probability $1.0 \times 0.8 \times 0.3 = 0.24$, or {S1, S2, S3, S5} with probability $1.0 \times 0.8 \times 0.7 = 0.56$. Therefore, 44 percent of cars eventually cause a transition through the revenue loss state. Of course, this model should be contrasted with an after-grace-period projected loss model, and a stochastic process describing the projected numbers of cars should be incorporated. Nevertheless, the model in Figure 6.6 can also be represented by a transition matrix as shown in Figure 6.7.

|       | Column j = |   |   |   |   |
|-------|---|---|-----|-----|-----|
|       | 1 | 2 | 3   | 4   | 5   |
| Row i = 1 | 0 | 1 | 0   | 0   | 0   |
| 2     | 0 | 0 | 0.8 | 0.2 | 0   |
| 3     | 0 | 0 | 0   | 0.3 | 0.7 |
| 4     | 1 | 0 | 0   | 0   | 0   |
| 5     | 1 | 0 | 0   | 0   | 0   |

**Figure 6.7** Revenue loss transition matrix.

In the transition matrix, $i$ refers to the number of the current state while $j$ refers to the number of the next state. The transition probability of 0.8 in row $i = 2$, column $j = 3$ is associated with transitioning from S2 to S3 and can be read as $P(2,3) = 0.8$. $P(i,j)$ is the probability of a particular transition. Note also that each row $i$ of the transition matrix must sum to 1.0. Using these models, we can plot a profile of the revenue streams and losses over time while employing different traffic loads for the system model. Let us now look at a different modeling methodology that can also help assess the impact of differing traffic loads on the system.

**Queuing Simulations** Of significant concern is the length of the ramp that ordinary drivers use to pay cash or credit for the toll. If traffic is busy enough and there are too many drivers not using Easy Pass, the queue of cars at the toll booth could potentially get so long that traffic on the main highway becomes disrupted. So the questions to answer are how long will the queue get during various times of the day and various days of the year, and what is the trade-off in loss of revenue versus length of ramp if we have to stop charging tolls when the traffic volume is too high for the pay toll booths to operate safely. A discrete-event *queuing simulation* is a methodology that can answer these questions.

Let us tackle the question of queue length by first examining data about the flow of traffic at the proposed site of the Easy Pass system. Our data collection team installed traffic counters for several months and found that the traffic during the day before the Thanksgiving Day holiday was by far the busiest day of the year. The executives decided that this was a statistical outlier that would have to be handled by intermittent no toll collection periods. However, they noted that the next busiest day of the year was equivalent to a sustained weekday rush hour interval and would like a weekday analysis as well as a heavy rush hour load analysis. From the traffic counters, we have the

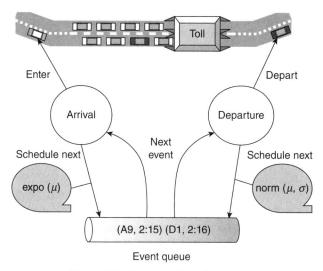

**Figure 6.8** Queuing simulation model.

changing rates of traffic flow for 2-h periods throughout the day averaged over many weekdays and, not surprisingly, the times between car counts are statistically similar to random draws from an exponential distribution having the traffic flow rate averages. Figure 6.8 captures the queuing simulation model for the system elements identified.

The model shows the two-lane toll booth with two separate queues, which we assume to be serviced by two separate toll booth cashiers. For simplicity, our model simulates just one queue and server combination. The server is the toll booth operator for the indicated lane. We make another assumption that cars entering the ramp will mostly choose the shortest queue and so the length of each queue on average will be the same. A final assumption is that the service time per car will be 2 min plus or minus an error of a few seconds provided by a normal distribution with a mean $\mu = 3$s and a standard deviation $\sigma = 4.5$. We already know the traffic rates per 2-h period, and the inverse of these will translate to interarrival (length of time between cars) averages for our random draws from an exponential distribution. For example, if the traffic rate is 50 cars per hour, this translates to a $1/50 = 0.02$h or 1.2 min between cars, but since we have two lanes, we can divide the rate in two to get $1/25 = 0.04$h or 2.4 min between cars entering into one lane of the toll ramp. Therefore, our interarrival distribution is exponential with $\mu = 2.4$.

This time the focus of this declarative model is on the arrival and departure events shown as the large circles in Figure 6.8. An arrival event signals the entry of the next car into this queuing simulation aspect of the Easy Pass system. A new entity arriving will enter the queue if one already exists or immediately begin being serviced. The entry of this entity will immediately cause a new arrival to be scheduled. An entity or car that has completed paying the toll will then depart the system causing a new departure to be scheduled for the next entity beginning service. Not shown in the figure is the fact that if there are no cars already in the system, a new arrival will not be queued but rather will go straight into service and therefore would also trigger a departure to be scheduled in addition to the scheduling of the next arrival. Newly scheduled arrival or departure events are placed in the *event queue* element shown at the bottom of Figure 6.8. All events are tagged with a time stamp, which indicates when the event should occur in time. The event queue in the figure shows that the arrival event #9 (A9) will occur at time 2:15, while the departure event (D1) will occur at time 2:16. The numbers on the events also correspond to the specific entity or car that entered into the simulation. The departure time for D1 minus the arrival time A1 will tell us how much time car #1 spent in the system. Since we are interested in the length of the queue, we will keep track over the entire day of how many cars enter the system and how many cars leave the system. The difference between these will indicate what was the queue length at a particular point in time. The maximum value from a plot of this quantity will answer our question about how long will the queue get, which will translate to how much space on the ramp is needed. We can compare these results with a steady-state queuing theory formula for the average length of the queue using the arrival and service rates above.

By making further assumptions such as an exponential service time instead of a normal distribution, the approximate average length of the queue is 4.2 cars using a relatively simple queuing theory formula. This can serve as a sanity check for our simulation model. However, as we have determined, the traffic rates change about every 2 h, while the queuing theory equations are for theoretically infinite runs. Our queuing simulation model will provide better estimates that are tailored to the Easy Pass system parameters. After having determined the queue length, we note that we also need to accommodate enough space for deceleration as cars come off the main highway and onto the toll ramp where they will come to a stop. We will examine this situation using a different type of model.

### Functional Models

Let us consider that in addition to knowing the average and maximum queue lengths for the toll ramp, we need to know that there is adequate stopping distance between the end of the queue and a car that has exited the main highway. So given the speed limit for this stretch of road and assuming good weather conditions that permit traveling at the speed limit, what is the stopping distance that is needed. The following diagram shows the situation and the parameters of interest. The stopping distance is given as "$d$" and the initial velocity as a vehicle enters the toll ramp is $v_0$. The resistive force due to friction after brakes are applied is given as $F_f$, which is equal to the coefficient of friction $\mu$ multiplied by the force applied by the weight of the vehicle, that is,

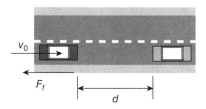

**Figure 6.9** Easy Pass deceleration diagram.

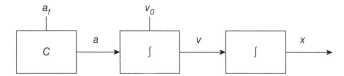

**Figure 6.10** Easy Pass deceleration block model diagram.

$F_f = \mu m g$ where $m$ is mass and $g$ is the acceleration due to gravity. Since this force is applied in the opposite direction to the movement of the car, the result is actually a deceleration (Fig. 6.9).

We refer back to physics-based relationships known as *first principles* to obtain our governing equations based on velocity, deceleration, and distance. The well-known equations $a = dv/dt$ and $v = dx/dt$ indicate that acceleration is the derivative of velocity with respect to time, and velocity is the derivative of distance with respect to time. With this information, we can create our functional model and simulate it to obtain our results. In Figure 6.10, we create a block model with these relationships. A *block model* is a connected network of components that have transfer functions with different capabilities depending on the type of component a block or connectable icon is representing. Inputs to a block will produce outputs via an internal equation. The integrator blocks shown by the integration symbol "∫" provides the derivative relationships described by the physics-based equations above.

Besides the integrator blocks, the figure also shows a constant block that will output the constant deceleration value obtained by multiplying the coefficient of friction with the value for $g$, which is equal to 9.8 m/s². If we assume that the friction coefficient on a nice weather day is 0.75, the deceleration $a_f$ will be equal to $\mu g = 0.75(9.8)$ or 7.35 m/s². Also, based on the speed limit, the initial velocity $v_0$ will be approximately 25 m/s. Configuring our block model with these initial values and simulating for 5 s, we get the following results.

Figure 6.11 shows two plots in which both horizontal axes display time. The plot on the left in Figure 6.11 shows the output of the middle velocity block in the block model. Note that the plot starts initially from 25 m/s on the vertical axis and consistently decreases until zero velocity. This deceleration occurs in about 3.4 s. The right plot is the output of the distance block in the block model

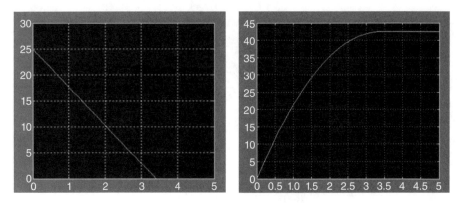

**Figure 6.11** Deceleration time and distance results.

(Fig. 6.10). The vertical axis here is distance traveled in meters. So, over the same amount of time, it can be seen that the car travels about 42.5 m, which answers our question about the stopping distance needed for cars entering the toll ramp from the main highway. The next section looks at this same issue but from a different perspective.

## Constraint Models

As mentioned previously, *constraint models* are similar to functional models except that the directed nature of the connected components, as can be seen from the block model, is less important than the physics-based relationships among system components and the balancing of those physical properties. Here we note that conservation of energy is involved. The law of conservation of energy tells us that the total energy in a closed system remains the same although some or all of the energy may be transferred from one form of energy to another. Therefore, we have a constraint that allows us to balance one form of energy with another. Such a concept is useful when looking at multidomain systems as different forms of energy are likely to involve elements where different physics-based equations apply. For instance, a light bulb transforms electrical energy into light and heat energy. All three domains have their own energy equations but can be balanced to show the transfer and relationships necessary to perform a proper analysis of the system.

Our Easy Pass stopping distance problem can be approached in this manner. First, the kinetic energy associated with the moving car is equal to $½mv^2$ where $m$ is mass and $v$ is velocity. When the car comes to a complete stop, its velocity is zero and so this form of energy goes to zero. Much of this energy loss is dissipated as heat. The change in energy here is equal to the work done in stopping the car, which essentially is the force due to friction $F_f$ applied in the opposite direction over the stopping distance. We learned earlier that the friction force $F_f$ is equal to $\mu mg$, so our constraint model is

$$\tfrac{1}{2}mv^2 = F_f d$$

or

$$\tfrac{1}{2}mv^2 - \mu mgd = 0.$$

Solving for the stopping distance $d$ we get

$$d = \tfrac{1}{2}v^2/\mu g.$$

The initial velocity we used for our car was 25 m/s and so $\tfrac{1}{2}v^2$ is equal to $625/2 = 312.5$. From before, we know $\mu g$ to be equal to 7.35 m/s$^2$. So, our stopping distance $d$ is equal to $312.5/7.35 = 42.517$ m, which is equivalent to what we obtained using our functional block model. The power associated with this energy transfer is equal to energy transferred per unit time. Assuming a 1088-kg vehicle and remembering that the vehicle was stopped in approximately 3.4 s, the power associated with this system is $1088(312.5)/3.4 = 100,000$ J/s or 100 kW.

In our Easy Pass system example, these relationships were kept within a single domain, but if we were using a motor to crank a generator to in turn create electricity that powers a lantern where our objective was to measure the heat from the lantern, these relationships would span many domains. The energy balance across the domains would create constraints that enable the needed analyses. Bond graphs are tools that often used to allow this multi-domain modeling. Bond graphs are based on a generic model of power transfer in various domains and can span electrical, mechanical, hydraulic, thermal, and chemical systems. The generic model focuses on effort and flow where the product of the two is power. Effort in the mechanical domain we have modeled is equivalent to force, while flow is equivalent to velocity. They are called bond graphs because the elements in the graph specify the relationships between effort and flow on the connections or bonds in the graph.

## Spatial Models

Spatial models are multielement models where either the spaces in the system are the elements or the entities within the spaces are the elements. An additional trait of these models is that elements are treated in a somewhat regularized manner so that rules of updating spaces or the behavior of entities within the spaces are uniform for many, if not all, of the elements. A common modeling method in which the space is the element that is focused on is the cellular automaton. In general, cellular automata are regularized matrices of cells in which each cell has the same set of rules to apply in order to update the information within itself. A given cell may also utilize the current information in neighboring cells but may not change information in other cells. The game of life is a popular implementation associated with cellular automata. The rules are very simple:

(1) A cell is either alive or dead.
(2) A cell stays alive if there are two or three other live cells around it.
(3) If there are less than two live cells around, it dies of loneliness.
(4) If there are more than three live cells around, it dies of overcrowding.
(5) If a cell is dead and there are three live cells around, it becomes alive.

The cells in the grid have a different color depending on whether they are alive or dead. As the rules are applied, the colors in the grid change and shift sometimes dramatically and sometimes none at all. Depending on the initial marking of alive and dead cells, various patterns can emerge that could be used to model real-life phenomena.

For our Easy Pass system, we will draw upon rules governing the transfer of heat using a different spatial model but also dividing up our system into elements. We will utilize a lot of assumptions for this one. The executives are concerned that the coefficient of friction we are using will change due to the heat from the car tires when pulling off onto the toll ramp and applying brakes. We will assume that brakes are applied when the car is traveling in one direction and that the maximum heat is transferred at the moment that brakes are applied. Also, we will look at a small one-dimensional segment of road and treat it as an isolated and homogeneous section of material. So we have the following figure.

Figure 6.12 is our section of road where a tire has applied one unit of heat to the center of the section. The assumed uniform material will conduct heat in both directions and so we want to study how this heat will dissipate over time to eventually determine based on time between vehicle arrivals whether or not this heat will linger to affect a change to our coefficient of friction used in the previous evaluations. To start, we divided our material into nine equal parts as shown in the figure, so this is indeed a multielement model and because we will use a partial differential equation (PDE), the heat equation, to model the relationships between elements; this is also a finite element

**Figure 6.12** Idealized section of the road affected by friction heat.

| 1 | 2 | 3 | 4 | 5 | 6 | 7 | 8 | 9 |
|---|---|---|---|---|---|---|---|---|
| 0 | 0 | 0 | 0 | 1 | 0 | 0 | 0 | 0 |
| 0 | 0 | 0 | 0.25 | 0.5 | 0.25 | 0 | 0 | 0 |
| 0 | 0 | 0.06 | 0.25 | 0.38 | 0.25 | 0.06 | 0 | 0 |
| 0 | 0.016 | 0.094 | 0.234 | 0.313 | 0.234 | 0.094 | 0.016 | 0 |
| 0.004 | 0.031 | 0.109 | 0.219 | 0.273 | 0.219 | 0.109 | 0.031 | 0.004 |
| 0.010 | 0.044 | 0.117 | 0.205 | 0.246 | 0.205 | 0.117 | 0.044 | 0.010 |
| 0.016 | 0.054 | 0.121 | 0.193 | 0.226 | 0.193 | 0.121 | 0.054 | 0.016 |
| 0.021 | 0.061 | 0.122 | 0.183 | 0.209 | 0.183 | 0.122 | 0.061 | 0.021 |
| 0.026 | 0.066 | 0.122 | 0.175 | 0.196 | 0.175 | 0.122 | 0.066 | 0.026 |

**Figure 6.13** Idealized section of the road affected by friction heat.

model (FEM).* The heat PDE specifies the rate of temperature change with respect to time and position, while ordinary differential equations define rates of change only with respect to time. Here we define the heat equation in one dimension,

$$\frac{\partial T}{\partial t} = \frac{\partial^2 T}{\partial x^2},$$

where $T$ is temperature, $t$ is time, and $x$ is position.

Estimating this one-dimensional heat equation by the finite difference method gives us the following finite difference equation:

$$T(x, t+\Delta t) = T(x,t) + \frac{\Delta t \{T(x+\Delta x, t) - 2T(x,t) + T(x-\Delta x, t)\}}{\Delta x^2},$$

where $\Delta x$ is the incremental change in position and $\Delta t$ is the incremental change in time.

Applying this equation to our Easy Pass idealized road section while using $\Delta x = 1$ and $\Delta t = 0.25\,\text{s}$, the next value for the road element at position 5 will be $T(5, 0.25) = 1 + 0.25(0 - 2 \times 1 + 0)/1 = 1 - 0.5 = 0.5$ after the initial application of 1 unit of heat. Values at $x$ positions 4 and 6 at the first time step are $T(4, 0.25) = 0.25$ and $T(6, 0.25) = 0.25$, respectively. Figure 6.13 shows the change in heat over 2 s of incrementing by 0.25-s time steps.

This analysis shows that after 2 s, the amount of heat transferred to a particular element dissipates to less than 20 percent (0.196) of its original value. We can use these results to someday determine whether or not to change our friction coefficient in our previous analysis.

*Adapted from Stoughton J. ECE 605 Lecture Notes, Old Dominion University, Fall 1999.

As mentioned in the beginning of this section, cellular automata are spatial models where the spaces in the system are the elements that are treated with regularized rules of updating the spaces. Spatial models also focus on systems where the entities within the spaces are the elements with regularized behavior for updating the states of entities. Multiagent modeling is a common modeling method that focuses on the entities within the space. Agents are software processes or objects that typically represent an element of the real-world system that provides a useful capability within the system. An agent's actions are governed by rules of behavior that are uniform for many, if not all, agents of the same type. When a system is primarily modeled by several agents acting together to accomplish a common goal, this system model is called a multiagent model.

Probably the most interesting characteristic of agents is their autonomous nature. Their rules of behavior allow them to know when and how to act on the system as necessary. When a global overall behavior occurs from the individual autonomous activities of single agents working disparately or cooperatively, the behavior is said to be *emergent* especially if the behavior was unexpected.

Multiagent models are used extensively in modeling human behavior such as in crowd models or in military battlefield scenarios, where a mathematical model of the whole system would be too complex with too many simplifications needed. As a result, validation methods for these models are a subject of much discussion and research. For our Easy Pass system, agents could obviously be the cars themselves, but they could also be the drivers as well. In fact, depending on the question being asked, our entity-based multielement Easy Pass spatial model could include other agents such as the camera, sensor, toll booth, toll booth operators, and so on.

### Multimodels

*Hybrid models* play an important role in most real-world systems. As we have seen from our Easy Pass system, several perspectives can be gained where each can answer a different question of import. To fully answer the range of questions one may have for a system, a combined group of model types may be the answer. *Multimodels* are composed of several models carefully connected in a network or graph where the models could easily be of more than one type. Such a hybrid model can employ a number of abstraction perspectives and can address a wider variety of questions. Additionally, large complex systems may have innate phases where different types of models are needed to define the activity occurring at those different stages.

Fishwick emphasizes the potential differences within multimodels by describing the relationships between aggregation and abstraction [1]. With abstraction, simulation of the model occurs with possibly every abstracted element of the system and since one element of the system may be abstracted to a greater degree than another, care must be taken when passing information from one modeled component to another. An important point is that more abstraction means more loss of information, while more refinement means an increase in information. Therefore, a datum in a more refined component will lose information when transferred to the more abstracted component, and a record transferred in the other direction will need to be supplemented with additional data. With decomposition, the multimodel is hierarchical with the actual simulation of the model occurring at the lowest decomposed level. Hence, no information is passed from the simulated (lowest) level to the higher levels. The higher levels serve only to arrange lower-level components in a design-friendly and semantically intuitive manner.

## ANALYSIS OF MODELING AND SIMULATION (M&S)

In addition to communicating the main qualities and capabilities of the target system, the raison d'être for system models is to learn something about the actual system. *Analysis* is the process we use to obtain the information. However, when we learn that something, steps should have been taken to ensure that the model has a high probability of being correct. Therefore, before this final analysis, we must verify that the implemented model is what we conceived would be needed to answer our questions about the system, and we must ensure that this model is providing valid reliable output similar to the real system or at least similar to a comparable system if the real one does not exist yet. These procedures are called verification and validation. In addition to ensuring that the conceived model was implemented, verification determines whether the correct variables are available and able to be measured, which may be needed to answer specific questions about the model. The most elegant implementation is useless if the necessary data sources are not available. Following, validation determines if the output data generated by the model is close enough to those generated by the real system so that we can say that there is no statistical difference between the model and the system when examining the required variables. This brings up a very important point in that often we do not need to model the whole system to learn what we want to know and being parsimonious with our modeling saves time and money. Since these savings begin when we are conceiving the model, we will first discuss tools for creating models ready for analysis followed by the analysis of the models themselves.

### Formal Modeling

Even before verification and validation, analysis is driven by choices made in the design, methods, and tools used to create the model. As we have learned, models can represent high-level system views where details are few or low-level system views where there are many details. These different abstractions do lead to intrinsic ambiguities that may or may not lead to

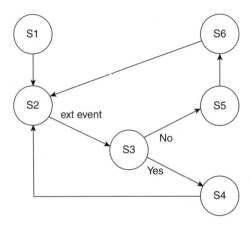

**Figure 6.14** Simplified FSA from Easy Pass account system object behavior.

difficulties depending on the questions being asked of the model. Nevertheless, when we know that a model will be subject to significant analyses, we may want to start out with modeling methods that promote specific rules on how to properly construct the model, so that these rules can be used in subsequently interpreting and analyzing the model. Methodologies that use precise rules about modeling and simulating are called *formalisms*. A model created using a formalism is a formal model. This confers the connotation that the model is unambiguous and mathematically manageable often allowing automated analyses.

Since utilizing rigorous characterizations of system elements provides the basis for system model analytics, we will first discuss system model description techniques. A system is described formally as a set or tuple with variables such as time, inputs, outputs, possible states, state transition functions, and output functions. From Zeigler et al., we get the following notation [4]:

$$\text{System tuple} = <X, S, Y, \delta_{int}, \delta_{ext}, \lambda, ta>,$$

where $X$ is the system model set of inputs, $S$ is the set of states that the system model can achieve, $Y$ is the system model output set, $\delta_{int}$ is the internal transition function that describes how the states within the system transition due to internal behavior, $\delta_{ext}$ is the external transition function that describes how the states transition due to external input events, $\lambda$ is the output function that describes what output a state will generate, and $ta$ is the time advance that describes how the system changes state with regard to time—for example, either discretely or continuously.

Using this notation, we can describe our previous Easy Pass FSA from Figure 6.4. This time we will map the state names to numbered state labels so we will not need the start state, and we will also leave off the end state to create Figure 6.14. The state name to state label mapping is below (Table 6.2).

Table 6.2 State label mapping

| | |
|---|---|
| S1 | Initializing accounts |
| S2 | Waiting for input |
| S3 | Checking accounts |
| S4 | Giving green light |
| S5 | Giving red light |
| S6 | Sending ticket |

Note that many of the event labels were left off the diagram except one, which is an external event triggered by a passing car and two internal events. The name of the external input event from Figure 6.4 is "compare license plate" and is represented in Figure 6.14 simply as "ext event." The other two labeled events refer to whether or not the driver of the car has an Easy Pass account—"no" refers to "account not found" and "yes" refers to "account credited." All other events that cause state transitions for the FSA are internal events that can be labeled "processing completed"; however, we will leave this label off since the unlabeled arrow means the same thing, which is to transition when processing in the state has completed.

Using our systems theory notation above, we can now formally describe this system. The set of inputs for the model is $X$ = {ext event, yes, no, processing completed}. The state set for the system is $S$ = {S1, S2, S3, S4, S5, S6}. The set of outputs is $Y$ = {Green Light, Red Light, Ticket, Bank.processTransaction}. For our external transition function we have

$$\delta_{ext}(S2, \text{ext event}) = S3;$$
$$\delta_{ext}(Sn|n \neq 2, \text{ext event}) = Sn.$$

The first statement means that when the system model is in state S2 and the external input event is triggered, the system will transition to S3. The last statement simply means that the external transition function does not have an effect on other states besides S2. For our internal transition function, the definitions are $\delta_{int}$(S1, processing completed) = S2; $\delta_{int}$(S2, <any>) = S2; $\delta_{int}$(S3, yes) = S4; $\delta_{int}$(S3, no) = S5; $\delta_{int}$(S4, processing completed) = S2; $\delta_{int}$(S5, processing completed and $ta > 2$) = S6; and $\delta_{int}$(S6, processing completed) = S2.

Note that any other inputs that occur other than those shown above will result in the system model remaining in its current state. Also, the definition $\delta_{int}$(S2, <any>) = S2 is provided to show that internal inputs do not affect a transition out of S2; <any> means any event. Finally, the output function $\lambda$ is based on the output of the current state whose output will be one of the members of set $Y$, and the time advance function $ta$ we will say is based on the set of positive integers and is used at least in S5 to satisfy the condition that the red light is on for at least 2 s before transition to S6.

Zeigler's formalism to capture state-event declarative models such as this Easy Pass FSA is called the *Discrete-Event System Specification* (DEVS).

There are other systems theory formalisms for other model types and methods. Functional models may be better described by the Discrete Time System Specification (DTSS) and constraint models such as bond graphs by the Differential Equation System Specification (DESS).* For DEVS, the basic component is an *atomic model*, which contains states and events as we have been discussing. Additionally, the atomic model is encapsulated with external communication only through input ports and output ports. For our Easy Pass FSA model, the input port would receive the external input that senses a car while there would be output ports for our identified external outputs (Green Light, Red Light, and Ticket). We may or may not consider Bank. processTransaction an internal output, which may go to a different element in the overall Easy Pass model. DEVS uses these input and output ports to connect atomic models together to compose *coupled models*. Therefore, coupled models contain sets of lower-level (more detailed) models and sets of their own input and output ports, as well as the coupling specifications between its own internal models and to other external coupled models.

## Semiformal Modeling

The advantage of formal models is that they provide a rigorous mathematical approach for representing dynamic systems that allow unambiguous understanding of characteristic system behavior and thorough analyses of system capabilities, properties, limits, and constraints. The disadvantage of formal models is that they are complex and somewhat tedious to design and implement. *Semiformal representation* methods are a trade-off that allows ease of use with less unambiguity, but with still a degree of analytic capability. One such representation method is the UML. As mentioned previously, UML is a standardized, semiformal, object-oriented, graphic notation that is widely used to model system and software structure and behavior.

UML is unified because at the time it was conceived, it brought together several prominent object-oriented design and analysis methodologies by Grady Booch, Jim Rumbaugh, and later by Ivar Jacobson—also known as the three amigos. The new unified modeling methodology supports the full life cycle of software-based systems from requirements to deployment with many design artifacts to ease update and maintenance. We will discuss several of its features including use case, activity, class, sequence, and statechart diagrams.

**Use Cases** *Use cases* are very important tools for linking the needs of the user with the design of the system. They are textual descriptions that describe a particular capability of the system. The typical use case begins with an external input event, which may be as simple as the user pushing a button or a car tripping a sensor and ends with an external output such as an update

---

*See Zeigler et al. [4] for descriptions of DTSS and DESS.

to a graphic display or the issuance of a ticket. The textual description of the use case then provides the conditions under which the use case can be activated, the dynamic internal interactions that take place to carry out the use case activity, any alternate threads or exceptions within the use case activity, and the conditions for ending the use case. The activities within the use case are associated with capabilities that should be tied to corresponding requirements that were identified for the system. Recall our text-based conceptual model of our Easy Pass system. We can write the requirements for this system as follows:

(1) The system will be placed on a two-lane highway.
(2) There shall be sufficient space to construct a toll ramp.
(3) The toll ramp shall contain a toll area that will provide cash and credit toll payment.
(4) The lanes of the main highway shall contain the Easy Pass electronic toll collection capability.
(5) Drivers enrolled in the Easy Pass system will remain on the main highway.
(6) Easy Pass drivers' accounts shall automatically be collected as their cars drive by using the constructed highway sensors.
(7) The Easy Pass account system shall compare passing license plate numbers with those of registered Easy Pass drivers.
(8) Registered Easy Pass Drivers on the main highway shall be given a green light.
(9) Nonregistered drivers shall be given a red light for at least 2 s and automatically issued a ticket to be received via the postal service.
(10) The bypass ramp toll lanes shall have sufficient space for non-Easy Pass users to queue while paying the toll.

Here, we use the word "shall" to emphasize that it is important to exactly meet this requirement, and we use the word "will" to say that this is most likely the way we want this requirement done. For example, requirement #1 leaves room for the circumstance that the system might be placed on a four-lane highway. Next, we have our use case as follows:

*Use Case Name.* Easy Pass Issues a Ticket

*Use Case Description.* When the license plate camera captures a license plate, it will send this information to the accounting component of the Easy Pass system, which will process this information as described here. Requirements 4, 6, 7, 8, and 9 are covered by this use case.

*Entry Conditions.* This use case begins when a car is detected by the license plate sensor and the licensed plate is acquired, thereby, providing a license plate to be compared.

*Main Thread.* The license plate number received is compared with the database of records in the accounting system. If a record is not found, the driver is given a red light and a loud buzzer sounds. Then, the accounting system records that a ticket must be issued to the driver.
*Alternate Threads.* If the record is found, the accounting system will credit the driver's account and the driver is given a green light.
*Exceptions.* The acquired record of the license plate may be empty indicating that it was not acquired properly. In this case, nothing will be done except wait for the next license plate notification.
*Exit Conditions.* This use case is completed when either a green light has been given to the driver or a record of the toll runner has been made and tagged for ticketing.

The use case name is "Easy Pass Issues a Ticket," which passes a rule of thumb that use case names should be at least five words long.* This is to prevent software engineers from thinking too quickly ahead and giving class specification names to use cases, which can become confusing. Note that the use case covers only requirements 4, 6, 7, 8, and 9, so there may be other use cases to cover other requirements such as requirement #3. Requirements 1, 2, 3, 4, 5, 6, and 10 have hardware/construction connotations, which means that requirements 3, 4, and 6 have both hardware and software considerations. One other thing to notice is the extra capability of a loud buzzer added in the main thread. This might have seemed like a useful and normal thing to do, but it was not required by the user as evidenced by a missing requirement for it. The addition of this capability is called requirement creep and can easily generate cost overruns in time and money. If it seems like the right thing to do, simply go back to the user and ask if it is wanted or needed so that it can be added as a requirement and budgets can be adjusted as necessary.

When the use case is finally graphically represented on a use case diagram, an oval encircling the name of the use case is used. The "Easy Pass Issues a Ticket" use case will be placed along with other use cases on the diagram with any relationships among them shown on the diagram. Also shown on the diagram will be icons representing actors or external entities that interact with the use cases and provide external input or an external outlet. For example, as in our concept map of the Easy Pass system, an actor representing a car will be placed on the diagram to indicate the outside of the system trigger for the use case. The next step is to convince ourselves that we have captured the required capabilities of the system and then create dynamic representations that elaborate the use case using design artifacts placed within interaction diagrams.

*This is the author's rule of thumb.

***Activity Diagrams*** Among other things, activity diagrams are useful in providing a high-level view of the behavior of the system from the time the system is brought online until it is no longer in use. These diagrams show how capabilities captured by use cases fit into the overall operation of the system. Consequently, they are often utilized to ensure that there are no missing capabilities not captured by implemented use cases. Activity diagrams are in essence decision flowcharts containing control, decision, and action elements that graphically describe the flow of the system throughout its life cycle.

***Class Diagrams*** As with many modeling tools, UML is concerned with capturing both the structure and the behavior of systems. Structure is concerned with how elements within systems are connected, while behavior is concerned with the dynamic way these elements interact. *Class diagrams* capture the specifications of classes and the static structural relationships among them. As previously mentioned, a class defines the general structure and functionality (methods) for objects that share commonalities. For instance, the class "automobile" can be the general name given to many instances or car objects that can be uniquely identified by their license plates and color and so on, but they are all generally built the same way and perform the same function.

Besides the expression of the attributes and operations that specifies classes, class diagrams show how each class is related to other classes in the system model. Since this is an object-oriented model, these specifications and relationships are often referred to as an object model. Relationships that can be represented include inheritance, aggregation, multiplicity, and generic association. Inheritance signifies that child classes can include or inherit from a parent class all of the attributes and methods belonging to that parent, while aggregation expresses a whole/part relationship where the class, which is the whole component, is composed of one or more other connected classes. *Multiplicity* identifies how many of one class is generally associated with how many of another. When a special relationship such as inheritance or aggregation does not exist, the connectivity is generally called an association and can be labeled explicitly with text to define the role of the relationship.

***Sequence Diagrams*** *Sequence diagrams* are a form of interaction diagram that expresses the dynamic interactions among objects and also frequently with actors. These representations elaborate the use cases by linking the object model defined during the creation of the class diagrams with the sequential activities defined in the use case. Therefore, sequence diagrams display objects and actors along with events (method calls and external inputs) occurring between them in a timeline fashion with time starting at the top of the diagram and advancing toward the bottom of the diagram. Figure 6.15 illustrates a sequence diagram based on our "Easy Pass Issues a Ticket" use case described earlier.

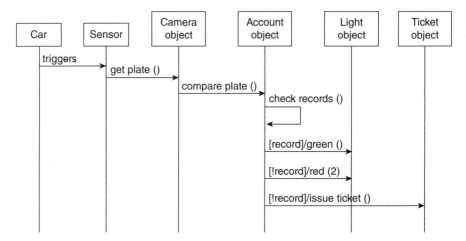

**Figure 6.15** Sequence diagram for Easy Pass issues a ticket use case.

In the diagram, the first two rectangles are actors that interact with the system and would be portrayed in the use case. These represent hardware components that would not have software associated with them, while the third rectangle represents a software object that would additionally have an associated hardware component. An interaction from the sensor invokes a method on the camera object that in turn invokes a method on the account object causing the comparison of the license plate record with the database of Easy Pass account holders. The comparison is performed by the account object. Arrows point to the objects that would carry out the specified behavior labeled on the arrow.

Note that the account object has an arrow pointing to itself. This represents an internal method invocation and can be placed on the sequence diagram to show that it is happening and also to indicate a performance requirement such as finding the record within 30 ms. Alternate paths on the sequence diagram are created using conditions of either finding the record [record] or not finding the record [!record] with the "!" symbol representing not. Then, the method invocations are given as actions or events after the "/" symbol. An example method invocation on the light object is LightObject.green(). The red method invocation is given with a parameter value of 2, which presumably means to stay on for 2 s. Finally, the ticket object is asked to issue a ticket. Clearly, this technique is useful in understanding and communicating the behavior associated with use cases and of the system.

**Statecharts** David Harel *statechart* notation was incorporated into UML to represent the dynamic behavior of objects and is typically used on especially complicated objects in the system model [3]. We have previously described this notation and illustrated its use on a finite-state machine model of the Easy

Pass account system object behavior (Figure 6.4). Most UML tools allow the development of such state machines. Some of these tools also allow the generation of software source code based on the UML object model and associated dynamic representations. This is a significant advantage of these tools. However, a disadvantage may occur when keeping the model design in sync with the generated code, so some tools also have the capability to reverse engineer the source code into an object model with mixed success.

**Other UML Diagrams**  UML utilizes many other diagrams that are effective for different stages of development. There are a few other types of interaction diagrams than sequence diagrams, and there are diagrams that help design how to package the software into files and libraries. There are also diagrams that express the interfaces with other components and describe how to deploy the software onto networked computer systems. UML is indeed the most complete system and software modeling tool to date.

## Model and Simulation Analyses

When we analyze a model, we consider the robustness and completeness of the model itself as well as the behavior of the model under dynamic execution. After conceiving the model, the first form of analysis is verification. During *verification*, we ensure that the requirements of the conceptual model have been properly translated in the executable system model and that the model is able to be run with no unforeseen errors and with reliable results. This can be done with automated debugging tools, manual code reviews, and model stress tests. Automated debugging tools provide syntax and some logic support in finding and fixing bugs, but finding semantic errors can be difficult. Manual code reviews are performed by sitting around a table and explaining the code to peers that can critique and find problems. These can be quite useful provided ample time is available and everyone can remain awake. Model stress tests are performed by running or executing the model with inputs set to their maximum and minimum boundary conditions to ensure that the model performs as expected.

*Validation* is mainly concerned with making sure our robust model provides useful results by being a close enough approximation to the target system. This form of analysis is done by performing statistical tests on the difference between the output of the model and the output of the target system. If the target system does not exist, a similar system may be used or an expert knowledgeable about the system or similar systems can examine the model output and make a decision as to the models' validity based on his or her past experiences. This type of subjective validation is called face validation. After validation, the model is ready to answer the question for which it was created to answer.

When we run the software program that executes the model, we say that we are simulating the model. When we simulate the model under a variety of input conditions, we say that we are experimenting with the model. It is through this form of analysis that we finally get our questions about the system answered. But, believe it or not, we still need to do a bit of design—experimental design. Experimental design is concerned with reducing the time and effort associated with simulating by identifying the information needed to be gathered from each simulation replication, how many replications need to be made, and what model parameter changes need to be compared. If the model contains stochastic processes and many parameter change considerations, millions of simulation replications could be needed. A replication is an execution of the model from start to finish in typically faster than real time. If the model is very complex, it may take several hours or days between runs. An experimental design strategy can save time and money when dealing with complex system models. One way it does this is by identifying and ruling out the change of parameters that may not be contributing to the desired output variables. Another way is by assigning high and low values to candidate parameters and running simulations using combinations of the high and low values to see which ones are having an effect. Candidate parameters are those that are likely to have an effect on the response of the output variables for which we have an interest.

## OR METHODS

When the system is well understood but contains many factors and many constraints, an optimal solution is difficult to find. OR is the discipline of using mathematical models applied to complex systems in order to optimize various factors that influence the performance of the system and to make business decisions that maximize profit or minimize cost. The solution space for an optimal or near-optimal solution is extremely large, so the art of OR is to find mathematical relationships that prune the solution space. Steps involved in finding the maxima or minima needed for the optimal solution are first to formulate the problem, next to develop a mathematical model to represent the system, and finally to search for a solution to the problem by solving the mathematical model. *Linear programming* is a deterministic technique often used to find optimal solutions to OR problems by solving linear equations of system variables including objective functions that describe what must be maximized or minimized for the system subject to several inequality constraints. OR methods can also use stochastic methods such as Markov chains and queuing theory. However, let us look at a linear programming solution to our Easy Pass system.

The executives would like to provide a park and pool service next to the Easy Pass system toll ramp. This service offers to customers the ability to park in a parking lot next to the toll ramp and then take either an electric van (EV) or an electric car (EC) to two designated metropolitan subway stops on different subway lines—one nearby and one farther away. The electric vehicles

Table 6.3 Metro stop demand

| Demand | Near Metro Stop | Far Metro Stop |
|---|---|---|
| Electric vans | 60 | 20 |
| Electric cars | 40 | 30 |

are part of a green initiative named Easy Green Commute (EGC) that is intended to appeal to environmentally conscious patrons as well as to customers wanting to avoid the inconvenience and expense of downtown traffic and parking. Metro subway stop parking is typically $15 per day, and downtown parking can be even more. To utilize the service, registered clients simply reserve either a car or a van and choose any available from the parking lot. There will be designated parking spaces at the metro stops for the EGC vehicles, making access to the subway easy and convenient. There are already plans to use the ECs for a lunchtime downtown service, so the executives have agreed to guarantee a minimum of 20 ECs at the near metro stop.

The EGC parking lot at the Easy Pass toll ramp has room to support 100 EGC vehicles and clients' cars. The executives have estimates for customer demands for taking vehicles to the metro stops, which are shown in the following table.

Table 6.3 shows that the highest demand is for taking the EVs to the near metro stop as this is the most economical way to downtown if a commuter group is large enough. The demands for other options are as shown in the table.

EVs cost $200 a day to reserve for the near metro stop and $220 a day for the far metro stop, while EC's cost $100 a day for the near metro and $150 a day for the far metro based on maintenance and power charging costs for the vehicles. The executives want to know how many EVs and ECs should be ordered to maximize the profit from this enterprise and how many of each should be earmarked to service the metro stops. The objective function for the linear program then is

$$\text{maximize } f(x) = 200 \times EV\_NM + 220 \times EV\_FM + 100 \times EC\_NM + 150 \times EC\_FM,$$

where $EV\_NM$ is the # of EVs for the near metro, $EV\_FM$ is the # of EVs for the far metro, $EC\_NM$ is the # of ECs for the near metro, $EC\_FM$ is the # of ECs for the far metro, and $x$ is a vector of the four variables ($EV\_NM$, $EV\_FM$, $EC\_NM$, and $EC\_FM$).

The objective function above adds the revenues from the total number of vehicles going to a particular metro stop multiplied by the rental costs for the particular type of vehicle. Note that this is a linear function, so constructing a mathematical model using this technique is called linear programming. The objective function is subject to a number of inequalities that are gleaned from the information we know about the system. From the demand information, we get the following inequalities:

$$EV\_NM \leq 60,$$
$$EV\_FM \leq 20,$$
$$EC\_NM \leq 40,$$
$$EC\_FM \leq 30.$$

The additional inequality equation below is based on the maximum number of EGC parking spaces at the Easy Pass toll ramp:

$$EV\_NM + EV\_FM + EC\_NM + EC\_FM \leq 100.$$

There are different methods for solving linear programs. Essentially, they search for the optimum solution that occurs at either the maximum or the minimum point in the solution space. For our problem, we will utilize a solver that requires the solution to be a minimization of the objective function and inequalities to be specified as vectors and matrices so that

$$\text{solution } f(x) = \text{linear\_programming\_solver}(c, A, b, lower\_bounds),$$

where

$$c = [-200 - 220 - 100 - 150] = \text{coefficient vector of the objective function},$$
$$Ax \leq b,$$

$$A = \begin{bmatrix} 1 & 0 & 0 & 0 \\ 0 & 1 & 0 & 0 \\ 0 & 0 & 1 & 0 \\ 0 & 0 & 0 & 1 \\ 1 & 1 & 1 & 1 \end{bmatrix}, \quad x = \begin{bmatrix} EV\_NM \\ EV\_FM \\ EC\_NM \\ EC\_FM \end{bmatrix}, \quad b = \begin{bmatrix} 60 \\ 20 \\ 40 \\ 30 \\ 100 \end{bmatrix},$$

$$lower\_bounds = \begin{bmatrix} 0 \\ 0 \\ 20 \\ 0 \end{bmatrix}.$$

Since our solver requires an objective function that needs to be minimized, we simply multiply our objective function by −1 to transform maximization to minimization, which gives us the coefficient vector $c$ above with the negative-valued coefficients. The matrix $A$ contains all of the coefficients from the inequality equations that serve to constrain the solution space of the objective function. For instance, the first row of the matrix is equivalent to

$$1 \times EV\_NM + 0 \times EV\_FM + 0 \times EC\_NM + 0 \times EC\_FM \le 60$$

or

$$EV\_NM \le 60.$$

The equation $Ax \le b$ describes the set of equations that constrain the solution space given the values of $A$ and $b$ and the set of variables $x$, while the *lower_bounds* vector contains constraints on the lowest values the variables can assume. For instance, the 20 value in the vector refers to variable $EC\_NM$ and the constraint given earlier that 20 ECs will be guaranteed to be serving the near metro stop.

The outputs of the solver are the values of variables in $x$ that provide the optimum solution to $f(x)$ and the optimum value itself that was achieved for $f(x)$. Using our solver, the results found are

$$x = \begin{bmatrix} 60.0 \\ 20.0 \\ 20.0 \\ 0.0 \end{bmatrix}$$

and

$$f(x) = 18,400.$$

The results show that we should use all the demand for EVs at the near metro stop and zero of the ECs for the far metro stop, while the other variables are both 20. Using these values, our objective function indicates that the EGC endeavor would make $18,400 per day.

But the executives forgot to mention that patrons that utilize the EGC system are not required to pay toll, so there is a loss of revenue associated with the park and pool service. Specifically, the toll for all vehicles is $10 and the EVs can hold 10 passengers, so this is a loss of $100 that has to be reimbursed to the Easy Pass system. ECs can hold four passengers. We need to update our linear program to reflect the $10 per passenger loss of revenue. Our objective function becomes

$$f(x) = 200 \times EV\_NM + 220 \times EV\_FM + 100 \times EC\_NM + 150 \times EC\_FM \\ - 10 \times 10TEV - 4 \times 10TEC,$$

where $TEV$ is the total number of EVs and $TEC$ is the total number of ECs.

Our solver has to be updated also to account for the additional constraints. This time we will add some equality constraints that show the relationships among the different variables such as $TEV = EV\_NM + EV\_FM$ and $TEC = EC\_NM + EC\_FM$. We will also add one more inequality, $TEV + TEC \le 100$, which constrains the total of these two variables to be less than or equal to the number of parking spaces available. Our new solver and corresponding parameters are provided below:

solution $f(x) = \text{linear\_programming\_solver}(c, A, b, A_{eq}, b_{eq}, \text{lower\_bounds})$,

where

$c = [-200\ -220\ -100\ -150\ 100\ 40]$ with added $10 \times 10$ and $4 \times 10$,

$Ax \leq b$

$$A = \begin{bmatrix} 1 & 0 & 0 & 0 & 0 & 0 \\ 0 & 1 & 0 & 0 & 0 & 0 \\ 0 & 0 & 1 & 0 & 0 & 0 \\ 0 & 0 & 0 & 1 & 0 & 0 \\ 1 & 1 & 1 & 1 & 0 & 0 \\ 0 & 0 & 0 & 0 & 1 & 1 \end{bmatrix}, \quad x = \begin{bmatrix} EV\_NM \\ EV\_FM \\ EC\_NM \\ EC\_FM \\ TEV \\ TEC \end{bmatrix}, \quad b = \begin{bmatrix} 60 \\ 20 \\ 40 \\ 30 \\ 100 \\ 100 \end{bmatrix},$$

$A_{eq} x = b_{eq}$,

$$A_{eq} = \begin{bmatrix} -1 & -1 & 0 & 0 & 1 & 0 \\ 0 & 0 & -1 & -1 & 0 & 1 \end{bmatrix}, \quad b_{eq} = \begin{bmatrix} 0 \\ 0 \end{bmatrix},$$

$$\text{lower\_bounds} = \begin{bmatrix} 0 \\ 0 \\ 20 \\ 0 \\ 0 \\ 0 \end{bmatrix}.$$

For the equation $A_{eq} x = b_{eq}$, we moved all the variables to one side of the equation to get $TEV - EV\_NM - EV\_FM = 0$ and $TEC - EC\_NM - EC\_FM = 0$. Note how adding just two more variables swells the inequality coefficient matrix $A$ from 20 numbers previously to now 36 numbers almost doubling in size—proving that these problems can get very large and complex very quickly.

Our new solver yields the following results:

$$x = \begin{bmatrix} 30.0 \\ 20.0 \\ 20.0 \\ 30.0 \\ 50.0 \\ 50.0 \end{bmatrix}$$

and

$$f(x) = 9900.$$

Now the results indicate that we should appropriate 50 EVs and 50 ECs. Also, 30 EVs should be allocated to service the near metro terminal and 20 for the far metro stop, while the minimum 20 ECs should be used for the near metro stop and 30 ECs should be allocated to the far metro terminal. Given these new values, our objective function now correctly indicates that the EGC endeavor would maximally make $9900 per day.

As we can see, linear programming is a powerful technique to help answer the business decisions of OR, but these problems are only as powerful as the accuracy of the model developed as it is in the case of all modeling methodologies. Additionally, as models become more complex, keeping track of variables, solving large matrices, and avoiding nonlinearities become more difficult, which then require different solvers and maybe different approaches to these OR problems.

## CONCLUSION

Types of models or modeling flavors allow us to learn or communicate something about a system in different ways depending on the questions being asked. Conceptual models are used for communicating the overall functionality of a system, while other types of models are more often used for system analysis. We use different types of models that may be state-event focused, physics-based, multidomain, spatial models, or composed of a hybrid of one or more of these modeling flavors. Upon choosing a model type, we must be careful what tools are used to implement the model so that proper analyses and evaluation results may be achieved.

## REFERENCES

[1] Fishwick PA. *Simulation Model Design and Execution: Building Digital Worlds*. Upper Saddle River, NJ: Prentice Hall; 1995.

[2] Aubert JP, Baget JF, Chein M. Simple conceptual graphs and simple concept graphs. In *Lecture Notes in Computer Science: Conceptual Structures: Inspiration and Application*. Vol. 4068. Carbonell JG, Siekmann J (Eds.). Berlin/Heidelberg: Springer; 2006, pp. 87–101.

[3] Harel D. Statecharts: A visual formalism for complex systems. *Science of Computer Programming*, 8(3):231–274; 1987.

[4] Zeigler B, Praehofer H, Kim TG. *Theory of Modeling and Simulation*. 2nd ed. New York: Academic Press; 2000.

## FURTHER READINGS

Eriksson H-E, Penker M, Lyons B, Fado D. *UML 2 Toolkit*. Indianapolis, IN: Wiley; 2004.

van Harmelen F, Lifschitz V, Porter B (Eds.). *Handbook of Knowledge Representation*. Oxford, UK: Elsevier; 2008, pp. 213–237.

# 7

# VISUALIZATION

Yuzhong Shen

As the Chinese proverb "A picture is worth one thousand words" says, vision is probably the most important sense of the five human senses. The human visual system is a sophisticated system with millions of photoreceptors in the eyes connected to the brain through the optic nerves, bringing about enormous information processing capabilities unpaired by any other human sensory systems. Vision is an inherent ability of human beings, while many other cognitive abilities such as languages are acquired with age. To exploit the immense computing power of the human visual system, visualization is widely utilized as one major form of information representation for various purposes.

Visualization is a process that generates visual representations, such as imagery, graph, and animations, of information that is otherwise more difficult to understand through other forms of representations, such as text and audio. Visualization is an essential component in modeling and simulation and is widely utilized at various stages of modeling and simulation applications. Visualization is especially important and useful for conveying large amount of information and dynamic information that varies with time. The foundation of visualization is *computer graphics*; thus, understanding of fundamental theories of computer graphics is very important for developing effective and efficient visualizations. This chapter first introduces the fundamentals of

---

*Modeling and Simulation Fundamentals: Theoretical Underpinnings and Practical Domains*,
Edited by John A. Sokolowski and Catherine M. Banks
Copyright © 2010 John Wiley & Sons, Inc.

computer graphics, including computer graphics hardware, 3D object representations and transformations, synthetic camera and projections, lighting and shading, and digital images and texture mapping. It then discusses contemporary visualization software and tools, including low-level graphics libraries, high-level graphics libraries, and advanced visualization tools. This chapter concludes with case studies of two commonly used software packages, namely, Google Earth and Google Maps.*

## COMPUTER GRAPHICS FUNDAMENTALS

In modeling and simulation, the term *visualization* generally refers to the use of computer graphics for various purposes, such as visual representation of data for analysis and visual simulation of military exercises for training. At the core of visualization is computer graphics. Developing effective and efficient visualizations in modeling and simulation requires in-depth understanding of computer graphics fundamentals. Computer graphics is a subfield of computer science that is concerned with the generation or manipulation of images using computer software and hardware [1–5]. Computer graphics played an essential role in easing the use of computers and thus bringing about the omnipresence of personal computers and other computing devices. Now computer graphics is used in almost all computer applications, including, but not limited to, human–computer user interface, computer-aided design (CAD) and manufacturing (CAM), motion pictures, video games, and advertisement.

This section briefly introduces several fundamental topics that are important for users of computer graphics in order to develop effective and efficient visualizations. Some low-level computer graphics details, especially those realized by hardware, are not covered in this section. However, it is important to note that as the latest computer graphics hardware provides more programmability, the application programmers have more control of the hardware that was not accessible before. In order to take advantage of the latest graphics hardware, knowledge of low-level computer graphics is still needed.

### Computer Graphics Hardware

Generation of sophisticated and realistic images using computer graphics usually involves sophisticated algorithms and large amount of data, which in turn translate into requirements on both computational power and memory capacity in order to execute the algorithms and store the data. Early computer systems did not have specialized computer graphics hardware, and the central processing unit (CPU) performed all graphics-related computations. Due to

---

*Please note that color versions of the figures in this chapter can be found at the book's ftp site at ftp://ftp.wiley.com/public/sci_tech_med/modeling_simulation.

the limited computational power of early computing systems, displaying just a few simple images could bog down their performance substantially [1,3]. Dedicated graphics acceleration hardware first became available on expensive workstations in the mid-1980s, followed by 2D graphics accelerators for personal computers in the early and mid-1990s, thanks to the increasingly widespread use of Microsoft Windows operating system. In the mid- and late 1990s, 3D graphics accelerations became commonplace on personal computers. These graphics cards contained what are now called fixed *graphics pipeline* since the graphics computations performed by these graphics cards such as geometric transformations and lighting could not be modified by the application programmers. Starting early 2000s, programmable graphics pipelines were introduced that allow application programmers to write their own shading programs (vertex and pixel shaders) to perform their customized graphics computations [6–8].

The dedicated hardware circuit to accelerate computer graphics computations is now commonly referred to as graphics processing unit or GPU for short. GPUs are highly parallel, high-performance computer systems themselves by any benchmark measure, and they contain much more transistors and have more computational power than traditional CPUs. The graphics hardware can be implemented as a dedicated graphics card, which communicates with the CPU on the motherboard with an expansion slot such as the Peripheral Component Interconnect (PCI) Express. The dedicated graphics card has its processing unit (GPU) and memory, such as the one illustrated in Figure 7.1. Graphics hardware can also be integrated into the motherboard, which is referred to as integrated video. Although integrated graphics hardware has its own processing unit (GPU), it does not have dedicated graphics memory and instead it shares with the CPU the system memory on the motherboard to store graphics data, such as 3D models and texture images. It is not surprising that dedicated graphics cards offer superior performance than integrated videos. With advances in very large-scale integrated circuit (VLSI) technology and thanks to mass production to satisfy the consumer market, the performance of graphics cards is improving rapidly while their costs have been

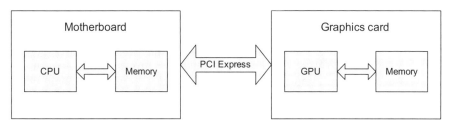

**Figure 7.1** Dedicated graphics card. The CPU is located on the motherboard, while the GPU is on the graphics card. Each has its own memory system. CPU and GPU communicate via the PCI Express interface.

continually falling. Now new high-performance graphics cards can be purchased for less than $100.

Previously, application programmers did not pay much attention to the GPU because the functions on the GPU were fixed, and the application programmers could not do much about it. Now that GPUs are providing more programmability and flexibility in the forms of programmable vertex shaders and pixel shaders, the application programmers need to know more about GPU architectures in order to develop effective and efficient visualizations [6–8]. It is critical for the modeling and simulation professionals to understand the contemporary computer graphics system architecture illustrated in Figure 7.1.

Because modern GPUs provide massively parallel computing capabilities, it is desirable to perform general-purpose computations on GPUs in addition to the traditional computer graphics applications. General-purpose computing on graphics processing unit, or GPGPU for short, addresses such issue and is a very active research area that studies the methods and algorithms for solving various problems, such as signal processing and computational fluid dynamics using GPUs [7]. Early GPGPU applications used specialized computer graphics programming languages such as GPU assembly language, which was very low-level and difficult to use. High-level shading languages such as High Level Shading Language (HLSL) and Open Graphics Library (OpenGL) Shading Language (GLSL) were released later. However, they were designed for computer graphics computations, and it was not convenient to represent general problems using such languages. The situation changed since NVIDIA, currently the world leader on GPU market, released Compute Unified Device Architecture (CUDA), which is an extension to the C programming language [9,10]. The users do not need to have in-depth knowledge of programmable shaders in order to use CUDA for general computations. Thus, the availability of CUDA significantly reduced the difficulties using GPU for general-purpose computations. To facilitate parallel computing on different GPUs or even CPUs, the Open Computing Language (OpenCL) has been released for various platforms [11].

## 3D Object Representations and Transformations

A computer graphics system can be considered as a black box, where the inputs are objects and their interactions with each other and the outputs are one or more 2D images that are displayed on output devices, such as liquid crystal display (LCD) and cathode ray tube (CRT) monitors. Modern 3D computer graphics systems are based on a *synthetic camera* model, which mimics the image formation process of real imaging devices, such as real cameras and human visual systems [1–5]. Various mathematical representations are needed in order to describe different elements, such as 3D objects, camera parameters, lights, and interactions between 3D objects and lights.

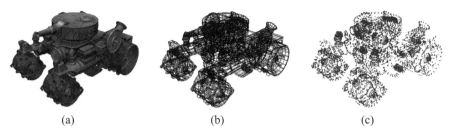

(a)　　　　　　　　　　　　(b)　　　　　　　　　　　(c)

**Figure 7.2** The tank model: (a) surface representation; (b) mesh (or wireframe) representation; (c) vertex representation (model courtesy of Microsoft XNA).

Among the various mathematical tools used in computer graphics, linear algebra is probably the most important and fundamental instrument.

Modern graphics hardware is optimized for rendering convex polygons. A convex object means that if we form a line segment by connecting two points in the object, then any point on that line segment is also in the same object. Convexity simplifies many graphics computations. For this reason, 3D objects are represented using convex polygonal meshes. As illustrated in Figure 7.2, the object (tank) consists of many polygons and each polygon is formed by several vertices. The most popular form of polygonal mesh is the triangular mesh, since triangles are always convex and all three vertices of a triangle are guaranteed to be on the same plane, which simplifies and facilitates many graphics computations, such as interpolation and rasterization. There are two fundamental mathematical entities used to represent 3D object locations and *transformations*: points and vectors. A point represents a location in space, and a mathematical point has neither size nor shape. A vector has direction and length but no location. In addition to points and vectors, scalars such as real numbers are used in computer graphics to represent quantities such as length. Most computer graphics computations are represented using matrix operations, such as matrix multiplication, transpose, and inverse. Thus, in-depth understanding and grasp of matrix operations is essential in order to develop effective and efficient computer graphics applications.

For 3D computer graphics, both points and vectors are represented using three components of real numbers, corresponding to the $x$, $y$, and $z$ coordinates in the Cartesian coordinate system. Both points and vectors can be represented as column matrices (matrices that have only one column) and row matrices (matrices that have only one row). This chapter utilizes column matrices to represent points and vectors as this notation is more commonly used in linear algebra and computer graphics literature. In practice, all points and vectors are internally represented by four-component column matrices in computer memory, which are called *homogeneous coordinates*. All affine transformations such as translations and rotations can be represented by matrix multiplications using a homogeneous coordinate system, which greatly

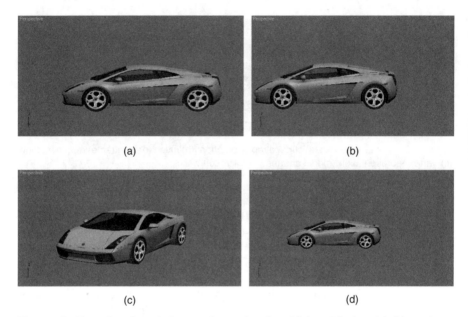

**Figure 7.3** Examples of translation, rotation, and scaling: (a) the original model; (b) translated model (to the left of the original location); (c) rotated model (rotation of 45 degrees about the vertical axis); (d) scaled-down model.

simplifies hardware implementation and facilitates pipeline realization and execution. In the homogeneous coordinate system, a point $P$ and a vector $\mathbf{v}$ can be represented as follows:

$$P = \begin{bmatrix} x \\ y \\ z \\ 1 \end{bmatrix}, \quad \mathbf{v} = \begin{bmatrix} x \\ y \\ z \\ 0 \end{bmatrix}. \tag{7.1}$$

It should be noted that the fourth component of the point $P$ is 1 while that of the vector $\mathbf{v}$ is 0. This is true for all points and vectors. Visualizations usually involve dynamic objects, which can change location, size, and shape with time, for example, a running soldier, a flying plane, and so on. Many of these dynamic changes can be represented using affine transformations, which preserve the collinearity between points and ratios of distances. Affine transformations are essentially functions that map straight lines to straight lines. The three most commonly used affine transformations are translation, rotation, and scaling, as illustrated in Figure 7.3.

Both translations and rotations are rigid body transformations; that is, the shape and the size of the object are preserved during the transformations.

COMPUTER GRAPHICS FUNDAMENTALS    187

Translation moves the object by a displacement vector $\mathbf{d} = [T_x\ T_y\ T_z\ 0]^T$, and the corresponding transformation matrix can be represented as

$$\mathbf{T}(\mathbf{d}) = \begin{bmatrix} 1 & 0 & 0 & T_x \\ 0 & 1 & 0 & T_y \\ 0 & 0 & 1 & T_z \\ 0 & 0 & 0 & 1 \end{bmatrix}. \tag{7.2}$$

Assume that a point $P = [x\ y\ z\ 1]^T$ is on the object. After applying translation $\mathbf{T}$ to the object, the point $P$ is translated to the point $P' = [x'\ y'\ z'\ 1]^T$, which can be computed as a matrix multiplication $P' = \mathbf{T}P$ as follows:

$$P' = \begin{bmatrix} x' \\ y' \\ z' \\ 1 \end{bmatrix} = \mathbf{T}P = \begin{bmatrix} 1 & 0 & 0 & T_x \\ 0 & 1 & 0 & T_y \\ 0 & 0 & 1 & T_z \\ 0 & 0 & 0 & 1 \end{bmatrix} \begin{bmatrix} x \\ y \\ z \\ 1 \end{bmatrix} = \begin{bmatrix} x + T_x \\ y + T_y \\ z + T_z \\ 1 \end{bmatrix}. \tag{7.3}$$

The inverse transformation of $T(\mathbf{d})$ is $T(-\mathbf{d})$, that is,

$$\mathbf{T}(\mathbf{d})^{-1} = \mathbf{T}(-\mathbf{d}) = \begin{bmatrix} 1 & 0 & 0 & -T_x \\ 0 & 1 & 0 & -T_y \\ 0 & 0 & 1 & -T_z \\ 0 & 0 & 0 & 1 \end{bmatrix}. \tag{7.4}$$

The 3D rotations are specified by a rotation axis (or vector), a fixed point through which the rotation axis passes, and a rotation angle. Here, we only describe three simple rotations in which the fixed point is the origin and the rotation axes are the axes of the Cartesian coordinate system. More complex rotations can be achieved through composite transformations that concatenate simple rotations. Rotations about $x$-, $y$-, and $z$-axes by an angle of $\phi$ can be represented by

$$\mathbf{R}_x(\phi) = \begin{bmatrix} 1 & 0 & 0 & 0 \\ 0 & \cos\phi & -\sin\phi & 0 \\ 0 & \sin\phi & \cos\phi & 0 \\ 0 & 0 & 0 & 1 \end{bmatrix},\ \mathbf{R}_y(\phi) = \begin{bmatrix} \cos\phi & 0 & \sin\phi & 0 \\ 0 & 1 & 0 & 0 \\ -\sin\phi & 0 & \cos\phi & 0 \\ 0 & 0 & 0 & 1 \end{bmatrix},$$

$$\mathbf{R}_z(\phi) = \begin{bmatrix} \cos\phi & -\sin\phi & 0 & 0 \\ \sin\phi & \cos\phi & 0 & 0 \\ 0 & 0 & 1 & 0 \\ 0 & 0 & 0 & 1 \end{bmatrix}. \tag{7.5}$$

Rotations about an arbitrary rotation axis can be derived as products of three rotations about the $x$-, $y$-, and $z$-axes, respectively. Any rotation matrix $\mathbf{R}(\phi)$ is always orthogonal, that is,

$$\mathbf{R}(\phi)\mathbf{R}^T(\phi) = \mathbf{R}^T(\phi)\mathbf{R}(\phi) = \mathbf{I}, \tag{7.6}$$

where $\mathbf{R}^T(\phi)$ is the transpose of $\mathbf{R}(\phi)$ and $\mathbf{I}$ is the identity matrix. A point $P$ will be at the location $P' = \mathbf{R}P$ after it is transformed by the rotation $\mathbf{R}$. The inverse transformation of rotation $\mathbf{R}(\phi)$ is $\mathbf{R}(-\phi)$, that is,

$$\mathbf{R}(\phi)^{-1} = \mathbf{R}(-\phi). \tag{7.7}$$

Scaling transformations are not rigid-body transformations; that is, the size of the object can be changed by scaling, as is the shape of the object. A scaling transformation scales each coordinate of a point on the object by different factors $[S_x \, S_y \, S_z]^T$. Its matrix representation is

$$\mathbf{S}(S_x, S_y, S_z) = \begin{bmatrix} S_x & 0 & 0 & 0 \\ 0 & S_y & 0 & 0 \\ 0 & 0 & S_z & 0 \\ 0 & 0 & 0 & 1 \end{bmatrix}. \tag{7.8}$$

Thus, a point $P = [x \, y \, z \, 1]^T$ is transformed into the point $P' = [x' \, y' \, z' \, 1]^T$ by the scaling operation $\mathbf{S}$ as follows:

$$P' = \begin{bmatrix} x' \\ y' \\ z' \\ 1 \end{bmatrix} = \mathbf{S}P = \begin{bmatrix} S_x & 0 & 0 & 0 \\ 0 & S_y & 0 & 0 \\ 0 & 0 & S_z & 0 \\ 0 & 0 & 0 & 1 \end{bmatrix} \begin{bmatrix} x \\ y \\ z \\ 1 \end{bmatrix} = \begin{bmatrix} S_x x \\ S_y y \\ S_z z \\ 1 \end{bmatrix}. \tag{7.9}$$

The inverse transformation of scaling $\mathbf{S}(S_x, S_y, S_z)$ is $\mathbf{S}(1/S_x, 1/S_y, 1/S_z)$, that is,

$$\mathbf{S}(S_x, S_y, S_z)^{-1} = \mathbf{S}\left(\frac{1}{S_x}, \frac{1}{S_y}, \frac{1}{S_z}\right) = \begin{bmatrix} 1/S_x & 0 & 0 & 0 \\ 0 & 1/S_y & 0 & 0 \\ 0 & 0 & 1/S_z & 0 \\ 0 & 0 & 0 & 1 \end{bmatrix}. \tag{7.10}$$

One common use of scaling is for unit conversion. Modeling and simulation applications contain many sophisticated assets, such as 3D models, textures, and so on. They are usually developed by different artists, and different units may be used. For example, one artist might use metric units, while another used English units. When the assets with different units are used in the same

# COMPUTER GRAPHICS FUNDAMENTALS

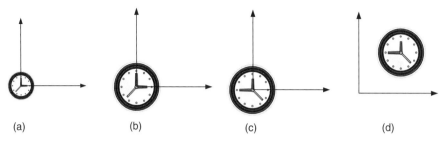

**Figure 7.4** Common sequence of transformations. The most common sequence of transformations from a local coordinate system to a world coordinate system is scaling, rotation, and translation: (a) original clock, (b) scaled clock, (c) the scaled clock in (b) is rotated for 90 degrees, and (d) the scaled and rotated clock in (c) is translated into a new location, which is the final location of the clock.

application, their units should be unified and scaling is used for unit conversion. Scaling can also be used to change object size and to generate special effects.

The 3D models are usually generated by artists in a convenient coordinate system that is called local coordinate system or object coordinate system. For example, the origin of the local coordinate system for a character can be located at the midpoint between the character's feet with $x$-, $y$-, and $z$-axes pointing right, up, and back, respectively. In applications involving many objects, the objects' local coordinates are first converted to the world or global coordinate system that is shared by all objects in the scene. The most common sequence of transformations is scaling, rotation, and translation, as shown in Figure 7.4.

## Synthetic Camera and Projections

As mentioned earlier in this chapter, modern 3D computer graphics is based on a synthetic camera model that mimics real physical imaging systems, such as real cameras and human visual systems. The inputs to the synthetic camera include 3D objects and camera parameters such as camera location, orientation, and viewing frustum. The output of the synthetic camera is a 2D image representing the projections of objects onto projection plane of the synthetic camera (corresponding to the film plane of a real camera). Figure 7.5 shows a car and a camera on the left and the image of the car captured by the camera on the right. It can be seen that different images of the same car can be formed by either moving the car or changing the camera settings.

Similar to using a real camera, the application programmer must first position the synthetic camera in order to capture the objects of interests. Three parameters are used to position and orient the camera: camera location $P_C$, object location $P_O$, and camera up direction $\mathbf{v}_{up}$. The camera location can

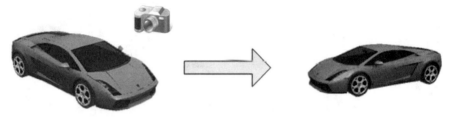

**Figure 7.5** A synthetic camera example. Shown on the left are a car and a camera and on the right is the image of the car captured by the camera. Different images of the same car can be captured by either moving the car or adjusting the camera.

be considered as the center of the camera or the camera lens; the object location is the target location, or where the camera is looking at; the camera up direction is a vector from the center of the camera to the top of the camera. The camera location and object location are 3D points, while the camera up direction is a 3D vector. A camera coordinate system can be constructed based on these three parameters, and the three axes of the camera coordinate system are represented as **u**, **v**, and **n** in the world coordinate system. That is,

$$\mathbf{u} = \begin{bmatrix} u_x \\ u_y \\ u_z \\ 0 \end{bmatrix}, \quad \mathbf{v} = \begin{bmatrix} v_x \\ v_y \\ v_z \\ 0 \end{bmatrix}, \quad \mathbf{n} = \begin{bmatrix} n_x \\ n_y \\ n_z \\ 0 \end{bmatrix}. \quad (7.11)$$

The transformation from the world coordinate system to the camera coordinate system is

$$\mathbf{M} = \begin{bmatrix} u_x & u_y & u_z & -u_x P_{C,x} - u_y P_{C,y} - u_z P_{C,z} \\ v_x & v_y & v_z & -v_x P_{C,x} - v_y P_{C,y} - v_z P_{C,z} \\ n_x & n_y & n_z & -n_x P_{C,x} - n_y P_{C,y} - n_z P_{C,z} \\ 0 & 0 & 0 & 1 \end{bmatrix}. \quad (7.12)$$

Since the final rendered images are formed by the lens of the camera, all world coordinates are transformed into camera coordinates in computer graphics applications. A point with world coordinate $P_{\text{world}}$ is transformed into its camera coordinate $P_{\text{camera}}$ as follows:

$$P_{\text{camera}} = \mathbf{M} P_{\text{world}}. \quad (7.13)$$

COMPUTER GRAPHICS FUNDAMENTALS 191

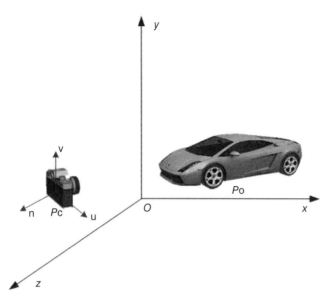

**Figure 7.6** Relationship between the world coordinate system and the camera coordinate system. The three axes of the camera coordinate systems are represented as **u**, **v**, and **n** in the world coordinate system. All objects' world coordinates are first converted to camera coordinates in all computer graphics applications.

Figure 7.6 shows the relationship between the world coordinate system and the camera coordinate system.

When using a real camera, we must adjust the focus of the camera after positioning the camera before taking a picture. Similarly, the focus of the synthetic camera needs to be adjusted too. The focus of the synthetic camera can be specified by defining different projections. There are two types of projections: *perspective projections* and *parallel projections*, which can be differentiated based on the distance from the camera to the object. The camera is located within a finite distance from the object for perspective projections, while the camera is at infinity for parallel projections. In perspective projections, the camera is called the center of projection (COP), and the rays that connect the camera and points on the object are called projectors. Projections of these points are the intersections between the projectors and the projection plane, as illustrated in Figure 7.7.

Perspective projections can be defined by a viewing frustum, which is specified by six clipping planes: left, right, bottom, top, near, and far (Fig. 7.7). Note that all the clipping planes are specified with respect to the camera coordinate system (in which the camera is the origin), while the camera's location and orientation are specified in the world coordinate system. A special but very commonly used type of perspective projections is symmetric perspective

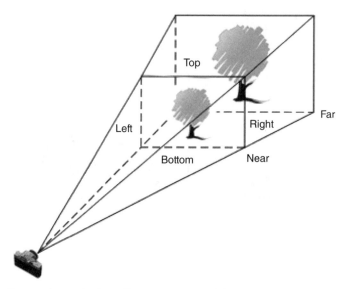

**Figure 7.7** Perspective projections. The camera is the COP. Projectors are rays connecting the COP and points on the objects. The intersections between the projection plane and the projectors are the projections (or images) of the points on the object.

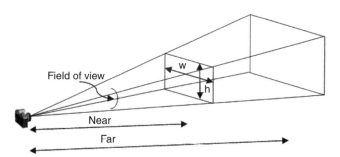

**Figure 7.8** Symmetric perspective projective. The most common type of perspective projection is symmetric perspective projection, which can be specified by four parameters: FOV, aspect ratio (w/h), near plane, and far plane. All the parameters are defined in the camera coordinate system in which the camera is the origin.

projection, which can be specified by four parameters: field of view (FOV), the aspect ratio between the width and height of the near clipping plane, and near and far clipping planes (Fig. 7.8). Again all the parameters are specified with respect to the camera coordinate system. Perspective projections are characterized by foreshortening, in which far objects appear smaller than close objects (Fig. 7.9).

**Figure 7.9** Perspective projections. Perspective projections are characterized by foreshortening, that is, far objects appear smaller than near objects (image courtesy of Laitche).

Parallel projections can be considered as special cases of perspective projections in which the camera is located infinitely far from the object. Thus, there is no COP in parallel projections. Instead, direction of projection (DOP) is used. Parallel projections can be classified into orthographic projections and oblique projections based on the relationship between the DOP and the projection plane. If the DOP is perpendicular to the projection plane, the parallel projection is called orthographic projection; otherwise, it is called oblique projection. Parallel projections are defined by specifying the six clipping planes (left, right, bottom, top, near, and far) and the DOP. Orthographic projections are frequently used in architectural design, CAD and CAM, and so on, because they preserve distances and directions along the projection plane, thus allowing accurate measurements.

In 3D modeling software, it is very common to have multiple perspective and orthographic views of the same object. The frequently used orthographic views are generated from the top, front, and side. Figure 7.10 shows four views of the model Tank in Autodesk Maya, a leading software package for 3D modeling, animation, visual effects, and rendering solution. Internally, all projections are converted into an orthographic projection defined by a canonical viewing volume through various transformations, such as scaling and perspective normalization [1–5].

**194** VISUALIZATION

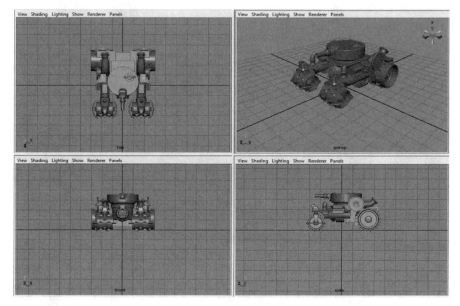

**Figure 7.10** Screen capture of Autodesk Maya. It is common to have multiple views of the same object in 3D modeling software and other modeling and simulation software. Shown here are three orthographic views (top, front, and side) and one perspective view (persp).

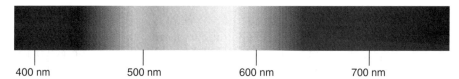

**Figure 7.11** Electromagnetic spectrum. Visible light occupies the range from 380 to 750 nm in the electromagnetic spectrum.

## Lighting and Shading

***Color*** Sources of excitation are needed to form images in physical imaging systems. For human visual systems, the source of excitation is visible light, which is one form of electromagnetic radiation with wavelengths in the range of 380–750 nm of the electromagnetic spectrum (Fig. 7.11). Although the visible light band covers the continuous range from 380 to 750 nm, human visual systems are most sensitive to three primary frequencies because of the physiological structure of human eyes. The retina on the back of human eye is responsible for forming images of the real world, and it has two types of photoreceptors cells: rods and cones. Rods are mainly responsible for night

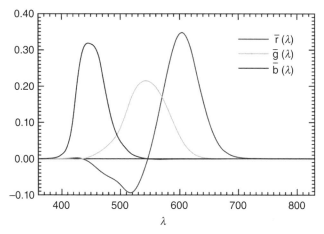

**Figure 7.12** The CIR 1931 RGB color matching functions. The color matching functions are the amounts of primary colors needed to match any monochromatic color with single wavelength (image courtesy of Marco Polo).

vision or low-intensity light, while cones for color vision. There are three types of cones that are most sensitive to red, green, and blue colors, respectively.

The *tristimulus theory* was developed to account for human eyes' physiological structure, and it states that any color can be represented as a linear combination of three monochromatic waveforms with wavelengths of 700, 546.1, and 438.1 nm, roughly corresponding to red, green, and blue colors, respectively. Figure 7.12 shows the color matching functions, which describe the amount of primary colors needed to match any monochromatic light of any single wavelength. Note that the red color matching function has some negative values, which represent the amount of red color needed to be added to the target color so that the combined color would match the sum of blue and green primary colors.

Display hardware devices, such as CRT and LCD, need only generate the three primary colors, and the final color is a linear combination of them. However, display devices cannot generate negative coefficients, and as a result, display devices cannot represent the entire visible spectrum. The range of colors that can be generated by a display device is called its color gamut, which is the range of colors that can be generated by adding primary colors. Different devices have different color gamut. In terms of color representations, any color can be described by its red, green, and blue components, denoted by R, G, and B, respectively. This representation is commonly referred to as the color cube, as shown in Figure 7.13. In the remainder of this chapter, we will not differentiate between the three primary colors, and all the processing and computations will be applied to each primary color separately, but in the same way.

**Figure 7.13** The color cube. Display devices use three primary colors: red, green, and blue to generate a color. Thus, a color can be represented as a point inside the cube determined by the red, green, and blue axes.

**Lights** In a dark room without any light, we cannot see anything. The same is true for computer graphics applications. In order to generate visualizations of the object, we first need to specify light sources that mimic natural and man-made light sources, be it the sunlight and a bulb inside a room. Different types of light models (*lighting*) were developed in computer graphics to simulate corresponding lights in real life [1–5]. The most commonly used light models in computer graphics applications are ambient light, point light, spotlight, and directional light [1–5].

(1) *Ambient lights* provide uniform lighting; that is, all objects in the scene receive the same amount of illumination from the same ambient light, independent of location and direction. However, different objects or even different points on the same object can reflect the same ambient light differently and thus appear differently. Ambient lights are used to model lights that have been scattered so many times that their directions cannot be determined. Ambient lights are commonly used to provide environment lighting.

(2) A *point light* is an ideal light source that emits light equally in all directions, as shown in Figure 7.14(a). A point light is specified by its color and location. The luminance of a point light attenuates with the distance from the light, and the received light at point $Q$ can be computed as follows:

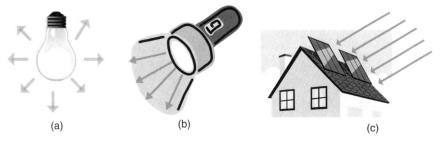

**Figure 7.14** Different types of lights: (a) point light; (b) spotlight; (c) directional light.

$$I(Q) = \frac{1}{a + bd + cd^2} I_0, \tag{7.14}$$

where $I_0$ is the point light intensity, $d$ is the distance between the light and the point $Q$, and $a$, $b$, and $c$ are attenuation coefficients, called constant, linear, and quadratic terms, respectively. The use of three attenuation terms can reduce the harsh rendering effects of using just the quadratic attenuation. Also, ambient lights can be combined with point lights to further reduce high-contrast effects.

(3) *Spotlights* produce cone-shaped lighting effects and are specified by three parameters: location, direction, and cutoff angle, as shown in Figures 7.14(b) and 7.15. The shape of the spotlight is a cone, which is determined by the light's location $P$ and the cutoff angle $\theta$. The spotlight does not produce any luminance outside the cone. The light intensity inside the cone varies as a function of the parameter $\phi$ (Fig. 7.15), which is the angle between the direction of the spotlight and a vector connecting the location of the spotlight (i.e., the apex of the cone) and a point on the target object. The light intensity is a decreasing function of the angle $\phi$, and the attenuation with $\phi$ is usually computed as $\cos^e \phi$, where $e$ is a parameter that can be used to adjust the tightness of the spotlight. In addition, spotlights also attenuate with distance, so Equation (7.10) should be applied to spotlights as well.

(4) *Directional lights*, or distant lights, are used to model light sources that are very far from the object, such as the sunlight. A directional light has constant intensity and is specified by its direction, as the one shown in Figure 7.14(c). Directional lights are assumed to be located at infinity; thus, they do not have a location.

**Reflection Models** After describing several common light models used in computer graphics, it is now time to discuss the interactions between lights and objects: *reflection*. It is these interactions that make non-self-illuminating objects visible. For the sake of simplicity, here we only consider opaque, non-

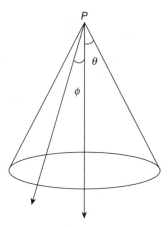

**Figure 7.15** Spotlight parameters. A spotlight is determined by three parameters: location, direction, and cutoff angle. There is no light outside the cone determined by the light location and the cutoff angle. Inside the cone, the light intensity is a decreasing function of the angle $\phi$, usually computed as $\cos^e\phi$.

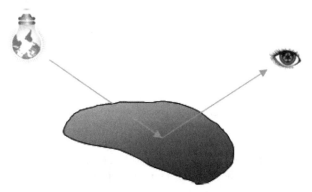

**Figure 7.16** Interactions between lights and objects. The light rays emitted by the light source are first reflected by the object surface; then, the reflected light rays reach human eyes and form the image of the object.

self-illuminating objects whose appearances are totally determined by their surface properties. (Self-illuminating objects can be modeled as lights directly.) The light rays emitted by light sources are first reflected by the object surface; then, the reflected light ray reaches human eye and forms the image of the object, as shown in Figure 7.16.

The interactions between lights and objects are very complicated, and various models have been proposed. Global illuminations are physics-based

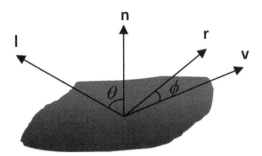

**Figure 7.17** The vectors used in lighting calculation are **l**: light direction from the object to the light source; **n**: surface normal; **r**: direction of reflection; and **v**: viewer direction from the object to the viewer.

modeling that simulates multiple interactions between light sources and objects and interactions between objects, while local illuminations only consider the interactions between light sources and objects. This chapter only considers local illumination models. Several reflection models have been developed to simulate a variety of interactions between light sources and object surfaces. These models utilize several vectors as illustrated in Figure 7.17. The vector **l** indicates the direction of light source, **n** is the surface normal, **r** is the direction of reflection, and **v** is the viewer direction. All four vectors are dependent on the point position on the object and thus can change from point to point.

The simplest reflection is ambient reflection. Because ambient light sources provide uniform lighting, the reflected light or brightness is not dependent on surface normal or viewer locations. Thus, ambient reflection can be represented as follows:

$$I(Q) = I_a k_a, \qquad (7.15)$$

where $I_a$ is the intensity of the ambient light and $k_a$ is the ambient reflection coefficient at point $Q$, which ranges from 0 to 1. It is important to note that different points or surfaces can have different ambient reflection coefficients and thus appear differently even illuminated with the same ambient light source. Figure 7.18(a) shows an example of ambient reflection. Because computation of ambient reflection does not involve any vectors or directions, the object appear flat and does not have any 3D feel.

Many objects have dull or matte surfaces, and they reflect light equally in all directions. Such kinds of objects are called Lambertian objects, and their reflections are called Lambertian reflection or diffuse reflection. The light

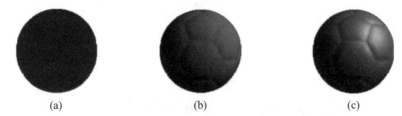

**Figure 7.18** Surface reflections. The figure shows different surface reflections of the same object (a soccer ball): (a) ambient reflection, (b) diffuse reflection, and (c) specular reflection.

reflected by Lambertian objects is determined by the angle between the light source and the surface normal as follows:

$$I(Q) = I_d k_d \cos\theta = I_d k_d \mathbf{l} \cdot \mathbf{n}, \tag{7.16}$$

where $I_d$ is the light intensity, $k_d$ is the diffuse reflection coefficient, and $\theta$ is the angle between the light direction $\mathbf{l}$ and the surface normal $\mathbf{n}$. When $\theta > 90°$, $\cos\theta$ would have a negative value, which is not reasonable because the object cannot receive a negative light intensity. Thus, Equation (7.16) can be further revised to accommodate this fact as follows:

$$I(Q) = I_d k_d \max(\cos\theta, 0) = I_d k_d \max(\mathbf{l} \cdot \mathbf{n}, 0). \tag{7.17}$$

Considering distance attenuation, Equation (7.17) can be further rewritten as

$$I(Q) = \frac{1}{a + bd + cd^2} I_d k_d \max(\cos\theta, 0) = \frac{1}{a + bd + cd^2} I_d k_d \max(\mathbf{l} \cdot \mathbf{n}, 0). \tag{7.18}$$

As Equation (7.18) indicates, the diffuse reflection is not dependent on view directions. Thus, the same point on the object appears the same to two viewers at different locations. Figure 7.18(b) shows an example of diffuse reflection from which it can be seen that the object (soccer) appears dull although it does have 3D appearance.

Shiny objects often have highlighted spots that move over the object surface as the viewer moves. The highlights are caused by specular reflections, which are dependent on the angle between the viewer direction and the direction of reflection, that is, angle $\phi$ in Figure 7.17. For specular surfaces, the light is reflected mainly along the direction of reflection. The specular reflections perceived by the viewer can be modeled as follows:

$$I(Q) = I_s k_s \cos^e \phi = I_s k_s (\max(\mathbf{r} \cdot \mathbf{v}, 0))^e, \tag{7.19}$$

where the exponent $e$ is called the shininess parameter of the surface and a larger $e$ represents smaller highlighted spot. Figure 7.18(c) shows an example of specular reflection. It can be seen that the object appears more realistic and has stronger 3D feel.

Most object surfaces have all the reflection components discussed above. The total combined effect of ambient reflection, diffuse reflection, and specular reflection can be described by the Phong illumination model as follows:

$$I(Q) = I_a k_a + \frac{1}{a + bd + cd^2}\left(I_d k_d \max(\cos\theta, 0) + I_s k_s (\max(\mathbf{r} \cdot \mathbf{v}, 0))^e\right). \quad (7.20)$$

It is important to note that Equation (7.20) should be calculated for each component (red, green, and blue) of each light source, and the final results are obtained by adding all the light sources. The Phong illumination model is implemented by the fixed graphics pipeline of OpenGL and Direct3D.

**Shading Models** Vertices are the smallest units to represent information for polygonal meshes. All the information of the polygonal mesh is defined at the mesh's vertices, such as surface normal, color, and texture coordinates (to be discussed in the next section). *Shading models* come into play when we need to determine the shade or the color of the points inside a polygon. Here we briefly discuss three commonly used shading models: flat shading, smooth shading, and Phong shading.

Recall that four vectors are used to calculate the color at a point: light direction, surface normal, direction of reflection, and viewer direction. These vectors can vary from point to point. However, for points inside a flat polygon, the lighting calculations can be greatly simplified. First, the surface normal is constant for all points inside the same flat polygon. If the point light is far away from the polygon, the light direction can be approximated as constant. The same approximations can be made for viewer direction if the viewer is far away from the polygon. With these approximations, all the points inside a polygon have the same color, and we need to perform the lighting calculation only once and the result is assigned to all points in the polygon. Flat shading is also called constant shading. However, polygons are just approximations of underlying curved smooth surface, and flat shading generally does not generate realistic results. Figure 7.19(a) shows an example of flat shading of a sphere.

The individual triangles that constitute the sphere surface are clearly visible and thus is not a good approximation of the original sphere surface. In smooth shading, lighting calculation is performed for each vertex of the polygon. Then, the color of a point inside the polygon is calculated by interpolating the vertex colors using various interpolation methods, such as bilinear interpolation. Gouraud shading is one type of smooth shading in which the vertex normal is calculated as the normalized average of the normals of neighboring polygons sharing the vertex. Figure 7.19(b) shows an example of Gouraud shading of the sphere. It can be seen that Gouraud shading achieves better

**202** VISUALIZATION

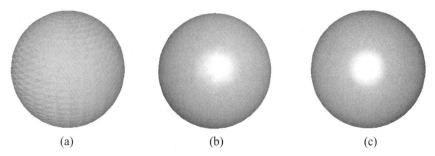

**Figure 7.19** Shading models: (a) flat shading; (b) smooth shading; (c) Phong shading.

results than the flat shading, but still has some artifacts such as the irregular highlighted spot on the sphere surface. The latest graphics cards support Phong shading, which interpolates vertex normals instead of colors across a polygon. The lighting calculation is performed at each point inside the polygon and thus achieves better rendering effects than smooth shading. Phong shading was not supported directly by the graphics cards until recently. Figure 7.19(c) shows an example of Phong shading of the same sphere, and it can be seen that it achieves the best and most realistic rendering effects.

### Digital Images and Texture Mapping

As mentioned previously, vertices are the smallest units that can be used to specify information for polygonal meshes. If the object has complex appearance, for example, the object contains many color variations and a lot of geometric details, many polygons would be needed to represent the object complexity because colors and geometry can be defined only at the polygon vertices. Even though the memory capacity of graphics cards has been increasing tremendously at a constant pace, it is still not feasible to store a large number of objects represented by high-resolution polygonal meshes on the graphics board. *Texture mapping* is a revolutionary technique that enriches object visual appearance without increasing geometric complexity significantly. Vertex colors are determined by matching the vertices to locations in digital images that are called texture maps.

Texture mapping can enhance not only the object's color appearance but also its geometric appearance. Texture mapping is very similar to the decoration of a room using wallpaper. Instead of painting the wall directly, pasting wallpaper onto the wall greatly reduces the efforts needed and increases visual appeal at the same time. Also, the wallpaper of the same pattern can be used repeatedly for different walls. The "wallpapers" used in texture mapping are called texture maps, and they can be defined in 1D, 2D, and 3D spaces. The most commonly used texture maps are 2D textures and digital images are the major form of 2D textures, although textures can be generated

(a) (b)

**Figure 7.20** Color image. (a) A color image of size 753 × 637 and (b) an enlarged version of the toucan eye in (a) where the pixels are clearly visible in (b).

(a) (b) (c)

**Figure 7.21** Each color image or pixel has three components and shown here are the (a) red, (b) green, and (c) blue components of the image in Figure 7.20.

in other ways, such as automatic generation of textures using procedural modeling methods.

***Digital Images*** The 2D digital images are 2D arrays of pixels, or picture elements. Each pixel is such a tiny square on display devices that human visual system can hardly recognize its existence, and the entire image is perceived as a continuous space. Figure 7.20(a) shows a color image whose original size is 753 × 627 (i.e., 753 columns and 627 rows), and Figure 7.20(b) shows an enlarged version of the toucan eye in which the pixels are clearly visible.

Each pixel has an intensity value for grayscale images or a color value for color images. As a result of the tristimulus theory, only three components (red, green, and blue) are needed to represent a color value. It is very common to use 1 byte (8 bits) to represent the intensity value for each color component, with a range from 0 to 255. Figure 7.21 shows the RGB components of the image in Figure 7.20(a).

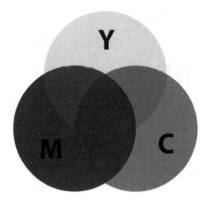

**Figure 7.22** CMYK color space.

Although the RGB format is the standard format for color image representation used by display hardware, other color spaces have been developed for different purposes, such as printing, off-line representations, and easy human perception. Here we briefly describe several of them, including CMYK, $YC_bC_r$, and HSV. The CMYK color space is a subtractive color space used for color printing. CMYK stands for cyan, magenta, yellow, and black, respectively. Cyan is the complement color of red; that is, if we subtract the red color from a white light, the resulting color is cyan. Similarly, magenta is the complement of green, and yellow is the complement of blue. The CMYK color space is shown in Figure 7.22.

The $YC_bC_r$ color space is commonly used in encoding of digital images and videos for different electronic devices and media, for example, digital camera, digital camcorders, and DVDs. $YC_bC_r$ represents luminance, blue-difference (blue minus luminance), and red-difference (red minus luminance). The $YC_bC_r$ color space is more suitable for storage and transmission than the RGB space, since the three components of the $YC_bC_r$ color space are less correlated. The HSV color space contains three components: hue, saturation, and value, and it is often used for color specification by general users since it matches human perception of colors better than other color spaces. Representations in different color spaces can be easily converted to each other; for example, the $YC_bC_r$ color space is used to store digital color images on the hard disk, but it is converted into the RGB color space in the computer memory (and frame buffer) before it becomes visible on the display devices.

Digital images can be stored on hard drives in uncompressed formats or compressed formats. Image compression refers to the process of reducing image file size using various techniques. Image compression or, more generally, data compression can be classified into lossless compression and lossy compression. Lossless compression reduces the file size without losing any information, while lossy compression reduces the file size with information

(a)  (b)  (c)

**Figure 7.23** JPEG standard format. JPEG is the standard format used for image compression. It can achieve compression ratios from 12 to 18 with little noticeable distortion: (a) the compression ratio is 2.6:1 with almost no distortion, (b) the compression ratio is 15:1 with minor distortion, and (c) the compression ratio is 46:1. Severe distortion is introduced, especially around sharp edges (image courtesy of Toytoy).

loss but in a controlled way. Lossy compression is commonly used for multimedia data encoding, such as images, videos, and audio, since some minor information loss in such data is not noticeable to humans. Image compression is necessary mainly because of two reasons: effective storage and fast file transfer. Even though the capacity of hard drives is increasing rapidly every year, the amount of digital images and video generated outpaces the increases of hard drive capacity, due to the ubiquitous digital cameras, increasing image resolution, and new high-resolution standards, for example, high-definition TV (HDTV). On the other hand, image compression can greatly reduce the time needed for transferring files between computers and between CPUs and peripheral devices, such as USB storage. Various image compression algorithms have been developed and among them Joint Photographic Experts Group (JPEG) is the most widely used format. JPEG can achieve compression ratios in the range from 12 to 18 with little noticeable loss of image quality (see the examples in Fig. 7.23). Several techniques are used in the JPEG standard, including discrete cosine transform (DCT), Huffman coding, run-length coding, and so on. The latest JPEG format is JPEG 2000, which uses discrete wavelet transform (DWT) and achieves better compression ratios with less distortion.

*Texture Mapping*   Texture mapping matches each vertex of the polygonal mesh to a texel (texture element) in the texture map. Regardless of the original size of the digital image used for the texture map, textures always have a normalized texture coordinates in the range [0, 1], represented by $u$ and $v$ for horizontal and vertical directions, respectively. Each vertex of the polygonal mesh is assigned a pair of texture coordinates, for example, [0.32, 0.45], so that the vertex's color is determined by the texel at the location specified by the texture coordinates in the texture map. The texture coordinates for points

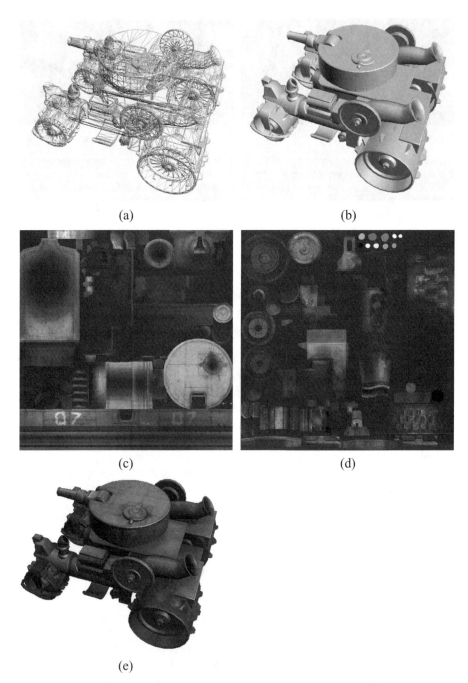

**Figure 7.24** Texture mapping: (a) the polygonal mesh of the model tank, (b) the model rendered without texture mapping, (c) and (d) are two texture maps used for the model, and (e) the final tank model rendered using texture mapping.

inside a polygon are determined by interpolating the texture coordinates of the polygon vertices. Figure 7.24(a) and (b) shows the polygonal mesh for the model tank and the surface of the tank rendered without texture mapping. It can be seen that the tank has only a single color and does not appear very appealing. Figure 7.24(c) and (d) shows the two texture maps used for the tank. The final tank model rendered with texture mapping shown in Figure 7.24(e) has a much richer visual complexity and appears much more realistic.

Various methods have been developed for determining the texture coordinates for each vertex of the polygonal mesh. If the object can be represented by a parametric surface, the mapping between a point on the surface and a location in the texture map is straightforward. Each point $P(x, y, z)$ on the parametric surface can be represented as a function of two parameters $s$ and $t$ as follows:

$$\begin{cases} x = x(s, t) \\ y = y(s, t) \\ z = z(s, t). \end{cases} \quad (7.21)$$

If the parameters $s$ and $t$ can be restricted to the range $[0, 1]$, then a mapping between $s, t$ and the texture coordinates $u, v$ can be established:

$$(s, t) \leftrightarrow (u, v). \quad (7.22)$$

Thus, a mapping between each point location and the texture coordinates can be established as well:

$$(x, y, z) \leftrightarrow (u, v). \quad (7.23)$$

For objects that cannot be represented by parametric surfaces, intermediate parametric objects, such as sphere, cylinder, and cube, are utilized. A mapping between each point on the original surface and the parametric surface can be established through various projections. The texture coordinates for a point on the parametric surface is used as the texture coordinates for its corresponding point on the original surface.

In addition to increasing color variations, texture mapping has also been used to change the geometric appearance either directly or indirectly. Displacement mapping changes the vertex positions directly based on a displacement or height map. Bump mapping does not change the object geometry directly, but changes the vertex normals based on a bump map so that the rendered surface appears more rugged and thus more realistic. Figure 7.25(a) shows an example of bump mapping. The plane appears rusty because of the effects of bump mapping. Texture mapping can also be used for modeling perfectly

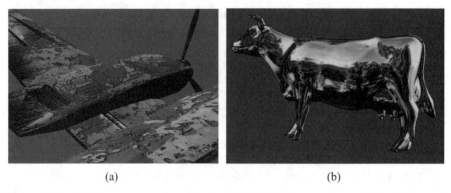

(a)          (b)

**Figure 7.25** Other texture mapping methods include (a) bump mapping and (b) environment mapping (courtesy of OpenSceneGraph).

reflective surfaces. Environment mapping uses images of surrounding environment as textures for the object so that the object appears perfectly reflective. Various environment mapping methods have been developing, such as spherical maps and cubic maps. Figure 7.25(b) shows an example of environmental mapping.

## VISUALIZATION SOFTWARE AND TOOLS

Various levels of software and tools have been developed to facilitate visualization development. These software and tools form a hierarchical or layered visualization architecture as illustrated in Figure 7.26. At the bottom of the hierarchy is the computer graphics hardware, which was discussed earlier in this chapter. Computer graphics hardware is constructed using VLSI circuits, and many graphics computations are directly implemented by hardware, such as matrix operations, lighting, and rasterization. Hardware implementations greatly reduce the time needed for complex computations, and, as a result, we see less nonresponsive visualizations. On top of graphics hardware are device drivers, which control graphics hardware directly and allow easy access to graphics hardware through its function calls. Low-level graphics libraries, such as *OpenGL* and *Direct3D*, are built on top of device drivers, and they perform fundamental graphics operations, such as geometry definition and transformations. The low-level graphics libraries are foundations of computer graphics, and they provide the core capabilities needed to build any graphics applications. However, it still takes a lot of effort to build complex applications using the low-level graphics libraries directly. To address this issue, high-level libraries were developed to encapsulate low-level librar-

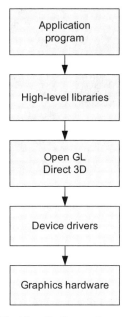

**Figure 7.26** Visualization system architecture.

ies so that it takes less time to develop visualizations. In addition, high-level libraries provide many advanced functionalities that are not available in low-level libraries. Finally, application programmers call the high-level libraries instead of the low-level libraries. This section first introduces the two most important low-level libraries, namely, OpenGL and Direct3D. It then discusses several popular high-level libraries. Finally, several case studies are described.

### Low-Level Graphics Libraries

Low-level graphics libraries are dominated by two application programming interfaces (APIs), namely, OpenGL and Direct3D. OpenGL is a high-performance cross-platform graphics API that is available on a wide variety of operating systems and hardware platforms, while Direct3D mainly works on Microsoft Windows operating systems and hardware, for example, Xbox 360 game console, and it is the standard for game development on Microsoft Windows platforms. Most graphics card manufacturers provide both OpenGL and Direct3D drivers for their products.

**OpenGL**  OpenGL [12] was originally introduced by Silicon Graphics, Inc. in 1992, and it is now the industry standard for high-performance professional

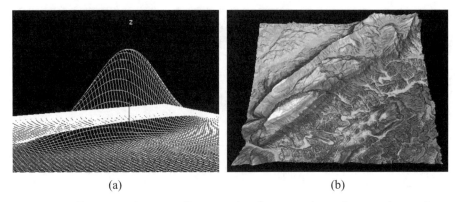

(a)                                                                                    (b)

**Figure 7.27** Example assignments. Two examples of programming assignments in a graduate course in Visualization offered at Old Dominion University: (a) visualization of 2D Gaussian distribution and (b) visualization of real terrain downloaded from U.S. Geological Survey (USGS). OpenGL was used to develop visualizations in the (chapter) author's course, *Visualization I*.

graphics, such as scientific visualization, CAD, 3D animation, and visual simulations. OpenGL is a cross-platform API that is available on all operating systems (e.g., Windows, Linux, UNIX, Mac OS, etc.) and many hardware platforms (e.g., personal computers, supercomputers, cell phones, PDAs, etc.). OpenGL provides a comprehensive set of functions (about 150 core functions and their variants) for a wide range of graphics tasks, including modeling, lighting, viewing, rendering, image processing, texture mapping, programmable vertex and pixel processing, and so on. OpenGL implements a graphics pipeline based on a state machine and includes programmable vertex and pixel shaders via GLSL. The OpenGL Utility Library (GLU) is built on top of OpenGL and is always included with any OpenGL implementation. GLU provides high-level drawing functions that facilitate and simplify graphics programming, such as high-level primitives, quadric surfaces, tessellation, mapping between world coordinates and screen coordinates, and so on. OpenGL is a graphics library only for rendering, and it does not support windows management, event processing, and user interactions directly. However, any user interfacing libraries that provide such capabilities can be combined with OpenGL to generate interactive computer graphics, such as Window Forms, QT, MFC, and GLUT. OpenGL is also the standard tool used for computer graphics instruction, and many good references are available [1,13–16]. Figure 7.27 shows screen captures of two programming assignments in the course Visualization I offered by the Modeling and Simulation graduate program at Old Dominion University. OpenGL is used as the teaching tool in this course.

***Direct3D*** Direct3D is a 3D graphics API developed by Microsoft for its Windows operating systems and Xbox series game consoles. Direct3D was first introduced in 1995, and early versions of Direct3D suffered from poor performance and usability issues. However, Microsoft has been continuously improving Direct3D, and now it has evolved into a powerful and flexible graphics API that is the dominant API for game development on Windows Platforms. Direct3D is based on Microsoft's Component Object Model (COM) technology, and it has a graphics pipeline for tessellation, vertex processing, geometry processing, texture mapping, rasterization, pixel processing, and pixel rendering. Direct3D is designed to exploit low-level, high-performance graphics hardware acceleration, and one of its major uses is for game development on personal computers. This is different from OpenGL, which is a more general-purpose 3D graphics API that is used by many professional applications on a wide range of hardware platforms. The vertex shader and pixel shader in the Direct3D graphics pipeline are fully programmable using the HLSL, and Direct3D does not provide a default fixed function pipeline. Direct3D supports two display modes: full-screen mode and windowed mode. In full-screen mode, Direct3D generates outputs for the entire display at full resolution, and the windowed mode generates outputs that are embedded in a window on the display. Similar to OpenGL, Direct3D is a constantly evolving graphics API and new features are introduced continually. At the time of writing of this chapter, the latest version is Direct3D 10 for Windows Vista. A third programmable shader, the geometry shader, is included in Direct3D 10 for geometric topology processing. Although Direct3D is less frequently used for computer graphics instruction, many good reference books are available [17–20].

### High-Level Graphics Libraries

Although both OpenGL and Direct3D are powerful and flexible graphics APIs, they are still low-level libraries, and it takes substantial effort to develop complex and advanced applications using these APIs directly. Thus, high-level graphics libraries have been developed that encapsulate low-level APIs so that sophisticated applications can be developed easily and quickly. Here we briefly describe several of them: OpenSceneGraph, XNA Game Studio, and Java3D.

***OpenSceneGraph*** *OpenSceneGraph* is an open source, cross-platform 3D graphics library for high-performance visualization applications [21]. It encapsulates OpenGL functionalities using object-oriented programming language C++ and provides many optimizations and additional capabilities. The core of OpenSceneGraph is a scene graph, which is a hierarchical data structure (graph or tree structure) for organization and optimization of graphics objects for the purpose of fast computation and rapid application development. Users of OpenSceneGraph do not need to implement and optimize low-level graphics functions and can concentrate on high-level content development rather than low-level graphics details. OpenSceneGraph is not simply an object-oriented encapsulation of OpenGL; it provides many additional capabilities

## 212 VISUALIZATION

**Figure 7.28** Screen captures of applications developed using OpenSceneGraph: (a) Vizard, courtesy of WorldViz LLC; (b) Priene the Greek Ancient City in Asia Minor, courtesy of the Foundation of the Hellenic World; (c) ViresVIG, courtesy of VIRES Simulatiuonstechnologie Gmbh; and (d) Pirates of the XXI Century, courtesy of !DIOsoft company.

such as such as view-frustum culling, occlusion culling, level of detail nodes, OpenGL state sorting, and continuous level of detail meshes. It supports a wide range of file formats via a dynamic plug-in mechanism for 3D models, images, font, and terrain databases. The two lead and most important developers of OpenSceneGraph are Don Burns and Robert Osfield, with contributions from other users and developers, including the author of this chapter. OpenSceneGraph is now well established as the leading scene graph technology that is widely used in visual simulation, scientific visualization, game development, and virtual reality applications. Figure 7.28 illustrates several applications developed using OpenSceneGraph.

***XNA Game Studio*** Microsoft XNA Game Studio is an integrated development environment to facilitate game development for Windows PC, Xbox 360 game console, and Zune media player. The target audience of XNA Game

Studio is academics, hobbyists, and independent and small game developers and studios. XNA Game Studio consists of two major components: XNA Framework and a set of tools and templates for game development. XNA Framework is an extensive set of libraries for game development based on the .NET Framework. It encapsulates low-level technical details so that game developers can concentrate on content and high-level development. XNA provides templates for common tasks, such as games, game libraries, audio, and game components. It also provides utilities for cross-platform development, publishing, and deployment. The games developed using XNA can be played on PC, Xbox 360, and Zune with minimal modifications. In addition, XNA provides an extensive set of tutorials and detailed documentations, which greatly reduce the learning curve and the time needed for complex game development. Microsoft also maintains and supports XNA Creators Club Online [22], a Web site that provides many samples, tutorials, games, utilities, and articles. Developers can sell games developed by them to Xbox Live, the world's largest online game community with about 17 million subscribers. Figure 7.29 shows several applications developed using XNA. A series of books on XNA have been published [23–25].

*Java 3D* Java is a revolutionary programming language that is independent of any hardware and software platforms; that is, exactly the same Java program

**Figure 7.29** Games developed using XNA: (a) racing, (b) role playing, (c) puzzle, (d) robot games (game courtesy of Microsoft XNA).

**214** VISUALIZATION

can run on different hardware and software platforms. The foundation of the Java technology is the Java virtual machine (JVM). Java source programs are first compiled into bytecode, which is a set of instructions similar to assembly language code. The JVM then compiles the bytecode into native CPU instructions to be executed by the host CPU. Java programs can run as stand-alone applications or applets in Internet browsers with minor modifications. *Java 3D* is a graphics API for the Java platform that is built on top of OpenGL or Direct3D. Java 3D was not part of the original Java distribution but was introduced as a standard extension. Java 3D is a collection of hierarchical classes for rendering of 3D object and sound. The core of Java 3D is also a scene graph, a tree data structure for organization of 3D objects and their properties, sounds, and lights for the purpose of fast rendering. All the objects in Java 3D are located in a virtual universe, and they form a hierarchical structure in the form of a scene graph. Java 3D also includes many utility classes that facilitate rapid visualization development, such as content loaders, geometry classes, and user interactions. Java 3D is widely used for game development on mobile platforms, such as cell phones and PDAs, and online interactive visualizations. Java 3D is now an open source project with contributions from individuals and companies [26]. Figure 7.30 shows the screen captures of several demos provided by Java 3D.

### Advanced Visualization Tools

Both the low-level and high-level graphics libraries discussed in the previous sections are intended for software development, and they require significant

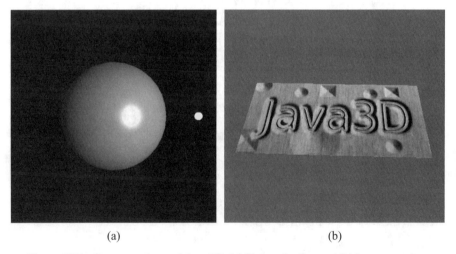

**Figure 7.30** Screen captures of Java 3D. (a) Phong shading and (b) bump mapping.

programming skills and experience. Since visualization is such a prevalent component in almost all applications, many software packages have been built so that users with less or no programming experience can make use of visualization for various purposes, such as 3D modeling, scientific visualization, and animations. This section briefly introduces three software packages that are widely used: MATLAB, Maya, and Flash.

***MATLAB*** MATLAB is a high-level computing language and interactive environment for numeric computation, algorithm development, data analysis, and data visualization. It is the most widely used tool for teaching science, technology, engineering, and mathematics (STEM) in universities and colleges and for research and rapid prototype development in the industry. The core of MATLAB is the MATLAB language, which is a high-level language optimized for matrix, vector, and array computations. It provides many toolboxes for various applications, including math and optimization, statistics and data analysis, control system design and analysis, signal processing and communications, image processing, test and measurement, computational biology, computational finance, and databases. The MATLAB desktop environment consists of a set of tools that facilitate algorithm development and debugging. Users of MATLAB can build applications with graphic user interfaces. MATLAB provides powerful visualization capabilities, and it has an extensive set of functions for matrix and vector visualization (such as line plot, bar graph, scatter plot, pie chart, and histogram), graph annotation and presentation, 3D visualization (such as mesh, surface, volume, flow, isosurface, and streamline), displaying images of various formats (such as jpeg, tiff, and png), as well as camera and lighting controls. The major advantage of using MATLAB for visualization is simplicity and rapid implementation. For many scientific and engineering applications, the visualization capabilities provided by MATLAB are sufficient. In other applications, MATLAB can be used to develop prototype visualizations in order to obtain quick understanding of the problem. Then, a full-featured stand-alone application can be developed, which does not need MATLAB in order to implement and run the visualizations. MATLAB is a product of MathWorks. Figure 7.31 shows two MATLAB visualization examples.

***Maya*** Maya is the industry standard for 3D modeling, animation, visual effects, and rendering. It is widely used in almost every industry that involves 3D modeling and visual effects, including game development, motion pictures, television, design, and manufacturing. Maya has a complicated but customizable architecture and interface to facilitate the application of specific 3D content workflows and pipelines. It has a comprehensive set of 3D modeling and texture mapping tools, including polygons, nonuniform rational B-spline (NURBS), subdivision surfaces, and interactive UV mapping. Realistic animations can be generated using Maya's powerful animation capabilities, such as key frame animation, nonlinear animation, path animation, skinning,

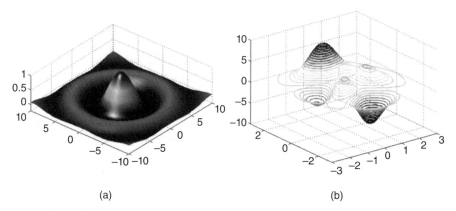

**Figure 7.31** Visualizations in MATLAB. (a) Visualization of the sinc function and (b) contour plot.

inverse kinematics, motion capture animation, and dynamic animation. Using advanced particle systems and dynamics, Maya can implement realistic and sophisticated visual effects and interactions between dynamic objects, such as fluid simulation, cloth simulation, hair and fur simulation, and rigid body and soft body dynamics. Maya includes a large collection of tools to specify material and light properties for advanced rendering. Programmable shaders using high level shading languages such as Cg, HLSL, and GLSL can be handily developed in Maya. In addition to Maya's graphic user interface, developers can access and modify Maya's existing features and introduce new features using several interfaces, including Maya Embedded Language (MEL), Python, Maya API, and Maya Python API. A Maya Personal Learning Edition (PLE) is available for learning purposes. Maya is a product of Autodesk. Figure 7.32 shows a screen capture of Maya. The building rendered in Maya is Virginia Modeling, Analysis, and Simulation Center (VMASC) in Suffolk, VA. VMASC also has laboratories and offices on the main campus of Old Dominion University in Norfolk, VA.

**Flash** *Flash* is an advanced multimedia authoring and development environment for creating animations and interactive applications inside Web pages. It was originally a product of Macromedia but was recently acquired by Adobe. Flash supports vector graphics, digital images and videos, and rich sound effects, and it contains a comprehensive set of tools for creating, transforming, and editing objects, visual effects, and animations for different hardware and software platforms. Animations can be developed quickly using optimized animation tools, including object-based animation, motion presets, motion path, motion editor, and inverse kinematics. Procedural modeling and filtering can be used to generate various visual effects more quickly and easily. 2D objects can be animated in 3D space using 3D transformation and rotation

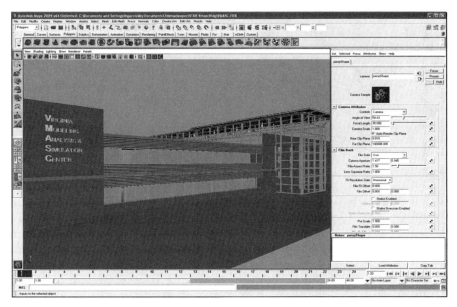

**Figure 7.32** A screen capture of Maya.

tools. In addition to the graphic authoring environment, Flash has its own scripting language, ActionScript, which is an object-oriented programming language that allows for more flexibilities and control in order to generate complex interactive Web applications. Flash applications can run inside Web pages using plug-ins, such as Flash player. Flash can also embed videos and audio in Web pages, and it is used by many Web sites, such as YouTube. Flash is currently the leading technology for building interactive Web pages, and it now integrates well with other Adobe products, such as Adobe Photoshop and Illustrator. Many interactive applications including games built using Flash are available on the Internet. Figure 7.33 shows a screen capture of Flash.

## CASE STUDIES

As the number 1 search engine on the Internet, Google has developed a series of products that make heavy uses of visualizations. Here we briefly introduce two popular Google products: Google Earth and Google Maps.

### Google Earth

*Google Earth* is a stand-alone application that is freely available from Google [27]. As its name indicates, it provides comprehensive information about

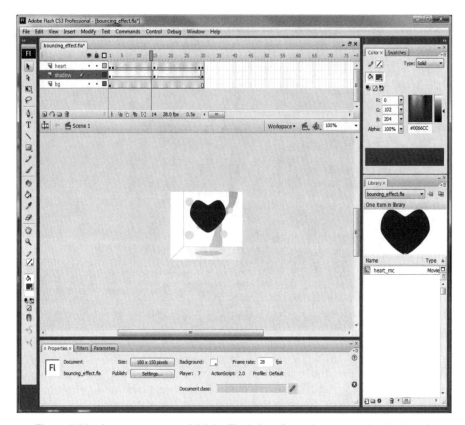

**Figure 7.33** A screen capture of Adobe Flash (one frame in an animation is shown).

Earth. It integrates satellite imagery, maps, terrain, 3D buildings, places of interest, economy, real-time display of weather and traffic, and many other types of geographic information [27]. Google Earth is a powerful tool that can be used for both commercial and educational purposes. Two versions of Google Earth are available: Google Earth and Google Earth Pro. Figures 7.34–7.37 shows several screen captures of Google Earth.

### Google Maps

*Google Maps* is an online application that runs inside Internet browsers [28]. In addition to road maps, it can also display real-time traffic information, terrain, satellite imagery, photos, videos, and so on. It can find places of interest through incomplete search and provide detailed driving directions. Users can conveniently control the travel route through mouse operations. Figures 7.38–7.41 illustrate several uses of Google maps.

CASE STUDIES 219

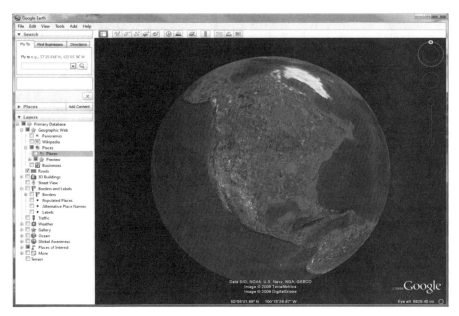

**Figure 7.34** Google Earth.

**Figure 7.35** Google Earth 3D display. Google Earth can display 3D buildings in many places in the world. Shown here are the buildings in lower Manhattan, NY.

**Figure 7.36** Google Earth display of 3D terrains (image is part of the Grand Canyon).

**Figure 7.37** Google Earth can display many types of real-time information. Shown is the traffic information in Hampton Roads, VA. Green dots represent fast traffic and red for slow traffic.

CASE STUDIES    221

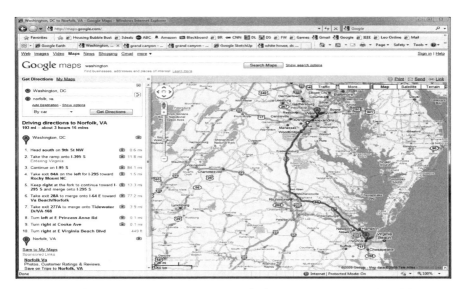

**Figure 7.38** Google Maps provides many types of information, such as detailed driving directions shown here.

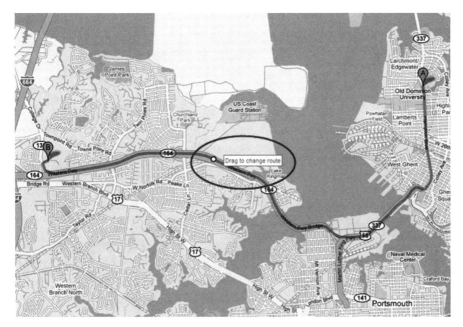

**Figure 7.39** Google Maps driving routes. Users of Google Maps can easily change driving route by dragging waypoints to the desired locations.

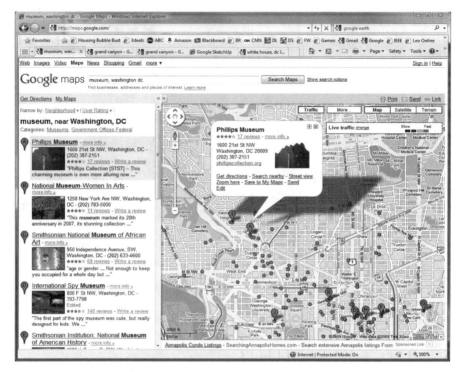

**Figure 7.40** Google Maps places of interest.

**Figure 7.41** Google Maps provides street views that enable users to have virtual tours of the streets that are constructed from real pictures of the streets.

## CONCLUSION

This chapter introduced fundamentals of computer graphics theories. GPUs have evolved into sophisticated computing architectures that contain more transistors and are more powerful than the CPUs. The increasing programmable capabilities provided by the latest GPUs require better understanding of GPU architectures. In addition, GPUs are increasingly being used for general-purpose computations. Transformations based on matrix computations are the foundation of computer graphics, and the use of homogeneous coordinate system unifies the representations for different transformations. The 3D computer graphics is based on the synthetic camera model that mimics real physical imaging systems. The synthetic camera must be positioned and oriented first, and different projections are developed to configure its viewing volume. Colors can be generated and represented using three primary colors. Various models have been developed for different lights and material reflection properties. Shading models are used to computehe color for points inside polygons. Texture mapping is an important technique in modern computer graphics, and it greatly increases visual complexity but with only moderate increase in computational complexity.

This chapter also discussed various types of software and tools to facilitate visualization development. OpenGL and Direct3D are the two dominant low-level graphics libraries. OpenGL is the industry standard for high-performance, cross-platform professional applications, while Direct3D is the tool of choice for game development on Microsoft Windows Platform and Xbox 360 game consoles. High-level graphics libraries encapsulate low-level libraries and provide additional functionalities and optimizations. OpenSceneGraph is an open source and cross-platform 3D graphics library, and it is the leading scene graph technology for high-performance visualization applications. XNA Game Studio is an integrated development environment to facilitate game development for Windows PC, Xbox 360, and Zune. Java 3D is also a scene graph, and it is widely used for mobile application and online applications. Advanced visualization tools that require less programming skills were also discussed in this chapter, including MATLAB, Maya, and Flash.

Highlighted were case studies of two popular Google software packages: Google Earth and Google Maps. Google Earth is a comprehensive visual database of Earth, while Google Maps provides many advanced capabilities for travel and planning purposes. Several examples were included to illustrate the uses of Google Earth and Google Maps. As can been from these two examples, visualization plays an increasingly important role in people's daily life. Visualization is especially important for modeling and visualization professionals, and in-depth understanding of visualization theories and techniques is essential in developing effective and efficient modeling and simulation applications.

## REFERENCES

[1] Angel E. *Interactive Computer Graphics: A Top-Down Approach Using OpenGL.* 5th ed. Boston: Pearson Education; 2008.

[2] Shirley P. *Fundamentals of Computer Graphics.* 2nd ed. Wellesley, MA: A K Peters; 2005.

[3] Foley JD, van Dam A, Feiner SK, Hughes JF. *Computer Graphics: Principles and Practice, Second Edition in C.* Reading, MA: Addison-Wesley Publishing Company; 1996.

[4] Lengyel E. *Mathematics for 3D Game Programming and Computer Graphics.* 2nd ed. Hingham, MA: Charles River Media; 2004.

[5] Van Verth JM, Bishop LM. *Essential Mathematics for Games and Interactive Applications: A Programmer's Guide.* Amsterdam: Morgan Kaufman Publishers; 2004.

[6] Fernando R. *GPU Gems: Programming Techniques, Tips and Tricks for Real-Time Graphics.* Boston: Addison-Wesley; 2004.

[7] Pharr M, Fernando R. *GPU Gems 2: Programming Techniques for High-Performance Graphics and General-Purpose Computation.* Boston: Addison-Wesley; 2005.

[8] Nguyen H. *GPU Gems 3.* Boston: Addison-Wesley; 2007.

[9] Fatahalian K, Houston M. GPUs: A closer look. *ACM Queue*, March/April 2008, pp. 18–28.

[10] Nickolls J, Buck I, Garland M, Skadron K. Scalable parallel programming with CUDA. *ACM Queue*, March/April 2008, pp. 40–53.

[11] Khronos Group. OpenCL: The Open Standard for Heterogeneous Parallel Programming. Available at http://www.khronos.org/developers/library/overview/opencl_overview.pdf. Accessed May 2009.

[12] OpenGL Architecture Review Board, Shreiner D, Woo M, Neider J, Davis T. *OpenGL Programming Guide Sixth Edition: The Official Guide to Learning OpenGL.* Version 2.1. Boston: Addison-Wesley; 2007.

[13] Wright RS, Lipchak B, Haemel N. *OpenGL SuperBible: Comprehensive Tutorial and Reference.* 4th ed. Boston: Addison-Wesley; 2007.

[14] Rost RJ. *OpenGL Shading Language.* 2nd ed. Boston: Addison-Wesley; 2006.

[15] Munshi A, Ginsburg D, Shreiner D. *OpenGL ES 2.0 Programming Guide.* Boston: Addison-Wesley; 2008.

[16] Kuehne R, Sullivan JD. *OpenGL Programming on Mac OS X: Architecture, Performance, and Integration.* Boston: Addison-Wesley; 2007.

[17] Luna FD. *Introduction to 3D Game Programming with DirectX 10.* Sudbury, MA: Jones & Bartlett Publishers; 2008.

[18] Luna FD. *Introduction to 3D Game Programming with DirectX 9.0c: A Shader Approach.* Sudbury, MA: Jones & Bartlett Publishers; 2006.

[19] Walsh P. *Advanced 3D Game Programming with DirectX 10.0.* Sudbury, MA: Jones & Bartlett Publishers; 2008.

[20] McShaffry M. *Game Coding Complete.* Boston: Charles River Media; 2009.

[21] OpenSceneGraph. Available at www.openscenegraph.org. Accessed May 2009.

[22] Microsoft. XNA Creators Club Online. Available at creators.xna.com. Accessed May 2009.
[23] Grootijans R. *XNA 3.0 Game Programming Recipes: A Problem-Solution Approach*. Berkeley, CA: Apress; 2009.
[24] Lobao AS, Evangelista BP, de Farias JA, Grootjans R. *Beginning XNA 3.0 Game Programming: From Novice to Professional*. Berkeley, CA: Apress; 2009.
[25] Carter C. *Microsoft XNA Game Studio 3.0 Unleashed*. Indianapolis, IN: Sams; 2009.
[26] Java 3D. Available at java3d.dev.java.net. Accessed 2009 May.
[27] Google Earth. Available at http://earth.google.com. Accessed May 2009.
[28] Google Maps. Available at http://maps.google.com. Accessed May 2009.

# 8

# M&S METHODOLOGIES: A SYSTEMS APPROACH TO THE SOCIAL SCIENCES

Barry G. Silverman, Gnana K. Bharathy, Benjamin Nye,
G. Jiyun Kim, Mark Roddy, and Mjumbe Poe

Most multiagent models of a society focus on a region's living environment and its so-called political, economic, social/cultural, and infrastructural systems. The region of interest might entail several states, a single state, and/or substate areas. Such models often support analysts in understanding how the region functions, and how changed conditions might alter its dynamics. However, these models do little or nothing to support analysts in answering critical questions regarding the region's key actors and between these key actors and the environmental elements they influence. And, these actors can be important. In some scenarios, they strongly influence the allocation of resources, flow of services, and mood of the populace. In other areas, such as those involving leadership survival, they often tend to dominate the situation.

Needless to say, personal interaction/behavioral modeling entails getting inside the head of specific leaders, key followers, and groups/factions, and bringing to bear psych–socio–cultural principles, rather than physics–engineering ones.

Recognizing the importance of behavioral modeling to crisis management, modeling and simulation (M&S) researchers are currently working to get a handle on sociocognitive modeling. There are many approaches that are being included:

---

*Modeling and Simulation Fundamentals: Theoretical Underpinnings and Practical Domains*,
Edited by John A. Sokolowski and Catherine M. Banks
Copyright © 2010 John Wiley & Sons, Inc.

(1) *Cognitive Modeling.* This approach is often attempted when studying an individual and his or her decision making. It is most often used to study microprocesses within the mind, though it has also been scaled to crew or team level applications. It can offer deep insights into what is driving a given individual's information collection and processing and how to help or hurt his or her decision cycle.

(2) *Ethnographic Modeling.* This is the main approach used in anthropology to study what motivates peoples of a given culture or group. This approach focuses on descriptive modeling of relations and relationships, morals and judgment, mobilization stressors, human biases and errors, and emotional activations such as in cognitive appraisal theories.

(3) *Social Agent Systems.* Sociological complexity theorists tend to use agent approaches to show how microdecisions of individual agents can influence each other and lead to the emergence of unanticipated macrobehaviors of groups, networks, and/or populations. Traditionally, the microdecision making of the agents is shallow, and it sacrifices cognitive and/or ethnographic modeling depths in order to compute macrobehavior outcomes.

(4) *Political Strategy Modeling.* In the rational actor theory branch of political science, classical game theory was successful in the Cold War era where two adversaries squared off in a conflict involving limited action choices, symmetrical payoff functions, and clear outcomes. To date, this approach has not borne fruit in trying to model net-centric, asymmetric games.

(5) *Economic Institution Modeling.* This is concerned with applying mathematical formalisms to represent public institutions (defense, regulators, education, etc.) and private sectors/enterprises (banking, manufacturing, farming, etc.), and to try and explain the economies and services of both developed and developing nations. Agents who make the producing, selling, distribution, consuming, and so on decisions are not modeled themselves, but rather a black box approach is the classical method where macrobehavior data are fit to regressive type curves and models. In this discipline, it is acceptable for institutional theories to be modeled with no evidence or observations behind them at all.

These are representative paradigms drawn from the major disciplines of what are normally considered the social sciences, that is, psychology, sociology, anthropology, political science, and economics, respectively [1]. Each of these disciplines typically has several competing paradigms accepted by researchers in those fields, in addition to those sketched above. These paradigms and disciplines each offer a number of advantages, but alone, they each suffer from serious drawbacks as well. The world is not unidisciplinary (nor uniparadigm), though it tends to be convenient to study it that way. The nature of scientific method (reductionism) over time forces a deepening and narrowness of focus; knowledge silos evolves and it becomes difficult for individuals in different disciplines (or even in the same discipline) to see a unifying paradigm.

This chapter focuses on research at the University of Pennsylvania on applying the systems approach to the social sciences. Our research agenda is to try and to synthesize the best theories and paradigms across all the social science disciplines, to provide a holistic modeling framework. There is no attempt to endorse a given theory but to provide a framework where all the theories might ultimately be tested. This is a new approach to social system modeling, though it makes use of tried and true systems engineering principles. Specifically, a *social system* is composed of many parts that are systems themselves. The parts have a functionality that needs to be accurately captured and encapsulated, though precision of a part's inner workings is less important than studying the whole. Provided a part's functionality is adequately captured, interrelation between the parts is of prime importance, as is studying the synergies that emerge when the parts interoperate. A challenge of social systems is that there are many subsystems that are themselves purposeful systems—many levels of functionality from the depths of the cognitive up to the heights of the economic institutions and political strategies—and one must find ways to encapsulate them in hierarchies, so that different levels may be meaningfully studied.

This chapter provides an overview of these important new developments. It begins by (1) elaborating on the goals of our behavioral modeling framework, (2) delving into the underpinning theory as well as its limitations, (3) providing some examples of games based on the theory, (4) describing implementation considerations, and (5) discussing how leader/follower models might be incorporated in or interfaced with comprehensive political, military, economic, social, informational (psyops), and infrastructure (PMESII) models in other chapters. We conclude with a wrap-up and way ahead.

## SIMULATING STATE AND SUBSTATE ACTORS WITH CountrySIM: SYNTHESIZING THEORIES ACROSS THE SOCIAL SCIENCES

We had three specific goals in developing both the underlying (FactionSim) framework and its country modeling application we describe here as *CountrySim*. One aim of this research is to provide a generic game simulator to social scientists and policymakers so that they can use it to rapidly mock up a class of conflicts (or opportunities for cooperation) commonly encountered in today's world. Simply put, we have created a widely applicable game generator (called *FactionSim*), where one can relatively easily recreate a wide range of social, economic, or political phenomenon so that an analyst can participate in and learn from role-playing games or from computational experiments about the issues at stake. Note that this game generator can be thought of as a kind of agent-based modeling framework. However, this is quite different from existing agent-based models because it is a framework that is designed for implementing highly detailed, cognitive agents in realistic social settings. This sociocognitive agent framework is called *PMFserv*. We have departed from the prevailing "Keep It Simple Stupid" (KISS) paradigm that is dominant in social science modeling because we see no convincing methodological or theoretical reasons why we should limit ourselves to simple agents and simple models, when interesting problems can be better analyzed with more complex models, that is, with realistic agents. We do understand the problems of complex models, and this issue will be discussed below.

Our second aim is to create plausible artificial intelligence (AI) models of human beings and, more specifically, leader and follower agents based on available first principles from relevant disciplines in both natural and social sciences. We want our PMFserv agents to be as realistic as possible so that they can help analysts explore the range of their possible actions under a variety of conditions, thereby helping others to see more clearly how to influence them and elicit their cooperation. A related benefit of having realistic agents based on evidence from video and multiplayer online games is that if the agents have sufficient realism, players and analysts will be motivated to remain engaged and immersed in role-playing games or online interactive scenarios. A "catch-22" of the first two aims is that, agent-based simulation games will be more valuable the more they can be imbued with realistic leader and follower behaviors, while the social sciences that can reliably contribute to this undertaking are made up of many fragmented and narrow specialties, and few of their models have computational implementations.

The third aim is to improve the science by synthesizing best-of-breed social science models with subject matter expert (SEM) knowledge so the country model merger can be tested in agent-based games, exposing their limitations and showing how they may be improved. In the social sciences and particularly in economics and, to a lesser extent, in political science, there seems to be an emerging consensus that a theory should be developed with mathematical rigor typically using a rational choice or some other approach (such as prospect or poliheuristic approaches) and tested using best available data (preferably large-N). It is also true that there is a resurgent interest in conducting experimental studies. Although we probably are not the first ones to point out this possibility, the idea of using realistic agent-based simulation to test competing theories in the social sciences looks like an attractive addition to these approaches. Especially when the availability of data is limited or the quality of data is poor, or when experimentations using human subject is either difficult or impossible, simulations may be the best choice. Simulators such as PMFserv can serve as virtual petri dishes where almost unlimited varieties of computational experimentations are possible and where various theories can be implemented and tested to be computationally proved (i.e., to yield "generative" proofs). These are only some of the virtues and possibilities of having a versatile simulator like PMFserv. In this discussion, we will limit these experiments to our CountrySim applications in Iraq and Bangladesh.

## Literature Survey

Our collection of country models, CountrySim, can best be described as a set of complex agent-based models that use hierarchically organized and cognitive–affective agents whose actions and interactions are constrained by various economic, political, and institutional factors. It is hierarchically organized in the sense that the underlying FactionSim framework consists of a country's competing factions, each with its own leader and follower agents. It is cognitive–affective in the sense that all agents are "deep" PMFserv agents with individually tailored and multi-attribute utility functions that guide a realistic decision-making mechanism. CountrySim, despite its apparent complexity, is an agent-based model that aims to show how individual agents interact to generate emergent macrolevel outcomes. CountrySim's user-friendly interface allows variables to be adjusted and results to be viewed in multiple ways. In this section, we briefly overview the field of agent-based modeling of social systems. For a more in-depth review, read the National Research Council's (NRC) *Behavioral Modeling and Simulation* or Ron Sun's *Cognition and Multi-Agent Interactions* [2,3].

*Agent-based modeling* is a computational method of studying how interactions among agents generate macrolevel outcomes with the following two features:

(1) Multiple interacting entities—from agents representing individuals to social groupings—compose an overall system
(2) Systems exhibit emergent properties from the complex interactions of various entities

˙Interactions are complex in the sense that the emergent macrolevel outcomes cannot be inferred by, for example, simply combining the characteristics of the composing entities. Since the 1990s, agent-based modeling has been recognized as a new way of conducting scientific research [4]. Agent-based modeling is based on a rigorous set of deductively connected assumptions capable of generating simulated data that is, in turn, amenable to inductive analysis. However, it does not rigorously deduce theorems with mathematical proofs or provide actual data that are obtained from reality.

Recently, agent-based modeling has received more attention thanks to books such as Malcolm Gladwell's hugely successful *Tipping Point*, where agent-based modeling was presented as the best available method for studying rare and important political and economic events such as riots and government and economic collapses [5]. Indeed, agent-based modeling is known to be particularly deft at estimating the probability of such unusual large-scale emergent events. Our CountrySim currently exhibits around 80 percent accuracy in retrodicting various events of interest (EOI) involving politico-economic instabilities.

Agent-based models can be categorized according to their respective structural features. For example, the NRC's commissioned study on behavioral M&S recently pointed out five dimensions along which agent-based models can be categorized. The five dimensions are:

(1) number of agents
(2) level of cognitive sophistication
(3) level of social sophistication
(4) the means of agent representation (rules versus equations)
(5) use of grid [2]

For the purpose of our review, we emphasize the first three dimensions. The distinction between the use of rules and the use of equations in agent representation seems to be increasingly blurred given the increasing proliferation of the combined use of rules and equations and given that equations can arguably be construed as a particular kind of rules. For example, our CountrySim uses both rules and equations for agent representation. The use of a grid also seems to be a distinction of limited significance given the predominance of grid-based models in the earlier stages of agent-based modeling development and given that modelers no longer have to make an either-or choice regarding grids. CountrySim, for example, is at the same time grid-based and not grid-based in that it uses a particular cellular automata, PS-I (Political Science Identity; [25]), to overcome computational constraints in terms of the number of agents. Hence, the first three dimensions identified by the NRC seem to provide the most germane distinctions. Figure 8.1 locates various well-known existing agent-based models along the three dimensions.

The first dimension is the level of cognitive sophistication, which shows significant variation from model to model. A famous model using agents with little cognitive sophistication is the residential segregation model devised by Nobel laureate *Thomas Schelling*. In Schelling's model, agents could have either black or white identities, and their decision making was limited to a single decision concerning whether or not to stay in a particular neighborhood, based on a simple rule concerning the neighborhood's color composition (i.e., when the percentage of neighbors of the opposite color exceeds certain threshold, move; otherwise, stay). Schelling's model can be easily implemented in agent-based modeling toolkits such as Swarm, Repast, and NetLogo. The agents that are used in these modeling toolkits are usually low on the cognitive sophistication dimension and tend to follow simple rules and use some sort of cellular automata.

In contrast, highly sophisticated cognitive agents can be modeled based on a computational implementation of one or another overarching theory of human cognition. This approach requires modeling the entire sequence of information-processing and decision-making steps human beings take from initial stimuli detection to responses via specific behavior. Two examples of

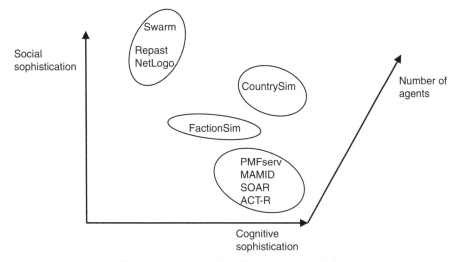

**Figure 8.1** Categories of agent-based models.

purely cognitive agents are *atomic components of thought or adaptive character of thought* (ACT-R) and *state, operator, and results* (SOAR) agents. ACT-R is currently one of the most comprehensive cognitive architectures that initially focused on the development of learning and memory and nowadays increasingly emphasizes the sensory and motor components (i.e., the front and back end of cognitive processing) [6–9]. SOAR is another sophisticated cognitive architecture that can be used to build highly sophisticated cognitive agents. SOAR is a computational implementation of Newell's unified theory of cognition and focuses on solving problems [10]. Agents using SOAR architecture are capable of reactive and deliberative reasoning and are capable of planning.

More recently, various efforts have been made to improve and complement these purely cognitive agents by including some aspects of the affective phenomena that are typically intertwined with human cognition. The effects of emotions on decision making and, more broadly, human behavior and the generation of emotion through cognitive appraisal are most frequently computationally implemented. Two examples are our own PMFserv and *methodology for analysis and modeling of individual differences* (MAMID). MAMID is an integrated symbolic cognitive–affective architecture that models high-level decision making. MAMID implements a certain cognitive appraisal process to elicit emotions in response to external stimuli and evaluates the effects of these emotions on various stages of decision making [11].

Our own PMFserv is a *commercial off-the-shelf* (COTS) human behavior emulator that drives agents in simulated game worlds and in various agent-based models including FactionSim and CountrySim. This software was developed over the past 10 years at the University of Pennsylvania as an architecture to synthesize many best available models and best practice theories of human behavior modeling. PMFserv agents are unscripted. Moment by moment, they rely on their microdecision-making processes to react to actions as they unfold and to plan out responses. A performance moderator function (PMF) is a micromodel covering how human performance (e.g., perception, memory, or decision making) might vary as a function of a single factor (e.g., event stress, time pressure, grievance, etc.). PMFserv synthesizes dozens of best available PMFs within a unifying mind–body framework and thereby offers a family of models where microdecisions lead to the emergence of macrobehaviors within an individual. For each agent, PMFserv operates its perception and runs its physiology and personality/value system to determine coping style, emotions and related stressors, grievances, tension buildup, and impact of rumors and speech acts, as well as various collective and individual action decisions, in order to project emergent behaviors. These PMFs are synthesized according to the interrelationships between the parts and with each subsystem treated as a system in itself. When profiling an individual, various personality and cultural profiling instruments are utilized. These instruments can be adjusted with Graphic User Interface (GUI) sliders and with data from Web interviews concerning parameter estimates from a country, leader, or area expert. PMFserv agents include a dialog engine and users can query the agents to learn their personality profiles, their feelings about the current situation, why they made various decisions, and what is motivating their reasoning about alliances/relations.

A significant feature of CountrySim is the way this first dimension, the level of cognitive sophistication, is intricately linked to the third dimension, the number of agents. In general, the more cognitively sophisticated the agents in an agent-based model, the smaller the number of agents the model can accommodate, given the computational constraints of processing multiple cognitively sophisticated agents in a timely manner [2]. For example, the sophistication of the agents used in ACT-R, SOAR, and MAMID means that these models are limited to no more than 10–20 agents. Indeed, no existing model except PMFserv has been used to build models of artificial social systems particularly to monitor and forecast various political and economic instabilities of interest to military and business end users. In FactionSim and CountrySim, PMFserv agents are skillfully sedimented to build various artificial social systems. Our cognitive–affective PMFserv agents are used to model leaders and other influential agents of a social system of interest using FactionSim and/or CountrySim, while follower agents are represented either by a couple of archetypical PMFserv agents or by numerous simple agents in a cellular automata (e.g., PS-I) that is intended to represent a physical and social landscape. In our latest version of CountrySim, the sophisticated leader agents from FactionSim and the simple follower agents from PS-I are dynamically interconnected so that leader agent decisions affect follower agents' actions and vice versa. Unless the processing speed of computers increases dramatically over the next few years, our method seems to be one of the more reasonable ways to get around the problem of using cognitively sophisticated agents to build artificial social systems.

The second dimension, the level of social sophistication, is also intricately linked to the third dimension, the number of agents, as well as to the first dimension, the level of cognitive sophistication. In general, the level of social sophistication is relatively low for small or large agent models while relatively high for midsized agent populations [2]. This relationship is intuitive given that sophisticated social behavior requires some level of cognitive sophistication, while cognitive sophistication beyond a certain level is limited by the aforementioned computational constraints. Agent-based models built using toolkits such as Swarm, Repast, and NetLogo generally tend to exhibit slightly higher social sophistication than models made using highly cognitively sophisticated agents based on ACT-R, SOAR, and MAMID. The former set of models is usually far less computationally constrained and can better represent spatial relationships among agents. It is certainly true that cognitively sophisticated agents can better represent multidimensional social behavior and network effects so long as they can be combined in large numbers to build large-scale artificial social systems. Thus, to maximize the dimension of social sophistication, the best options appear to be a midsized population model with moderately cognitively sophisticated agents or a hybrid model (using both cognitively sophisticated and simple agents) as in the case of FactionSim and CountrySim.

The NRC's review of agent-based modeling points out three major limitations on this third way of conducting research. The three limitations are found in the following three realms of concern:

(1) degree of realism
(2) model trade-offs
(3) modeling of actions [2]

With regard to the first of these (degree of realism), we believe that our particular approach to agent-based modeling, highlighted by our recent effort in building CountrySim, has achieved a level of realism that has no precedent. We use realistic cognitive–affective agents built by combining best available principles and conjectures from relevant sciences. The rules and equations that govern the interactions of these agents are also derived from the best available principles and practices, with particular attention to their realism. When populating our virtual countries with agents and institutions, we triangulated three sources:

(1) Existing country databases from the social science community and various government agencies and nongovernmental organizations
(2) Information collected via automated data extraction technology from the Web and various news feeds
(3) Surveys of SEMs

This approach has yielded the best possible approximations for almost all parameter values of our model [12].

Scholars in more tradition-bound research communities such as econometrics and game theory may point out that our approach violates the prevailing KISS paradigm and commits the sin of "overfitting." As mentioned previously, we simply depart from this prevailing paradigm because we see no convincing methodological or theoretical reasons for adhering to it. There is an equally important and convincing emerging paradigm named "Keep It Descriptive Stupid" (KIDS) that provides an alternative framework for building realistic and complex models of social systems [13]. This approach emphasizes the need to make models as descriptive as possible and accepts simplification only when evidence justifies it. The concern regarding overfitting is also misdirected given that, unlike in econometrics, data are not given a priori in agent-based modeling, and the addition of new model parameters or rules and equations increases the number of simulation outcomes that can be generated instead of simply fitting the model to the given data [2].

With regard to the second concern (regarding model trade-offs) and the third concern (regarding the modeling of actions), we again believe that our particular approach to agent-based modeling, highlighted by our recent effort in building CountrySim, has good theoretical and practical justification. It is true that using cognitively sophisticated agents limits the modeling of large-scale interactions among numerous agents with an entire range of possible interactions. Also, when using simple agents, we can only conduct high-level exploratory analyses at a high level of abstraction without gaining detailed insights or being able to evaluate specific impacts of particular actions as required by various PMSEII studies. We overcome this trade-off by simultaneously using both sophisticated and simple agents and by linking FactionSim and CountrySim to a cellular automata such as PS-I.

Concerning the modeling of actions, we overcome the need to model actions at a reductively abstract level (attack vs. negotiate) or a highly detailed level (seize a particular village using a particular type of mechanized infantry vs. provide a specific amount of money and years of education to the village's unemployed in return for them not joining a particular local extremist group) by building a model that can account for these two vastly different levels of implementable actions. Our solution is to combine a higher-level CountrySim and a lower-level VillageSim (also known as NonKin Village) that can be linked to each other in order to serve the particular needs of different audiences, from a military or a business perspective.

### Technical Underpinnings: Behavioral Game Theory

Game theory, analytic game theory, in particular, has been employed for many years to help understand conflicts. Unfortunately, analytic game theory has a weak record of explaining and/or predicting real-world conflict—about the same as random chance according to Armstrong [14] and Green [15]. In the field of economics, Camerer pointed out that the explanatory and predictive powers of analytic game theory are being improved by replacing prescriptions from rational economics with descriptions from the psychology of monetary judgment and decision making [16]. This has resulted in "behavioral game theory" that adds in emotions, heuristics, and so on. We pursue the same approach and believe the term "behavioral game theory" is broad enough to cover all areas of social science, not just economics.

In particular, the military, diplomatic, and intelligence analysis community would like for (behavioral) game theory to satisfy an expanding range of scenario simulation concerns. Their interest goes beyond mission-oriented military behaviors, to also include simulations of the effects that an array of alternative *diplomatic, intelligence, military, and economic* (DIME) actions might have upon the *political, military, economic, social, informational (psyops), and infrastructure* (PMESII) dimensions of a foreign region. The goal is to understand factional tensions and issues, how to prevent and end conflicts, and to examine alternative ways to influence and possibly shape outcomes for the collective good.

Our research is aimed at supporting this. Specifically, we focus on the following questions: How can an analyst or trainee devise policies that will influence groups for the collective good? And, what must a sociocultural game generator encompass?

## Political, Social, and Economic "Games"

Figure 8.1 attempts to portray a fairly universal class of leader–follower games that groups often find themselves in and that are worthy of simulation studies. Specifically, the vast majority of conflicts throughout history ultimately center around the control of resources available to a group and its members. This could be for competing groups in a neighborhood, town, region, nation, or even between nations. Further, it applies equally to social, political, and/or economic factions within these geographic settings. That is, this principle of resource-based intergroup rivalry does not obey disciplinary boundaries even though theories within single disciplines inform us about some aspect of the game. Analysts would need an appropriate suite of editors and a generator to help them rapidly mock up such conflict scenarios and analyze what outcomes arise from different courses of action/policies. We describe this game intuitively here and more fully in subsequent sections.

Specifically, the sociocultural game centers on agents who belong to one or more groups and their affinities to the norms, sacred values, and interrelational practices (e.g., social and communicational rituals) of those groups. Specifically, let us suppose there are $N$ groups in the region of interest, where each group has a leader archetype and two follower archetypes (loyalists and fringe members). We will say more about archetypes shortly, and there can certainly be multiple leaders and followers in deeper hierarchies, but we stick in this discussion to the smallest subset that still allows one to consider beliefs and affinities of members and their migration to more or less radical positions. There is an editable list of norms/value systems from which each group's

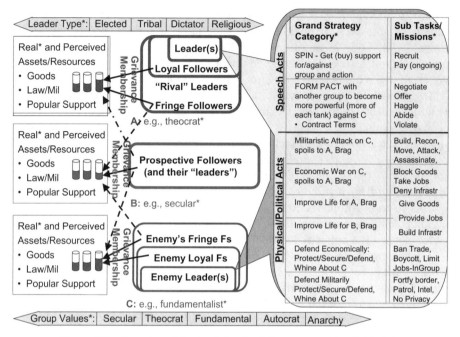

**Figure 8.2** Overview of the basic leader–follower game. *, editable list.

identity is drawn. The range across the base of Figure 8.2 shows an example of a political spectrum for such a list, but these could just as easily be different parties in a common political system, diverse clans of a tribe, different groups at a crowd event, sectors of an economy, and so on. Each entry on this list contains a set of properties and conditions that define the group, its practices, and entry/egress stipulations. The authority of the leader in each group is also indicated by a similarly edited list depicted illustratively across the top of Figure 8.2.

While a number of assumptions made by the classical analytic game theory are defensible (well-ordered preferences, transitivity), others are meant for mathematical elegance. Without assumptions doing most of the "heavy lifting," it is impossible to develop mathematically tractable models [17]. This is the "curse of simplicity." Simple or stylized game models are unable to encode domain information, particularly the depth of the social system. For example, human value systems are almost always assumed, hidden, or at the best, shrunk for the purpose of mathematical elegance. Yet, human behavior is vital to the conflict–cooperative game behavior.

While mathematical convenience is one explanation, there is more involved. Many modeling platforms would simply not allow value systems to be made explicit, and there is no modeling process that would allow one to revisit the values. As computational power increases to accommodate more complex models, social system modelers are beginning to address this curse of simplicity.

Even though such models cannot be solved mathematically, we can find solutions through validated simulation models with deep agents. If one could find clusters of parameters that pertain to a corresponding game model, we can also start talking about correspondence between game theoretic models and cognitively deep simulation models. There is room for a lot of synergy.

Now, let us return to the cognitively detailed game. The resources of each group are illustrated along the left side of Figure 8.2 and are summarized for brevity into three tanks that serve as barometers of the health of that aspect of the group's assets—(1) political goods available to the members (jobs, money, foodstuffs, training, health care etc.); (2) rule of law applied in the group as well as level and type of security available to impose will on other groups; and (3) popularity and support for the leadership as voted by its members. In a later section, we will see that many more resources can be modeled, but for the discussion here, we will start with this minimal set of three. Querying a tank in a culture game will return current tank level and the history of transactions or flows of resources (in/out), who committed that transaction, when, and why (purpose of transactional event).

To start a game, there are initial alignments coded manually, though these will evolve dynamically as play unfolds. Specifically, each group leader, in turn, examines the group alignments and notices loyal in-group (A), resistant out-group (C), and those "undecideds" in middle (B) who might be turned into allies. Also, if there are other groups, they are examined to determine how they might be enlisted to help influence or defend against the out-group and whatever alliance it may have formed. Followers' actions are to support their leader's choices or to migrate toward another group they believe better serves their personal value system. Actions available to leader of A are listed in the table on the right side of Figure 8.2 as either speech acts (spin/motivate, threaten, form pact, brag) or more physical/political acts. Of the latter, there are six categories of strategic actions. The middle two tend to be used most heavily by stable, peaceful groups for internal growth and development. The upper two are economic and militaristic enterprises and campaigns taken against other groups, while the lower two categories of actions are defensive ones intended to barricade, block, stymie the inroads of would be attackers. The right-hand column of the action table lists examples of specific actions under each of these categories—the exact list will shift depending on whether the game is for a population, organizational, or small group scenario. In any case, these actions require the spending of resources in the tanks, with proceeds going to fill other tanks. Thus, the culture game is also a resource allocation problem. Leaders who choose successful policies will remain in power, provide benefits for their followers, and ward off attackers. Analysts and trainees interacting with this game will have similar constraints to their policies and action choices.

The lead author spent much of 2004 assembling a paper-based version of Figure 8.2 as a role-playing diplomacy game and play-testing it with analysts [18]. The goal of the game is to help players to experience what the actual leaders are going through and, thereby, to broaden and deepen their understanding, help with idea generation, and sensitize them to nuances of influencing leaders in a given scenario. The mechanics of the game place the player at the center of the action, and play involves setting objectives, figuring out campaigns, forming alliances when convenient, and backstabbing when necessary. This is in the genre of the diplomacy or risk board games, though unlike diplomacy, its rapidly reconfigurable to any world conflict scenario.

After completing the mechanics and play-testing, three implementations of the game were created: (1) a software prototype called LeaderSim (or Lsim) that keeps world scenarios and action sets to the simplest possible, so that we can easily build and test all of the core ideas of the theory [19]; (2) a scaled-up version called Athena's Prism that has been delivered as a fully functioning computer game in mid-2005, though AI opponent features are continually being added [18]; and (3) a fleshed out version called FactionSim that adds public and private institutions (agencies) that manage the resources and run the services of a given faction or government minister [20]. This FactionSim version also includes group hierarchies and many more layers of leader and follower agents who decide on their own whether to do a given leader's bidding (e.g., go to war, work in a given sector, vote for his or her reelection, etc.). This last version is still under development, though we discuss elements of it in subsequent sections of this paper and have working examples of it that plug into third-party simulators to run the minds and behavior of agents in those worlds.

### Social Agents, Factions, and the FactionSim Test Bed

This section introduces FactionSim, an environment that captures a globally recurring sociocultural "game" that focuses upon intergroup competition for control of resources (security, economic, political, etc., assets). The FactionSim framework facilitates the codification of alternative theories of factional interaction and the evaluation of policy alternatives. FactionSim is a tool that allows conflict scenarios to be established in which the factional leader and follower agents all run autonomously; use their groups' assets, resources, and institutions; and freely employ their microdecision making as the situation requires. Macrobehaviors emerge as a result. This environment thus implements PMFserv within a game theory/PMESII campaign framework. One or more human players interact with FactionSim and attempt to employ a set of DIME actions to influence outcomes and PMESII effects.

To set up a FactionSim game one simply profiles the items overviewed in this section. Types of parameters for typical social system models in PMFserv entities are given below. These may be edited at the start, but they all evolve and adapt dynamically and autonomously as a game plays out. In addition, there are other parameters that are automatically generated (e.g., the 22 emotions of each agent, relationship levels, models of each other, etc.). Profiling includes:

(1) **Agents** (decision-making individual actors)
   - value system/goals, standards, and preference (GSP) tree: hierarchically organized values such as short-term goals, long-term preferences and likes, and standards of behavior including sacred values and cultural norms
   - ethno–linguistic–religious–economic/professional identities
   - level of education, level of health, physiological/stress levels
   - level of wealth, savings rate, contribution rate
   - extent of authority over each group, degree of membership in each group
   - personality and cultural factor sets (conformity, assertivity, humanitarianism, etc.)
(2) **Groups/factions**
   - philosophy, sense of superiority, distrust, perceived injustices/transgressions
   - leadership, membership, other roles
   - relationship to other groups (in-groups, out-groups, alliances, atonements, etc.)
   - barriers to exit and entry (saliences)
   - group level resources such as political, economic, and security strengths
   - institutional infrastructures owned by the group
   - access to institutional benefits for the group members (level available to group)
   - fiscal, monetary, and consumption philosophy
   - disparity, resource levels, assets owned/controlled
(3) **Region's resources**
   *Security model* (force size, structure, doctrine, training, etc.)
   - power-vulnerability computations [19]
   - Skirmish model/Urban Lanchester model (probability of kill)
   *Economy model* (dual sector—Lewis-Ranis-Fei (LRF) model) [21]
   - formal capital economy (Solow growth model)
   - undeclared/black market [22]
   *Political model* (loyalty, membership, voting, mobilization, etc.) [23]
   - follower social network [4,24,25]
   - info propagation/votes/small world theory [26]

(4) ***Institutions available to each group*** (public works, protections, health/ education, elections, etc.)
- capital investment, capacity for service, # of jobs
- effectiveness, level of service output
- costs of operation, depreciation/damage/decay
- level of corruption (indicates usage vs. misuse), group influence

Now, with this framework in mind, let us look at the different types of actors required to construct the kind of social system models we have built. Frequently, we create two different types of individual actors:

(1) Individually named personae, such as leaders, who could be profiled.
(2) Archetypical members of the society or of a particular group whose model parameters are dependent on societal level estimates.*

These individuals then have the following types of action choices (at the highest level of abstraction):

(1) Leader-actions (A) = {Leader-actions (target) = {Speak (seek-blessing, seek-merge, mediate, brag, threaten), Act (attack-security, attack-economy, invest-own-faction, invest-ally-faction, defend-economy, defend-security)}
(2) Follower-actions(target) = {Go on Attacks for, Support (econ), Vote for, Join Faction, Agree with, Remain-Neutral, Disagree with, Vote against, Join Opposition Faction, Oppose with Non-Violence(Voice), Rebel-against/Fight for Opposition, Exit Faction}}

Despite efforts at simplicity, stochastic simulation models for domains of this sort rapidly become complex. If each leader has nine action choices "on each of the other (three) leaders," then he or she has 729 (=$9^3$) action choices on each turn (and this omits considering different levels of funding each action). Each other leader has the same, so there are $729^3$ (~387 million) joint action choices by others. Hence, the strategy space for a leader consists of all assignments of his or her 729 action responses to each of the $729^3$ joint action choices by the other three. This yields a total strategy set with cardinal-

---

*For each archetype, what is interesting is not strictly the mean behavior pattern, but what emerges from the collective. To understand that, one expects to instantiate many instances of each archetype where each agent instance is a perturbation of the parameters of the set of PMFs whose mean values codify the archetypical class of agent they are drawn from. This means that any computerization of PMFs should support the stochastic experimentation of behavior possibilities. It also means that individual differences, even within instances of an archetype, will be explicitly accounted for.

ity 387 million raised to 729, a number impossibly large to explore. As a result, FactionSim provides an Experiment Dashboard that permits inputs ranging from one course of action to a set of parameter experiments the player is curious about. All data from PMFserv and the sociocultural game is captured into log files. At present, we are finalizing an after-action report summary module, as well as analytic capabilities for design of experiments, for repeated Monte Carlo trials, and for outcome pattern recognition and strategy assessment.

### The Economy and Institutional Agencies

This section overviews the version of the economic models implemented within FactionSim as of late 2007 beginning with a macroview and moving to individual institutions. At the macro-level, the framework of the previous section makes it fairly straightforward to implement ideas such as the Nobel Prize-winning LRF model or "dual sector theory." This argues that a developing nation often includes a small, modern technology sector (faction) run by elites. They exploit a much larger, poor agrarian faction, using them for near-free labor and preventing them from joining the elites. This gives rise to the informal economy faction, which provides black market income and jobs, and which may also harbor actor intent on chaos (rebellion, insurgency, coup, etc.). Whether or not there is malicious intent to overthrow the current government and elites, the presence of the informal sector weakens the formal economy (elite faction) by drawing income and taxes away from it, and by potentially bribing its institutions and actors to look the other way.

We set up many of our country models with these types of factions. In the balance of this section, we examine how the institutions of a single faction work and may be influenced. The discussion focuses on public institutions to keep it brief, but we also model private ones and business enterprises that the actors may manage, work at, get goods and services from, and so on. Also, we will examine how one can substitute more detailed, third-party models of these institutions and enterprises without affecting the ability of our cognitive agents to interact with them. Thus, the models discussed in this section are defaults, and one can swap in other models without affecting how the actors think through their resource-based, ethnocultural conflicts.

The economic system currently in FactionSim is a mixture of neoclassical and institutional political economy theories. Institutions are used as a mediating force that control the efficiency of certain services and are able to be influenced by groups within a given scenario to shift the equitableness of their service provisions. Political sway may be applied to alter the functioning of the institution, embedding it in a larger political–economy system inhabited by groups and their members. However, the followers of each group represent demographics on the order of millions of people. To handle the economic production of each smaller demographic, a stylized *Solow growth model* is employed [27]. The specific parameter values of this model depend on the status of the followers. Each follower's exogenous Solow growth is embedded inside a political economy, which endogenizes the Solow model parameters. Some parameters remain exogenous, such as savings rate, which is kept constant through time. As savings rates are modeled after the actual demographics in question and the time frame is usually only a few years, fixing the parameter seems reasonable.

Each follower demographic's production depends on their constituency size, capital, education, health, employment level, legal protections, access to basic resources (water, etc.), and level of government repression. These factors parameterize the Solow-type function, in combination with a factor representing technology and exogenous factors, to provide a specific follower's economic output. The economic output of followers is split into consumption, contribution, and savings. Consumption is lost for the purposes of this model. Savings are applied to capital to offset depreciation. Contribution represents taxation, tithing, volunteering, and other methods of contributing to group coffers. Both followers and groups have contributions, with groups contributing to any supergroups they belong to. Contributions are the primary source of growing groups' economy resources.

The unit of interaction is the institution as a whole, defined by the interactions between it with groups in the scenario. An institution's primary function is to convert funding into services for groups. Groups, in turn, provide service to members. Groups, including the government, provide funding and infrastructure usage rights. In turn, each group has a level of influence over the institution, which it leverages to change the service distribution. Influence can be used to increase favoritism (e.g., for one's own group), but it can also be used to attempt to promote fairness. The distribution of services is represented as a preferred allotment (as a fraction of the total) toward each group. Institutions are also endowed with a certain level of efficiency. Efficiency is considered the fraction of each dollar that is applied to service output, as opposed to lost in administration or misuse.

The institutions currently modeled as of end of 2007 are public works, health, education, legal protections, and elections. Public works provide basic needs, such as water and sanitation. Health and education are currently handled by a single institution, which handles health care and K-12 schools. Legal protections represent the law enforcement and courts that enforce laws. Their service is the expectation to protection of full rights under law, as well as to basic human rights. The electoral institution establishes the process by which elections are performed, and handles vote counting and announcement of a winner.

The electoral institution's function occurs only periodically, and favoritism results from tampering with ballot counting. Elections are implemented in tandem with PS-I, a cellular automata that allows incorporations of numerous followers and the geography of a particular country, which handles the district-level follower preference formation and transformation (see Lustick et al. [25]). The electoral institution receives the actual vote results for each party leader. The electoral institution handles electoral systems effects (variations of the first past the post, plurality, and hybrid systems), vote tampering (i.e., corruption), and districting effects (i.e., gerrymandering). We envision our later releases to include strategic AI leader agents that maximize their respective political power vis-à-vis other AI leaders and human agent (analyst) through the districting effects.

## Modeling Agent Personality, Emotions, Culture, and Reactions

Previous sections of this chapter presented a framework for implementing theories of political science, economics, and sociology within an agent-based game engine. The discussion thus far omitted treatment of the actors who populate these worlds—run the groups, inhabit the institutions, and vote and mobilize for change. These are more on the domain of the psychological and anthropological fields. In this section, we introduce PMFserv, a COTS human behavior emulator that drives agents in simulated game worlds. This software was developed over the past 10 years at the University of Pennsylvania as an architecture to synthesize many best-of-breed models and best practice theories of human behavior modeling. PMFserv agents are unscripted, but use their microdecision making, as described below, to react to actions as they unfold and to plan out responses.

A PMF is a micromodel covering how human performance (e.g., perception, memory, or decision making) might vary as a function of a single factor (e.g., sleep, temperature, boredom, grievance, etc.). PMFserv synthesizes dozens of best-of-breed PMFs within a unifying mind–body framework and thereby offers a family of models where microdecisions lead to the emergence of macrobehaviors within an individual. None of these PMFs are "homegrown"; instead, they are culled from the literature of the behavioral sciences. Users can turn on or off different PMFs to focus on particular aspects of interest. These PMFs are synthesized according to the interrelationships between the parts and with each subsystem treated as a system in itself.

The unifying architecture in Figure 8.4 shows how different subsystems are connected. For each agent, PMFserv operates what is sometimes known as an *observe, orient, decide, and act* (OODA) loop. PMFserv runs the agents perception (observe) and then orients all the entire physiology and personality/value system PMFs to determine levels of fatigues and hunger, injuries and related stressors, grievances, tension buildup, impact of rumors and speech acts, emotions, and various mobilizations and social relationship changes since the last tick of the simulator clock. Once all these modules and their parameters are oriented to the current stimuli/inputs, the upper-right module (decision making/cognition) runs a best response algorithm to try to determine or decide what to do next. The algorithm it runs is determined by its stress and emotional levels. In optimal times, it is in vigilant mode and runs an expected subjective utility algorithm that reinvokes all the other modules to assess what impact each potential next step might have on its internal parameters.

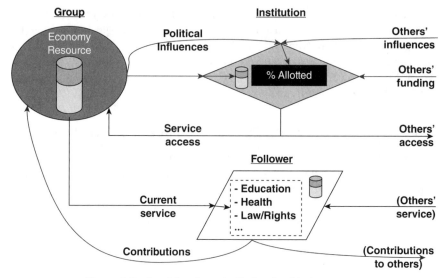

**Figure 8.3** Asset flow in an institutional political economy.

When very bored, it tends to lose focus (perception degrades), and it runs a decision algorithm known as unconflicted adherence mode. When highly stressed, it will reach panic mode, its perception basically shuts down and it can only do one of two things: (1) cower in place or (2) drop everything and flee. In order to instantiate or parameterize these modules and models, PMFserv requires that the developer profiles individuals in terms of each of the module's parameters (physiology, stress thresholds, value system, social relationships, etc.). Furthermore, the architecture allows users to replace any or all of these decision models (or any PMFs) with ones they prefer to use. PMFserv is an open, plug-in architecture.*

This is where an agent (or person) compares the perceived state of the real world to its value system and appraises which of its values are satisfied or violated. This in turn activates emotional arousals. For the emotion model, we have implemented one as described by Silverman et al. [28]. To implement a

---

* It is worth noting that because our research goal is to study best-of-breed PMFs, we avoid committing to particular PMFs. Instead, every PMF explored in this research must be readily replaceable. The PMFs that we synthesized are workable defaults that we expect our users will research and improve on as time goes on. From the data and modeling perspective, the consequence of not committing to any single approach or theory is that we have to come up with ways to readily study and then assimilate alternative models that show some benefit for understanding our phenomena of interest. This means that any computer implementation we embrace must support plug-in/plug-out/override capabilities, and that specific PMFs as illustrated in Figure 8.3 should be testable and validatable against field data such as the data they were originally derived from.

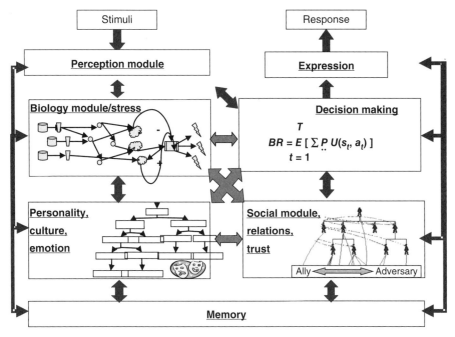

**Figure 8.4** PMFserv: an open architecture for agent cognitive modeling.

person's value system, this requires every agent to have GSP trees filled out. GSP trees are multi-attribute value structures where each tree node is weighted with Bayesian importance weights. A *preference tree* represents an agent's long-term desires for world situations and relations (for instance, no weapons of mass destruction, an end to global warming, etc.) that may or may not be achieved within the scope of a scenario. Among our agents, this set of "desires" translates into a weighted hierarchy of territories and constituencies.

As an illustration of one of the modules in Figure 8.3 and of some of the best-of-breed theories that PMFserv runs, let us consider "cognitive appraisal" (personality, culture, emotion module)—the bottom-left module in Figure 8.4 also expanded in Figure 8.5.

The *standards tree* defines the methods an agent is willing to employ to attain his or her preferences. The standard tree nodes that we use merge several best-of-breed personality and culture profiling instruments such as, among others, Hermann traits governing personal and cultural norms, standards from the GLOBE study, top-level guidelines related to the economic and military doctrine, and sensitivity to life (humanitarianism) [29,30]. Personal, cultural, and social conventions render the purely Machiavellian action choices inappropriate ("one should not destroy weak allies simply because they are currently useless"). It is within these sets of guidelines that

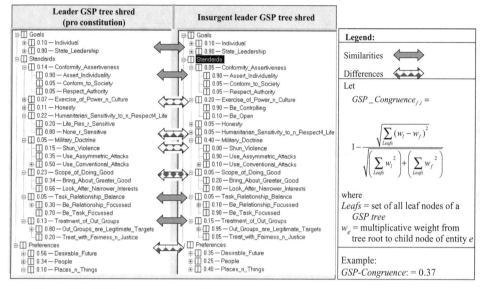

**Figure 8.5** GSP tree structure, weights, and emotional activations for opposing leaders.

many of the pitfalls associated with shortsighted AI can be sidestepped. Standards (and preferences) allow for the expression of strategic mindsets.

Finally, the *goal tree* covers short-term needs and motivations that drive progress toward preferences. In the Machiavellian- and Hermann-profiled world of leaders, the goal tree reduces to the duality of growing/developing versus protecting the resources in one's constituency [29,31]. Expressing goals in terms of power and vulnerability provides a high-fidelity means of evaluating the short-term consequences of actions (Fig. 8.5). For nonleader agents (or followers), the goal tree also includes traits covering basic Maslovian type needs.

This has been an abbreviated discussion of the internals of the cognitive layer, the PMFserv framework. The workings of each module are widely published and will not be repeated here. Elsewhere in other publications we have discussed how these different functions are synthesized to create the whole (PMFserv) [12,20,28,32]. For example, among other things, Silverman et al. reviewed how named leaders are profiled within PMFserv, and how their reasoning works to figure out vulnerability and power relative to other groups, to form/break alliances, and to manage their careers and reputations [32]. Likewise, it also describes the way in which archetypical follower agents autonomously decide things like emotional activations, social mobilization, group membership, and motivational congruence with a given leader or group. It explains how they attempt to satisfy their internal needs (physiological, stress, emotive, social, etc.), run their daily lives, carry out jobs and missions, and otherwise perform tasks in the virtual world. It also reviews the many

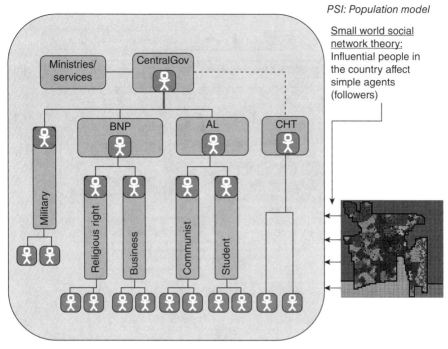

**Figure 8.6** Overview of the components of a CountrySim model: Bangladesh shown as an example of political science application. BNP, Bangladeshi National Party; AL, Awani League; CHT, Chittagong Hill Tract.

best-of-breed PMFs and models that are synthesized inside an agent to facilitate leader–follower reasoning.

The National Research Council of the National Academies indicated that as of 2007, there were no frameworks that integrate the cognitive with the social layer agent modeling [2]. Dignum et al. suggested several intriguing ideas for doing this, but so far have not completed that implementation [33]. So, the PMFserv–FactionSim symbiosis offers a unique innovation by itself. Further, we know of no environments other than CountrySim that attempt to bring the cognitive and social agent ideas together with a landscape agent model for modeling state and substate actors as we do in the CountrySim generator described in the next section (Fig. 8.6).

## Modeling Methodology

In the ensuing section, we will briefly outline how the models are built. In recent years, modeling methodologies have been developed that help to construct models, integrate heterogeneous models, elicit knowledge from diverse sources, and also test, verify, and validate models [34]. A diagrammatic representation of the process is given in Figure 8.7. The details of the process are

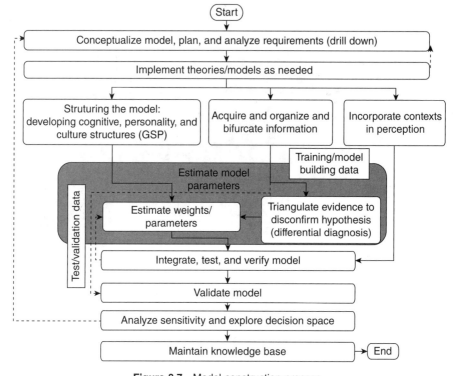

**Figure 8.7** Model construction process.

beyond the scope of this paper, but can be found elsewhere [12,18]. We recap the salient features briefly here.

These models are knowledge-based systems, and, to a significant extent, the modeling activity involves eliciting knowledge from SEMs as well as extracting knowledge from other sources such as databases and event data, consolidating the information to build a model of the social system.

We designed and tested the *knowledge engineering-based model* building process (KE process) to satisfy the following functional requirements:

(1) Systematically transform empirical evidence, tacit knowledge, and expert knowledge into data for modeling
(2) Reduce human errors and cognitive biases (e.g., confirmation bias)
(3) Verify and validate the model as a whole
(4) Maintain the knowledge base over time

***Conceptualize Model, Plan, and Analyze Requirements (Drill Down)*** The modeling problem is characterized based on the specific objective (type and purpose of the system envisaged) and the nature of the domain (how much and what information is available, as given by the typology based on, e.g., personality). In general, the objective of the modeling problem, along with the context, provides what needs to be accomplished and serves to define the method to go about it.

At this stage, we clarify the objectives, learn about the contexts surrounding the model, immerse ourselves in the literature, consult SEMs, and ultimately build a conceptual model of the modeling problem.

*Review/Implement Theories/Models* The basic theories necessary to describe the social systems are implemented in the framework. However, in some cases, additional theories may have to be incorporated, as has been the case with economic growth model implementing Solow in the current case [27]. We reviewed our framework to verify whether the framework is capable of describing the identified mechanism by allowing for the same pathways to exist in our model. It must be noted that while we implement theories, we do not hard code the dynamics of agent behavior into the framework. The latter is emergent.

*Structuring the Model* The cognitive structure of the agent world being modeled is represented with values (in turn consisting of GSP) and contexts and will consist of entities such as agents, groups/factions, and institutions. The generic structure has evolved over a long period of research, has been built collaboratively with an expert as well as using empirical materials, and does not change between countries.

For each country, the country experts select the configuration of actors, groups, and institutional parameters and provide values to those parameters.

*Acquire and Organize and Bifurcate Information* Once we have the conceptual model, we determine the general requirements of such a model as well as broad-brush data requirements. In a separate paper, we have discussed key issues involved in obtaining data from event databases and automated extraction techniques, as well as employing SEMs [12].

A number of databases contain surveys. There were two difficulties we faced in using these data for our purpose. First, it was hard to find a one-to-one correspondence between a survey questionnaire item and a parameter of, say, our GSP tree, when the surveys were not designed with our parameters in mind. The unit of analysis for these public opinion surveys were countries, while, for our joint sociocognitive PMFserv/FactionSim framework, the appropriate unit of analysis is at the faction level. Both these difficulties were not insurmountable; we selected survey questionnaire items that can serve as proxy measures for our parameters of interest. By cross-tabulating and sorting the data according to properties that categorize survey respondents into specific groups that match our interests, we also obtained what was close to faction level information for a number of parameters.

While the existing country databases and event data from Web scraping are good assets for those of us in the M&S community who are committed to using realistic agent types to populate our simulated world, they are useful as supplementary sources of information. A more direct source of parameter information is SMEs, who are experts in the countries they study. We, therefore, designed an extensive survey to elicit knowledge from SMEs. Through the survey, we elicited SMEs to provide this information in our preferred format for our countries of interest.

However, there are three main difficulties associated with using SMEs to elicit the information we need. First, SMEs themselves, by virtue of being human, have biases and can make mistakes and errors [3,35]. More importantly, being a country expert does not mean that one has complete and comprehensive knowledge; a country expert does not know everything there is to know about a country. Second, eliciting SME knowledge requires significant financial and human resources and limit the number of SMEs that can be employed on the same country. Third and finally, simply finding SMEs for a particular country of interest may by itself pose a significant challenge. This short supply of expertise, a high cost of employing SMEs, and potential SME biases and errors mean that SME knowledge itself requires verification. This verification of SME input may be provided by triangulating multiple SME estimates against each other as well as against estimates from databases and event data.

***Estimate Model Parameters*** After eliciting the expert input, we verify critical pieces of information by pitting against other sources of information such as database and event data. For this, we build an evidence table by organizing the empirical evidence or expert input by breaking statements into simpler units with one theme (replicate if necessary), adding additional fields (namely, reliability and relevance), and then sorting. The organized information is then assessed for reliability and relevance.

Any specific reliability info is used to identify and tag for further investigation and sensitivity analysis. The technique could be used in conjunction with other KE techniques even when an expert is involved in providing the information. In order to ensure separation of model building (training and verification) and validation data, the empirical materials concerned are divided into two different parts. One part is set aside for validation. The model is constructed and verified out of the remaining part.

With respect to consolidation of inputs from diverse sources and to determine the model parameters through rigorous hypothesis testing, essence of two techniques are employed. We describe these in terms of eliciting weights in GSP trees in the earlier Figure 8.4. However, the same approach is also used for eliciting all the parameters of the CountrySim:

(1) *Differential Diagnosis/Disconfirming Evidence.* Typically, a modeler would tend to build a model by confirming his or her evidence/data based on satisfying strategy. This is a cognitive bias in humans. Instead, a novel strategy or tool for disconfirming hypotheses embraces the

scientific process. In developing the GSP tree structure, the structure of the tree is considered as a hypothesis, and a paper spreadsheet-based tool is used to disconfirm the hypothesis (also known as differential diagnosis) against the evidence.

(2) *Determining the Weights.* The simplest way for SMEs to holistically assign the weights is based on intuitive assessment after reading a historical account. In this KE process, the weights of the nodes are semi-quantitatively assessed against every other sibling node at the same level, through a pair-wise comparison process. The assessment process itself is subjective and involves pair-wise comparison. Incorporation of pair-wise comparison caters to the fact that, at a given time, the human mind can comfortably and reliably compare only two attributes. This also helps eliminate inconsistent rankings within the same groups, provides more systematic processes for assessment of weights, and leaves an audit trail in the process. This process could be used with empirical evidences, expert input, or a combination of the techniques.

Finally, we construct the models of agents, factions, and institutions, and then integrate them all to make the consolidated model of the country. In summary, databases, Web scraped event data, and SMEs are each not entirely sufficient, but in unison, they can provide a significantly more accurate picture, provided a rigorous process is employed to integrate their knowledge together. For additional details, a stylized example of how model building is carried out has been given in Silverman et al. [12] and Bharathy [34].

**Incorporate Contexts in Perception** In our architecture, which implements situated ecological psychology, the knowledge about the environment and contexts is imbued in the environment (or contexts). The agents themselves know nothing a priori about the environment or the actions that they can take within that environment, but archetypical microcontexts (pTypes) are identified and incorporated in the environment. We mark up the context in which decisions are occurring through a semantic mark up. The details of the PMFserv architecture can be found in the published literature [28].

**Integrate, Test, and Verify Model** Since our intention is to model instability in countries, we define aggregate metrics or summary outputs of instability from default model outputs (such as decision by agents, levels of resources, emotions, relationships, membership in different factions, etc.). The direct (or default or base) outputs from the CountrySim model include decisions by agents, levels of emotions, resources, and so on. These parameters are tracked over time and recorded in the database. The aggregate metrics (summary outputs) are called EOIs. EOIs reveal a high-level snapshot of the state of the conflict.

Once the models were constructed, these were verified through a hierarchical and life cycle-based inspection, against the specifications. Over the training period, simulated EOIs were fit to real EOIs. Specifically, the weights in functions transforming indicators to EOIs were fitted for the training period and then employed to make out-of-sample predictions in the test period.

**Validate Model** The intention is to calibrate the model with some training data, and then see if it recreates a test set (actually validation). Considering that the decision space is path-dependent and the history is only a point in the complex space, other counterfactuals (alternative histories) might be expected to emerge. In carrying out a detailed validation process, we primarily aim to create correspondence with historical scenarios or higher-level outcomes with respect to:

(1) Descriptive and naturalistic models of human micro- or individual behavior to test if a model recreates a historical situation
(2) Low mutual entropy of emerging macrobehaviors in simulated worlds versus real ones
(3) Model alignment to confirm whether any outcomes of the existing, abstract, higher-level models relating to multistate could be reproduced or correlated by mutual entropy, provided another independent model could be found.

Some of the potential validation techniques are as follows:

(1) Concept validation and preliminary face validation exercises
(2) Detailed validation exercises, such as correspondence testing against an independent set of historical/literature evidence, model docking, a modified Turing test, cross-validation between experts, use of SMEs, or assessment or interrogation of human subjects for stylized cases, as appropriate.

In the ensuing example case study, we show statistical correspondence with testing data (independent set of data set aside for validation).

**Analyze Sensitivity and Explore Decision Space** In the Monte Carlo analysis, one uses domain knowledge and the evidence tables created for differential diagnosis to select a large subset of variables. Based on this initial list, one should carry out the sensitivity analysis with respect to those parameters to determine which have significant uncertainty associated with them, as well as those that were most significant for policy making.

**Maintain Knowledge Base** A country model must be maintained and the actions, institutions, and so on updated at key intervals. Additionally, we continuously improve (by design as well as by learning through feedback loops) strategies, instruments, and steps that are taken for the management of models, including refining and reusing, as well as monitoring through a spiral development.

This knowledge engineering-based modeling process has been tested by applying it to several real-world cases that we address.

## THE CountrySIM APPLICATION AND SOCIOCULTURAL GAME RESULTS

FactionSim and PMFserv have been, or currently are being, deployed in a number of applications, game worlds, and scenarios. A few of these are listed below. To facilitate the rapid composition of new casts of characters, we have created an *integrated development environment* (IDE) in which one knowledge engineers named and archetypical individuals (leaders, followers, suicide bombers, financiers, etc.) and assembles them into casts of characters useful for creating or editing scenarios.

Many of these previous applications have movie clips, tech reports, and validity assessment studies available at www.seas.upenn.edu/~barryg/hbmr. Several historical correspondence tests indicate that PMFserv mimics decisions of the real actors/population with a correlation of approximately 80 percent [32,36]. In 2008, we have applied the framework to model 12 representative countries across Asia (e.g., China, India, Russia, Bangladesh, Sri Lanka, Thailand, North Korea, etc.). We codified this into a generic application for generating country models that we call CountrySim. The CountrySim collection of country models can best be described as a set of complex agent-based models that use hierarchically organized and cognitive–affective agents whose actions and interactions are constrained by various economic, political, and institutional factors. It is hierarchically organized in the sense that the underlying FactionSim framework consists of a country's competing factions, each with its own leader and follower agents. It is cognitive–affective in the sense that all agents are "deep" PMFserv agents with individually tailored and multi-attribute utility functions that guide a realistic decision-making mechanism. CountrySim, despite its apparent complexity, is an agent-based model that aims to show how individual agents interact to generate emergent macrolevel outcomes. CountrySim's user-friendly interface allows variables to be adjusted and results to be viewed in multiple ways. It is very easy in CountrySim to trace inputs through to outputs or vice versa, examine an output and trace it back to the inputs that may have caused it.

For a given state being modeled, CountrySim uses FactionSim (and PMFserv) typically to profile 10s of significant ethnopolitical groups and a few dozen named leader agents, ministers, and follower archetypes. These cognitively detailed agents, factions, and institutions may be used alone or atop of

**Table 8.1 Past PMFserv applications**

| Domestic Applications | International Applications |
|---|---|
| (1) Consumer modeling<br>  Buyer behavior<br>  Ad campaign<br>(2) Pet world<br>  Pet behavior<br>(3) Gang members<br>  Hooligans<br>(4) Crowd scenes<br>  Milling<br>  Protesting<br>  Rioting<br>  Looting | (1) Models of selected countries (Bangladesh, Sri Lanka, Thailand, and Vietnam) including major factions, its leaders, decisions as well as summary conflict indicators such as rebellion, insurgency, domestic political crisis, intergroup violence, and state repression<br>(2) Intifadah recreation (leaders, followers)—Roadmap sim<br>(3) Somalia crowds—Black Hawk Down (males, females, trained militia, clan leaders)<br>(4) Ethnic village recreations (tribal, political, and economic factions at the street level for tactical DIME_PMESII training)<br>(5) Iraq DIME-PMESII sim—seven ethnic groups, parliament (leaders and 15,000 followers)<br>(6) Urban Resolve 2015—Sim-Red (multiple insurgent cell members/roles/missions)<br>(7) State instability modeling 12 representative nations across Asia<br>(8) Interstate, world diplomacy game (Athena's Prism); many world leaders profiled |

another agent model that includes 10,000s of lightly detailed agents in population automata called PSI. Figure 8.4 shows the architecture of a typical country model, in this case Bangladesh. We will describe its structure more fully in the Bangladesh section, but here let us focus on how the PMFserv agents are organized into FactionSim groups and roles. Further, there is a bridge to the PSI population substrate through which the cognitively detailed PMFserv agents pass on their DIME actions and decisions that affect 10,000s of simple agents in the landscape. This PSI landscape is the topic of several published papers, and we will describe it only from the viewpoint of the services it provides to CountrySim [25]. Specifically, PSI organizes the simple agents in a spatial distribution similar to how identities and factions are geographically oriented in the actual country. This provides detail about regime extent and reach, and about message propagation delays that FactionSim alone omits. The FactionSim and PSI landscape agents thus are bridged together, and a two-way interaction ensues in which FactionSim leaders, ministers, and influential follower archetypes tend to make decisions that affect the landscape agents. In the sociopolitical context of CountrySim, the landscape then propagates the impacts and returns simple agent statistics that FactionSim uses to update faction resources and memberships, count votes for elections, and in part determine some of the well-being and instability indicators used in our overall summary metric forecasts and computations.

CountrySim, as just described, offers a capability that is unique for analysts in at least three dimensions. It does of course support the exploration of possible futures and sensitivity experiments; however, that alone is not unique to our approach. In terms of novelty, CountrySim elicits the qualitative models of SMEs of a given nation and permits them to run a quantized version of their model. These SME models tend to differ from traditional statistical (or even AI models) models and often incorporate insights into the personality and underlying motivations of the leaders involved, insights about the cultural traits and ethnopolitical group cleavages, and local knowledge about the history of grievances and transgressions at play. Eliciting this permit us to better understand each SME's models, observe its performance, track its forecasts, and help to improve it over time. CountrySim offers a uniquely transparent drill-down capability where one can trace potential causalities by working backward from summary outcome EOIs to indicators and events that are summed up in those indicators. Further, one can find the agents that precipitated those events and query them through a dialog engine to inspect their rationale and motivations that lead them to the action choices they made. This is very helpful to analysts trying to diagnose potential causes and find ways that might better influence outcomes. Finally, CountrySim is able to integrate best-of-breed theories and practices from the social and behavioral sciences and engineering into the simulator components—in fact, components are built exclusively by synthesizing social science theories. The SME mental models are elicited as parameterizations of these best practice scientific theories/models. As such, we provide a pathway for studying the underlying social sciences including their strengths, gaps, and needs for further research.

## Case Study: CountrySim Applied to Iraq

During the spring of 2006 and well before the "surge" in U.S. troops, five student teams assembled a total of 21 PMFserv leader profiles across seven real-world factions so that each faction had a leader and two subfaction leaders. The seven factions—government (two versions—CentralGov and LoclGov), Shia (two tribes), Sunnis, Kurds, and Insurgents—could be deployed in different combinations for different scenarios or vignettes. The leader and group profiles were assembled from open source material and followed a rigorous methodology for collecting evidence, weighing evidence, considering competing and incomplete evidence, tuning the GSP trees, and testing against sample data sets [8]. This Iraqi CountrySim model did not include a PSI population layer. The PMFserv agents provided all the decision making, action taking, and opinion/voting feedback.

Validation testing of these models was run at a military organization for 2 weeks in May 2006. They assembled 15 SMEs across areas of military, diplomatic, intel, and systems expertise. Within each vignette, the SMEs attempted dozens of courses of action across the spectrum of possibilities (rewards, threats, etc.). A popular course of action of the diplomats was to "sit down" with some of the persuadable leaders and have a strong talk with them. This was simulated by the senior diplomat adjusting that leader's personality weights (e.g., scope of doing good, treatment of out-groups, etc.) to be what he or she thought might occur after a call from President Bush or some other influential leader. The SME team playing the multinational coalition presented their opinions at the end of each vignette. The feedback indicated that the leader and factional models corresponded with SME knowledge of their real-life counterparts. They accepted the profiling approach as best in class and invited us onto the team for the follow on.

Here we show an illustrative policy experiment on four factions initially organized into two weak alliances (dyads):

(1) CentralGov trying to be secular and democratic with a Shia tribe squarely in their alliance but also trying to embrace all tribes,
(2) a Shia tribe that initially starts in the CentralGov's dyad but has fundamentalist tendencies,
(3) a secular Sunni tribe that mildly resents CentralGov but does not include revengists,
(4) Insurgents with an Arab leader trying to attract Sunnis and block Shia control.

Each faction has a leader with two rival subleaders (loyal and fringe) and followers as in Figure 8.1—all 12 are named individuals, many are known in the United States. This is a setup that should mimic some of the factional behaviors going on in Iraq, although there are dozens of political factions there in actuality. Figure 8.3 summarizes the outcomes of three sample runs (mean of 100 trials each) over a 2-year window. The vertical axis indicates the normalized fraction of the sum across all security tanks in these factions, and thus the strip chart indicates the portion of the sum that belongs to each faction. Rises and dips correspond either to recruiting and/or to battle outcomes between groups. The independent variable is how much outside support is reaching the two protagonists—CentralGov and Insurgents. When CentralGov and Insurgents are externally supported (3A), CentralGov aids the Shia militia economically, while the Shia battle the Insurgents. Fighting continues throughout the 2-year run. A takeaway lesson of this run seems to be that democracy needs major and continuous outside help, as well as luck in battle outcomes and some goodwill from tribes for it to take root. When only the Insurgents are supported (3B), the CentralGov is crippled by Insurgent attacks and civil war prevails. When the borders are fully closed and no group receives outside support (3C), the insurgency ultimately fails, but the CentralGov becomes entirely reliant upon the Shia group for military strength—a puppet government. These runs suggest the elasticity of conflict with respect to outside support is positive, and with no interference, the country seems able to right itself, although we in the West might not like the outcome. Of course, these runs only include four of the many factions one could set up and run, plus due to page limits, we only displayed the effects of actions upon the security tank, and not other resources of the factions.

FactionSim, with the help of PMFserv, is able to help the analyst to generate and understand why (space limits prevent us from showing the drill-down diagrams, so we summarize them briefly here). The agents and factions in our runs fight almost constantly and are more likely to attack groups with which they have negative relationships and strong emotions. Relationship and emotions also factor into the formation of alliances. For example, across all runs, CentralGov has a friendly relationship toward the Shia, who are moderately positive back. This leads to CentralGov giving aid to the Shia and consistently forming an ally. Likewise, the Sunni secular has slight positive feelings toward the Insurgents and is more likely to assist them, unless others are more powerful. Finally, some action choices seem to have purely emotional payoffs. For example, from an economic perspective, the payoff from attacking an enemy with zero economy is zero—a wasted turn. Yet in run 2c, when the Insurgents fail, the Shia still occasionally attack them simply because the Insurgents are their enemy. This seems to be a case where emotional payoffs are at least as important as economic payoffs (Fig. 8.8).

## Case Study: CountrySim Applied to Bangladesh

The previous case showed a CountrySim model assembled by laypersons from open literature, which was shown to pass the validity assessment of a panel of experts who accepted it as similar to the personalities and ethnic factions they knew in the country. Since that time, we added a Web interview front end so that experts can fill in their own country models themselves. The Bangladesh model shown earlier in Figure 8.4 was input by an SME we contracted as a consultant for 12 h of his time. One can see his model has the government, military, the two major political groups (Bangladeshi National Party, Awami League) that have alternated being in power, and a minority ethnic group that formerly had threatened a rebellion but which is now appeased (i.e., Chittagong Hill Tract). This model is interesting since it quantizes the SMEs' qualitative model into FactionSim and PMFserv parameter sets. Thus, the SME had to fill in all the parameters of each group and leader and archetypical follower. We also separately contracted for a political scientist (Lustick and some assistants) to fill in the population layer within PSI.

As mentioned previously, CountrySim also includes viewers on the back end that help to summarize performance metrics and allow the user to drill in and trace outcomes back to the Web interview inputs. We present some example results here to illustrate how the model performs and how one may follow a thread from outcome back to input, and to see how it is validated.

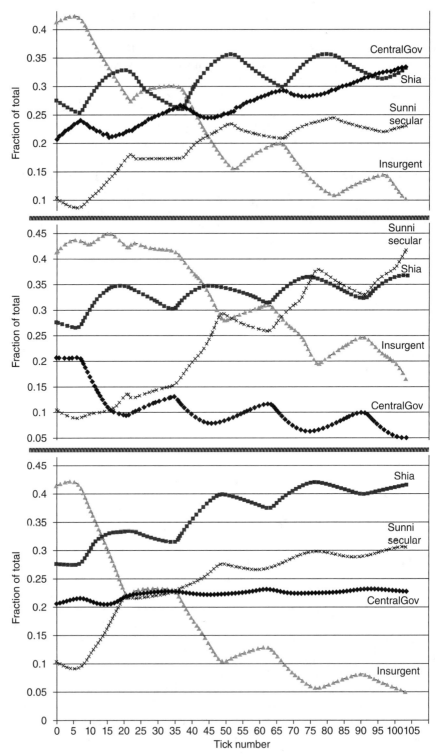

**Figure 8.8** Military power of Iraqi factions under alternate DIME actions (mean of 100 runs).

Since our intention for Bangladesh is to model instability, we define aggregate metrics or summary outputs of instability from default model outputs. The direct (or default or base) outputs from the CountrySim model include decisions by agents, levels of emotions, relationships, membership in different factions, levels of resources, and so on. These parameters are tracked over time and recorded in the output database.

All forecasts are aggregations of week-by-week activity in the model (i.e., 52 ticks/year over 3 years). Our country forecasts, in turn, aggregate these into quarterly statistics. To do this, raw events are summarized into indicators (e.g., all fightback decisions taken by members of the separatist faction in a given quarter are added into the rebellion indicator). Many such midlevel indicators then get aggregated into highest level performance metrics called EOIs. EOIs reveal a high-level snapshot of the state of the country. Specifically, CountrySim generates several EOI scores important to instability such as the four we now define, among others:

(1) *Rebellion* is an organized opposition whose objective is to seek autonomy or independence. (Secession, or substantial devolution of power, occurs when rebellion is successful.)

(2) *Insurgency* is an organized opposition by more than one group/faction, whose objective is to usurp power or change regime by overthrowing the central government by extra legal means.

(3) *Domestic political crisis* is the significant opposition to the government but not to the level of rebellion or insurgency.

(4) *Intergroup violence* is violence between ethnic or religious groups that is not specifically directed against the government and not carried out by the government.

In carrying out a detailed validation process, we primarily aim to create correspondence with historical scenarios or higher-level outcomes. The results of the likelihood of occurrence of EOIs were compared with EOIs obtained (with the same definitions) from Ground Truth from an independent data provider (University of Kansas [UK], 2008). The initial values of the Ground Truth are machine extracted and coded event set (built using a complex logistic regression model) obtained from the UK. These initial estimates were further augmented by human inspection (UK+) at the University of Pennsylvania. In the results shown below, we have compared the simulated output against the Ground Truth values for each of the EOI over the validation period (every quarter in 2004–2006 periods).

In a complex, stochastic system (such as a real country), a range of counterfactuals (alternate futures) are possible. Our simulated outputs are likelihood estimates and are shown as a band (with max, mean, and min) to account for counterfactuals resulting from multiple runs, while the Ground Truth values are shown as binary points (the diamonds indicated for each quarter).

Although we generate and display the multiple futures (from multiple runs), in metrics and calculations, we only employ the mean values across alternative histories. We cast mean likelihood estimates from multiple runs into a binary prediction by employing threshold systems, consisting of:

- a single threshold line (1Threshold)
- a double threshold system with upper and lower bounds (2Threshold)

In the figures, we display threshold values of 0.5 for single threshold system and 0.65 and 0.35 for double threshold systems. Based on these Ground Truths and Threshold Systems, we calculated our metrics such as precision, recall, and accuracy for multicountry, multiyear study.

As can be seen in Figure 8.9, there is a high degree of correlation between our prediction and that of the Ground Truths. The details are given in the figure for each EOI as follows:

(1) *Upper Left.* EOI rebellion has a very low likelihood of occurrence in both real as well as simulated outputs. The government forged a treaty agreement with the CHT tribe, once a separatist group in Bangladesh. There is nothing in the way of separatist conflict in Bangladesh today. Both Ground Truth and CountrySim agree with this estimate.

(2) *Upper Right.* Our model shows an increasing likelihood of coup or military takeover in Bangladesh circa 2006. Actual insurgency (i.e., military takeover) occurred in the first quarter of 2007. The simulated likelihood of EOI (after applying threshold) is a quarter off from the EOI. Although CountrySim model does not get the timing of the insurgency, the CountrySim agents are acting up, so some indications of this about to occur is reflected. Note that Ground Truth does not reveal any indications of events occurring.

(3) *Lower Left.* The likelihood of domestic political crisis is estimated to be high in Bangladesh (circa 2004, 2006), which corresponded to internal political tensions, horse trading in the country, and culminating in the military coup. Right after the forecast period, the domestic political crisis leads to riots and a military takeover in the first quarter of 2007.

(4) *Lower Right.* Intergroup violence also shows a limited occurrence for Bangladesh, except toward the end of 2005. During the latter part of 2005, the violent activities by religious extremists such as the JMJB Group against other factions and the government occurred.

With the double threshold system (with 2/3–1/3 thresholds), the likelihood estimates at or above the upper threshold are classified as 1, while those at or below the lower threshold are classified as 0. It must be acknowledged that when one imposes a 2Threshold System (with a conservative 1/3–2/3 band)

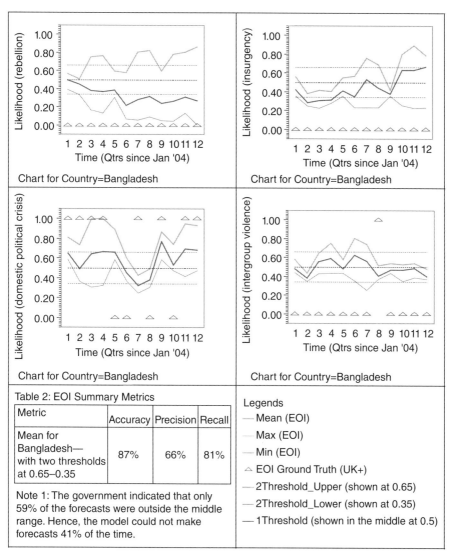

**Figure 8.9** CountrySim quarterly forecasts for Bangladesh (mean, upper, and lower bands on 12 Monte Carlo runs) compared with Ground Truth (UK+) statistical forecasts (triangle shapes) for 2004–2006.

upon the predictions, a number of likelihood estimates fall in the middle region. We have ignored all cases that might be classified as uncertain or in the middle band and then proceeded to calculate the above metrics. With 2/3–1/3 thresholds, our accuracies are at about 87 percent, the precision and recall are lower at about 66 and 81 percent, respectively, for Bangladesh. This shows that with the 2Threshold System, about 40 percent of CountrySim

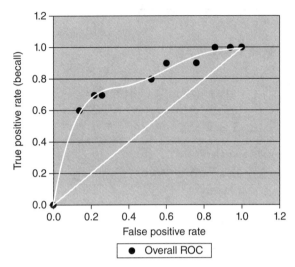

**Figure 8.10** Relative operating characteristic (ROC) curve for CountrySim (Bangladesh).

predictions fall in the middle range for Bangladesh. This could be simply interpreted as limited discriminatory power of the model for Bangladesh. One can improve the discriminatory power of the model by designing the EOI calculator to separate the EOI likelihood outcomes into two binary bands (i.e., use a single threshold system at 50 percent). Detailed consideration of all the threshold issues deserves a paper of its own.

In order to get a quantitative relationship between CountrySim and Ground Truth forecasts, we make use of a relative operating characteristic (ROC) curve. The ROC plots the relationship between the true positive rate (sensitivity or recall) on the vertical and the false positive rate (1-specificity) on the horizontal. Any predictive instrument that performs along the diagonal is no better than chance or coin flipping. The ideal predictive instrument sits along the $y$-axis.

The consolidated ROC curve for Bangladesh is plotted in Figure 8.10. In the two threshold forms presented, it was difficult to present the ROC curve for the model due to elimination of those cases that fell in the middle band of uncertainty. There were not enough recall and specificity data points to construct an ROC curve for Bangladesh using the two threshold systems. Instead, we present the ROC curve based on the single threshold system. This curve shows that the CountrySim largely agrees with the Ground Truth. In fact, its accuracy measured relative to Ground Truth is 80+ percent, while its precision and recall were listed at the base of Figure 8.10.

In closing, these results show that the agent approach offers nearly the same performance as statistical models, but brings to bear a greater transparency, explainability, and means to draw understanding of the underlying dynamics that are driving behaviors. This is possible since one can drill down to the events that each CountrySim agent participated in and then find that agent and interview him about what motivated him and why he did what he did (via the dialog engine).

## CONCLUSIONS AND THE WAY FORWARD

Four lessons are learned from our current efforts to build SimCountries using FactionSim/PMFserv in order to monitor and assess political instabilities in countries of interest. The first lesson is the need to speed up the maturation process of the social sciences so that there will be a sufficient set of theories that are close to being first principles that are widely accepted by social scientists. Our PMFserv's biology and physiology modules are based on proven first principles from the medical and natural sciences. However, our social, cognitive appraisal, and cognition modules in PMFserv and the leader–follower dynamics in FactionSim—to name just a few from an extensive list of implementations—are built by computationally implementing what we consider to be best-of-breed models based on recommendations from social science SMEs. Without a set of first principles, the best we can do is to rely on these recommendations. However, this constraint is something that is clearly beyond our control, and we are painfully aware of the possibility of never obtaining such a neat and tidy set of first principles from the social sciences.

A related second lesson that we learned is the need to expedite the process of developing our own more formalized and computationally implementable theories and conjectures in political science. This challenge is less acute with regard to economics, since this domain is already more mathematized. Albert Hirschman's exit, voice, loyalty framework was suggested by many as the best-of-breed model for us to capture and computationally implement the leader–follower dynamics in FactionSim. However, this framework was never a formalized model or a theory developed with computational implementation in mind. Further, it is only a small piece of the explanation of what drives loyalty. Hence, computational implementation required that we take additional steps, converting the theory into a computationally implementable one with the necessary formalizations and adding in other theories that complement and extend it (e.g., motivational congruence, mobilization, perceived injustice, etc.). We are eager to see social science theories become formalized as much as possible so that our new kind of political science—using simulators with realistic AI agents—can truly take off without being hampered by a slow process of formalization.

The third lesson that we learned has to do with the need to develop a state-of-the art toolset that would allow an analyst who uses an AI simulator like ours to construct realistic profiles for all the actors and issues at play in a given conflict region of interest. Currently, it is possible for an analyst who intimately understands a particular conflict scenario—with its key actors, factional member profiles, resource distributional factors (such as greed), and disputed issues and grievances—to use the available agent-based modeling editors to manually mock up a new scenario within a matter of days. However, we would like to speed and enhance this process with a state-of-the-art data extraction and model parameter generator that will query the analyst regarding his or her region of interest in order to zoom in on the particular questions to be investigated, and the specific corpus of texts, data sets, and Web sites to be scraped for the relevant data. This issue is an exciting new frontier for the M&S community, and our most up-to-date efforts are reported in another paper [20].

The fourth lesson that we learned is to increase the multidisciplinary nature of social sciences and to educate the new generation of sociopolitical analysts to be able to intelligently use these exciting toolsets that are being developed outside of the social sciences. We do not expect social scientists to be computer scientists and engineers and spend excessive time and energy in tasks they are not trained for, such as software development. Instead, we are suggesting that the new generation of sociopolitical scientists should receive the training they need to be informed users of these toolsets, just as they learn to use statistical software such as SPSS, R, or STATA to conduct regression analysis. We believe there is tremendous potential in our AI M&S technology for all aspects of political science. We are currently building virtual countries for a specific purpose of monitoring and assessing political instabilities. However, in so doing, we are required to construct a realistic social system of practical values to analysts complete with a minimal set of leader and follower agents, groups, institutions, and more. We have delved into the vast archive of political science literature in all conceivable areas. Consequently, we have created a tool that will be of interest and practical use to a very wide user base.

M&S hold great promise, especially when combined with toolsets that allow analysts and policymakers with a modicum of training in M&S software to conduct experiments that provide them with useful information. A fifth lesson learned is that our agent-based approach offers statistical performance nearly on a par with regression models, yet has the added benefit that it permits one to drill down into details of what is causative and what emerged from the action decisions of the stakeholders. It seems that if the world is expected to be unchanging and one needs no deeper insights, that regression models might be preferred as they give slightly more accurate forecasts. However, if the world is unstable, and avoiding and understanding potential surprises are important, then the agent-based approach holds the prospect that one can interview the agents, examine their grievances and motivations, and trace outcomes back through to input parameters that one can then experiment with to see how to improve and otherwise influence the society. This is a new capability that logistic regression does not support.

As suggested in our introduction, we believe that these new tools set a new standard for rigor and provide a new methodology for testing hypotheses. This new methodology is particularly useful with regard to problems that are mathematically intractable or difficult to research "empirically" because of the poor quality or unavailability of data. In addition, this kind of political science truly opens up a way to conduct counterfactual analyses. The ultimate value of this new approach lies in providing policymakers and analysts with a cutting-edge toolset that will improve their intelligence capability.

As a final thought, we conjecture about how business might also benefit from the types of agent-based models we offer here. Our joint CountrySim, FactionSim, and PMFserv framework has a large variety of military and business applications. Let us briefly present three of the more obvious ones. First, our CountrySim is arguably the best available country-level agent-based model for military users wishing to monitor, assess, and forecast various political developments of interest such as insurgencies, rebellions, civil wars, and other forms of intrastate political crises. In fact, CountrySim was built precisely for this purpose. It represents a synthesis of all the best available social science and area studies theories and information, expert inputs based on expert surveys, and state-of-the-art and cutting-edge agent-based modeling and systems approach methodologies. It has undergone multiple rounds of revision and improvement. As mentioned previously, our four existing virtual countries (Bangladesh, Sri Lanka, Thailand, and Vietnam) all have shown better than 80 percent accuracy in retrodicting the past political trajectories of these countries. We are now in the process of preparing for the challenge of forecasting the future political developments of these countries. It is of course impossible to predict the future with pinpoint accuracy. We emphasize that our goal is to generate reasonably accurate forecasts of possible political developments in our countries of interest: for example, the probability of a military coup in country X in the year 2010 expressed as a percentage, along the same lines as forecasting the percent chance of rain in Philadelphia tomorrow. Having the ability to monitor, assess, and ultimately forecast political developments should aid our military's capacity to anticipate and prepare for the futures that may be harmful to the national interest of the United States. Second, CountrySim is arguably the best available toolset for military users to conduct DIME-PMESII and other forms of computational counterfactual experiments. Sometimes political instabilities outside our borders can have significant impacts on our nation's well-being and require our intervention along diplomatic, informational, military, and economic lines. The need for timely and well-targeted intervention is particularly true in the age of globalization. In our virtual countries, our military and diplomatic users can computationally implement and experiment with specific kinds of interventions to aid their planning for future contingencies. When considering costly types of military and economic intervention, planning well and anticipating as many possibilities as possible using the best tools and the best available information become crucial, and we provide this crucial capacity. Third, we are increasingly aware that politics and business are inseparably interlinked, and this linkage is especially pronounced in certain parts of the world. It seems that the less economically developed a country, the greater the interdependence between these two realms. As the events of the past six months have shown, however, the same interdependence may also emerge in advanced economies: When the private sector falters, business turns to government for help. In many ways, government is the biggest business of any country. A toolset that allows one to monitor, assess, and forecast a country's political future is therefore a tremendous asset to investors and entrepreneurs doing business abroad. Our technology also makes it possible to build detailed virtual economies of a variety of countries around the world, tailoring them for business and economic applications. In sum, the military and business applications of our framework are limited only by our users' imagination.

## REFERENCES

[1] Cioffi-Revilla C, O'Brien S. Computational Analysis in US Foreign and Defense Policy. Paper presented at the First International Conference on Computational Cultural Dynamics. University of Maryland, College Park, MD, August 27–28, 2007.

[2] National Research Council. *Behavioral Modeling and Simulation: From Individuals to Societies*. Washington, DC: The National Academies Press; 2008.

[3] Tetlock P. *Expert Political Judgment: How Good Is It? How Can We Know?* Princeton, NJ: Princeton University Press; 2005.

[4] Axelrod R. Advancing the art of simulation in the social sciences. In *Simulating Social Phenomena*. Conte R, Hegselmann R, Terna P (Eds.). Berlin: Springer; 1997, pp. 21–40.

[5] Gladwell M. *The Tipping Point: How Little Things Can Make a Big Difference*. Boston: Little, Brown, and Company; 2000.

[6] Anderson JR. *The Architecture of Cognition*. Cambridge, MA: Harvard University Press; 1983.

[7] Anderson JR. *The Adaptive Character of Thought*. Hillsdale, NJ: Lawrence Erlbaum Associates; 1990.

[8] Anderson JR. *Rules of the Mind*. Hillsdale, NJ: Lawrence Erlbaum Associates; 1993.

[9] Anderson JR, Bothell D, Byrne MD, Douglass S, Lebiere C, Qin Y. An integrated theory of the mind. *Psychological Review*, 111(4):1036–1060; 2004.

[10] Newell A. *Unified Theories of Cognition*. Cambridge, MA: Harvard University Press; 1990.

[11] Hudlicka E. Modeling effects of behavior moderators on performance: Evaluation of the MAMID methodology and architecture. *Proceedings of the 2003 Conference on Behavior Representation in Modeling and Simulation (BRIMS)*. Scottsdale, AZ; 2003.

[12] Silverman BG, Bharathy GK, Kim GJ. Challenges of country modeling with databases, news feeds, and expert surveys. In *Agents, Simulation, Applications*. Uhrmacher A, Weyns D (Eds.). Boca Raton, FL: Taylor and Francis; 2009.

[13] Edmonds B, Moss S. From KISS to KIDS—an 'anti-simplistic' modelling approach. In *Lecture Notes in Artificial Intelligence*. Vol. 3415. Multi-Agent Based Simulation 2004. Davidsson P. et al. (Eds.). Springer; 2005, pp. 130–144.

[14] Armstrong JS. Assessing game theory, role playing and unaided judgment. *International Journal of Forecasting*, 18:345–352; 2002.

[15] Green KC. Forecasting decisions in conflict situations: A comparison of game theory, role playing and unaided judgment. *International Journal of Forecasting*, 18:321–344; 2002.

[16] Camerer C. *Behavioral Game Theory*. Princeton, NJ: Princeton University Press; 2003.

[17] de Marchi S. *Computational and Mathematical Modeling in the Social Sciences*. Cambridge, UK: Cambridge University Press; 2005.

[18] Silverman BG, et al. Athena's Prism—A diplomatic strategy role playing simulation for generating ideas and exploring alternatives. In *Proceedings of the First International Conference on Intelligence Analysis*. MacLean, VA: MITRE; 2005.

[19] Johns M. Deception and Trust in Complex Semi-Competitive Environments. Dissertation, University of Pennsylvania; 2006.

[20] Silverman BG, Bharathy GK, Nye B, Smith T. Modeling factions for 'effects based operations': Part II—Behavioral game theory. *Journal of Computational & Mathematical Organization Theory*, 14(2):120–155; 2008.

[21] Lewis WA. Economic development with unlimited supplies of labour. *Manchester School*, 28(2):139–191; 1954.

[22] Harrod RF. A second essay in dynamic theory. *Economic Journal*, 70:277–293; 1960.

[23] Hirschman AO. *Exit, Voice, and Loyalty*. Cambridge, MA: Harvard University Press; 1970.

[24] Epstein JM. Modeling civil violence: An agent-based computational approach. *Proceedings of the National Academy of Sciences of the United States of America*, 99(3):7243–7250; 2002.

[25] Lustick IS, Miodownik D, Eidelson RJ. Secessionism in multicultural states: Does sharing power prevent or encourage it? *American Political Science Review*, 98(2):209–229; 2004.

[26] Milgram S. The small world problem. *Psychology Today*, 1:60–67; 1967.

[27] Solow RM. A Contribution to the theory of economic growth. *Quarterly Journal of Economics*, 70(1):65–94; 1956.

[28] Silverman BG, Johns M, Cornwell J. Human behavior models for agents in simulators and games: Part I—Enabling science with PMFserv. *Presence*, 15(2):139–162; 2006.

[29] Hermann MG. Who becomes a political leader? Leadership succession, generational change, and foreign policy. Annual Meeting of the International Studies Association. Honolulu, HW; 2005.

[30] House RJ, et al. *Culture, Leadership, and Organizations: The GLOBE Study of 62 Societies*. Thousand Oaks, CA: Sage Publications; 2004.

[31] Machiavelli N. *The Prince*. Skinner Q, Price R (Eds.). Cambridge, UK: Cambridge University Press; 1988.

[32] Silverman BG, Bharathy GK, Nye B, Eidelson RJ. Modeling factions for 'effects based operations': Part I—Leader and follower behaviors. *Journal of Computational & Mathematical Organization Theory*, 13(4):379–406; 2007.

[33] Dignum F, Dignum V, Sonenberg L. Exploring congruence between organizational structure and task performance: A simulation approach. Available at http://people.cs.uu.nl/dignum/papers/ooop-dignum-final.pdf.

[34] Bharathy GK. Agent-Based Human Behavior Modeling: A Knowledge Engineering-Based Systems Methodology for Integrating of Social Science Frameworks for Modeling Agents with Cognition, Personality & Culture. PhD Dissertation, University of Pennsylvania; 2006.

[35] Heuer RJ Jr. *Psychology of Intelligence Analysis*. Washington, DC: Center for the Study of Intelligence, Central Intelligence Agency; 1999.

[36] Silverman BG, Bharathy GK, O'Brien K. Human behavior models for agents in simulators and games: Part II—Gamebot engineering with PMFserv. *Presence*, 15(2); 2006.

# 9

# MODELING HUMAN BEHAVIOR

Yiannis Papelis and Poornima Madhavan

*Human behavior* is the collective set of actions exhibited by human beings, either individually or in groups of various sizes and compositions. Modeling human behavior is an interdisciplinary field that aims at developing models that reflect or even replicate reality, given a set of appropriate initial conditions. The field is quickly gaining momentum and has wide ranging and diverse applications in civilian and military domains. Because of the widespread utilization of human behavior models, it is important to have a firm grasp on the theoretical underpinnings of the field and what such models can provide; it is also important to recognize the associated limitations. As human beings, we are familiar with our own and others' behavior and tend to confuse this familiarity with a reliable and generalizable scientific process. We are also accustomed to the predictive power of physics-based models and tend to presume that human behavior models possess similar abilities of predicting behavior, something that is rarely true [1]. It is thus important to realize the nature of the complexity of human behavior and try to address it when developing analytic models.

The issues that affect human behavior modeling vary greatly depending on the level of required behaviors and the overall size of the population being modeled. At the *physical level*, human behavior is driven by physiology and

---

*Modeling and Simulation Fundamentals: Theoretical Underpinnings and Practical Domains*,
Edited by John A. Sokolowski and Catherine M. Banks
Copyright © 2010 John Wiley & Sons, Inc.

**Table 9.1 Human behavior modeling issues**

|  | Individual | Group | Society |
|---|---|---|---|
| Strategic | High-level decision making based on intuition and emotion | Interaction dominant models, collective intelligence | Changes to culture, political situations |
| Tactical | Concrete decision making; short-term emotions affect decisions | Social influence, emotional reflection, collaboration | Communication-dominant models |
| Physical | Reactive models based on stimulus–response, physiology, motor skills | Physics-based interactions | Population dynamics |

**Table 9.2 Example applications**

|  | Individual | Group | Society |
|---|---|---|---|
| Strategic | Persistent virtual reality characters | Economics models, financial markets | Insurgency modeling, effect of sanctions |
| Tactical | Route selection; obstacle avoidance; procedural tasks | Crowd behaviors, traffic simulation, war games, computer-generated teams | Public opinion polling, consumer models, social simulation |
| Physical | Reaction time, skills assessment | Crowd movement simulation | Disease spreading, population aging |

automated processes, whereas at the *tactical level*, decision making and emotions are the primary drivers for our short-term actions. At the highest level, strategic behaviors involve longer planning and complex decision making according to intuition, emotions, and experience. At the same time, there are significant differences between modeling individuals or groups and even further differences that depend on the size of the group that is modeled. It is thus beneficial to define a taxonomy within which to classify behavioral modeling approaches and identify related challenges. Tables 9.1 and 9.2 present a classification of human behavior modeling organized according to the level of cognitive processing involved in the behavior as well as the size of the population being modeled, along with examples.

The issues associated with each classification vary greatly, as is the maturity of the related research. For example, models at the physical level are based on empirical research that can often be validated and thus have predictive value. Similarly, individual-based models are typically less challenging than group- or society-level models because of the complexity introduced by the interactions among individuals and the chaotic nature of evolutionary behaviors. On the other hand, tactical and strategic behaviors are harder to model due to the adaptive and unpredictable nature of human behavior. When incorporating larger populations, the complexity drastically increases to the point where such models are difficult, if not impossible to validate. It is interesting to note that despite these apparent limitations, such models are in widespread use in both civilian and military applications. The reasons for this popularity are many. The output of such models is easy to understand because simulations using such models often resemble real life. When presented through means that obscure the source of the results, it is often impossible to differentiate simulation from real life. For example, results of military simulations presented with the same tools used during actual operations obscure the source of the displayed data. Another reason for the popularity of tactical models is the ability to change input parameters and rerun the simulation obtaining a new answer for the what-if scenario. Iterative use of models in such a manner supports exploratory analysis that helps us gain an intuitive feel for the modeled situation. And even though such models elude validation on a strictly scientific basis, it is often possible to perform sensitivity analysis and identify broad trends as opposed to exact predictions. For example, a war-gaming simulation may help a planner realize that increasing the number of armored vehicles has a positive effect on the outcome of a scenario, but it cannot pinpoint the exact number of vehicles required to win the battle with a certain probability. By simulating models of human behavior, we seek to understand, not necessarily predict, the aggregate behavior of an inherently complex system for which we have no better model.

In summary, the issues and techniques used for implementing human behavior models vary greatly depending on the scope and size of population, and it is thus important to put each approach in the appropriate context. The remainder of this module summarizes material in the context of the taxonomy illustrated in Tables 9.1 and 9.2.

## BEHAVIORAL MODELING AT THE PHYSICAL LEVEL

The physical level of behavior provides the most opportunity for modeling using strict scientific and engineering approaches. By assuming simple and constant motivation functions, such as "seeking to a goal," "avoiding an obstacle," "operating a vessel," or "tracking a lane," models can predict measurable actions that are primarily physical in nature and depend on physiological or psychological principles [2]. Decisions are done at an instinctive or reactive level, and emotions have little impact on the process; instead, performance is governed by the level of workload, fatigue, situational awareness, and other similar factors. Some modeling efforts treat a human being as a limited capacity processor that can selectively parallel process but with specific penalties for splitting attention among many tasks [3]. Based on such assumptions, models can use laws of physics, experimental data, or other "hard" facts about human performance that yield predictive operational models of human behavior [4]. On a more practical basis, fuzzy logic is often used to simulate human control tasks. *Fuzzy logic* is a theory of developing controllers that utilize fuzzy variables, a concept derived from fuzzy sets [5,6]. A fuzzy variable takes on approximate as opposed to exact values; for example, when driving a vehicle, we may refer to the speed as fast or slow, in lieu of an exact number, that is, 53.5 mph. Even though fuzzy logic is often used as a means to obtaining better control (e.g., in a thermostat), it provides a good basis for mimicking human physical behavior because it simulates our perception of the world, which is approximate and not exact. As a contrast to that approach, it is also possible to build traditional control systems that simulate human control tasks, as shown by Gabay and others who developed an experimental approach to obtaining parameters for a traditional control system that mimics the performance of a human being in closed-loop tracking tasks [7]. Similar models exist that predict human performance while operating a vehicle, piloting a plane, or performing general tasks.

## BEHAVIORAL MODELING AT THE TACTICAL AND STRATEGIC LEVEL

The tactical level of behavior modeling focuses on human behavior when seeking short-term goals. The difference between physical and higher behavior levels is rooted in the complexity of the goal and duration of the activity being modeled. Whereas physical behavior models assume singular and/or simple goals and focus on human capabilities, higher-level behaviors have more complicated goals that take longer to achieve and involve decision making at various levels. For example, consider the task of driving. At the physical level, one would model the ability of a driver to track the lane of a road. At the tactical level, however, one would model a passing maneuver or a lane change, including the decision making involved in selecting the particular maneuver. Furthermore, a strategic level of behavior would focus on route selection, or even the choice between driving versus taking an alternative means of transport.

When modeling higher-level behaviors, there are two broad schools of thought. One considers human beings as rational operators that attempt to maximize a utility function or achieve a specific goal. Such models ignore the effect of emotion and instead concentrate on rational decision making using deterministic or stochastic approaches. The second school of thought considers human beings as *quasi-rational entities* that still pursue a specific utility function, but frequently make suboptimal decisions and exhibit actions that can act contrary to their stated goals and can even reach the level of self-destructive behavior [8,9].

Rational decision making is by far easier to model and simulate than quasi-rational behavior. Behaviors stemming from rational thought are by their nature algorithmic and procedural and often represent skills for which humans train, that is, driving an automobile, flying a plane, being a soldier, being a participant in a team sport, and so on. Techniques for developing such models include state machines, rule-based systems (RBSs), and pattern recognition, among others. Modifications that add stochastic variation to behavior can be blended into the baseline approaches and because such variation is not driven by emotions, traditional statistical approaches can be used to enhance and fine-tune rational decision models. It is also possible to incorporate limits that reflect how excessive demands may stress human mental and physical capabilities, effectively blending the physical and tactical levels of behavior into a singular model. The inclusion of physics- or psychology-based perception models can further enhance the overall behavior model by reducing the quality of data used for decision making. More advanced models utilize processes that attempt to simulate experience and intuition. *Intuition* in particular, which refers to the ability of a human being of recognizing something without a full understanding of how this recognition takes place, is a key characteristic of strategic decision making. Research has shown that such decision making is rooted in pattern matching based on prior knowledge, as a result, technical approaches to construct models that exhibit strategic decision making depend on knowledge databases, pattern matching using neural networks, hidden Markov models (HMMs), and similar techniques [10].

*Quasi-rational behavior models* are more challenging to develop. The processes that govern suboptimal decision making are an active topic of research and not yet fully understood. Such processes are complex and depend on a variety of current and expected outcome emotions, and on the context within which choices are made [8,11,12]. Lacking strict constraints that are based on optimizing functions, it is tempting to substitute stochastic decision making as a means to simulate quasi-rational behavior. However, suboptimal decisions are not random decisions; it is just that they are governed by processes that we do not yet fully understand and thus cannot model on a first-principle basis. Suboptimal decisions are also not intuitive decisions; in fact, as discussed earlier, it has been shown that intuitive behavior is rooted in fact-based experience and pattern matching. Nevertheless, it is possible to develop models that capture some of the better understood factors involved in quasi-rational behavior. For example, we can use models that simulate emotional state, with actions being a side effect or result of the current emotional state [13]. By simulating emotions and letting the emotional state control the actions exhibited by the model, behavior is by definition suboptimal as it is detached from goal seeking, yet driven by processes that conform to reality. In general, for models requiring fidelity at strategic levels of behavior, the challenge lies in simulating irrationality and other abstract and suboptimal decision-making processes that are not strictly driven by goal-seeking human behavior traits.

## Lumped versus Distributed Models

The issues that differentiate the physical, tactical, and strategic behavior models are further amplified when considering the incorporation of multiple entities into the simulation. The treatment so far has concentrated on models of an individual human; we now look at the issues involved when modeling multiple humans.

First, let us define the difference between lumped and distributed models in human behavioral modeling. The notion of a *lumped model* originates in electrical engineering and refers to studying a circuit assuming that each element is an infinitesimal point in space and all connections are perfect, thus their effect on the circuit need not be considered. The assumption here is that the effects of the connections are either small, or are taken into account in the element model. Lumped systems are typically modeled by using differential equations. In human behavior modeling, a lumped model is one that treats modeling of multiple entities as a unified model that ignores the specific interactions, or presumes that such interactions are factored into the singular model. As in other disciplines, such models typically involve differential equations that represent the change of system variables over time. Examples include population dynamics and disease spreading [14,15]. Lumped system models are powerful when considering that they operate independently of the size of the population they model. At the same time, their weakness lies in assumption that interpopulation interactions are captured by the basic model, something that is not always the case. In fact, lumped models of human behavior implicitly presume that motivation and thus behavior remain largely constant. For example, consider the limited capacity population equation developed by the eighteenth century mathematician Verhulst [16]:

$$p_{t+1} = r \cdot p_t \frac{K - p_t}{K}.$$

The equation provides a model that predicts the size of the population at a particular time ($p_{t+1}$) given the population at a prior time ($p_t$), the growth rate ($r$), and the maximum population size ($K$). Note that numerous factors such as reproductive choice, socioeconomic prosperity, state of medical care, and so on, are implicitly factored into the growth rate, thus making this a lumped model.

Let us contrast this with a distributed system. Much like lumped systems, the notion of a *distributed system* originates in engineering, where systems are represented by taking into account variations of elements not only over time but over other variables; in the case of electrical engineering, connections are imperfect having finite impedances; in the case of mechanical engineering, the twist of a rod or link along its length is not lumped into the model; instead, it is taken into account explicitly. In basic engineering modeling, distributed systems utilize partial differential equations to capture changes over time as well as changes over spatial variables. In human behavior modeling, one can approach distributed system modeling in two ways. One is by subscribing to Nagel's *reductionism theory*, which states that one way to understand the behavior of multiple entities is to understand a single entity and then coalesce that knowledge predicting the behavior of multiple entities [17]. Nagel proposed this theory to suggest that a generalized approach to science can be achieved by reducing complex theories into more fundamental theories, provided that there are explicit linkages between the two theories that allow us to transition the knowledge from the reduced to the more complex one.

This approach is parallel to augmenting the basic equations governing the aggregate behavior of a population with additional factors, constructing complex models by combining simpler ones. Surely, use of partial differential equations can be construed as one way to achieve this, as is the use of ad hoc modeling equations that incorporate additional factors into the model. Another school of thought, which contradicts Nagel's reductionism theory, is that superposition of behaviors is not adequate; instead, the interactions among entities themselves create fundamentally new rules and laws that apply to the aggregate behavior of the group and such laws do not exist when dealing with a single entity. As stated in *Artificial Intelligence*, "Often the most efficient way to predict and understand the behavior of a novel complex system is to construct the system and observe it" [18]. By using agent-based modeling, a topic that is covered in Chapters 1 and 8 in this book, we can address the limitations of lumped models in human behavior simulation and study the effect of individual behavioral models on the evolutionary behavior of populations of varying size. Of course, agent-based modeling can become computationally expensive as the size of the population grows. For studying phenomena that depend on population sizes that exceed current computing capacity, lumped models may be the only practical approach. It is important to note that the taxonomy presented in Table 9.1 differentiates between *group-level* and *society-level* modeling, that is, a reflection of practical limitations that make lumped system models the only viable alternative for simulating entities whose size approaches whole societies (i.e., millions of entities or more). It is conceivable that computational advances in the near future will support practical agent-based modeling or arbitrarily large populations, fusing the group- and society-level classifications.

The remainder of the chapter summarizes various techniques utilized in building human behavior models. These techniques apply to individual or multiagent simulations.

## TECHNIQUES FOR HUMAN BEHAVIOR MODELING

### Fuzzy Logic Control

As indicated earlier, the use of fuzzy logic is targeted primarily at building controllers that can better handle control tasks for which traditional control approaches fail. This typically occurs in nonlinear or adaptive systems, which, due to their dynamic nature, violate the assumptions of linearity on which classical control approaches depend. Fuzzy logic controllers on the other hand, respond better to such systems because they mimic human adaptability associated with control tasks. Fuzzy controllers depend on fuzzy logic, which in turn is based on multivalued logic supporting variables that can take values between conventional evaluations. For example, consider a binary variable representing "yes" or "no"; it can have only two valuations 0 and 1 (or any other pairs of numbers); however, there is no way to represent "maybe." A fuzzy variable on the other hand, can have any value between "yes" and "no," including "maybe," "possibly," "not likely," and so on. The ability to classify variables in humanlike terms is beneficial, if for no other reason for it makes sense. For example, if one is asked the question: "What is the room temperature," the most typical answers are qualitative; for example, one may answer: "It is fine," or "it is cold" or "it is pretty hot"; rarely does one reply by stating: "The temperature is 79.4 degrees Fahrenheit." If our goal is to build a temperature controller that acts humanlike, then using fuzzy variables provides a natural way to deal with valuations that conform to human interpretation. At the same time, using such variables provides little value unless a corresponding theory allows manipulation and effective use of such variables when solving a problem. It is the development of the theory of fuzzy sets [5] that provided the mathematical tools necessary to handle fuzzy variables and operationalize control strategies, and in turn can facilitate modeling human behavior.

In the same way that traditional set theory defines set operations (union, difference, etc.), fuzzy set theory provides associated operators that allow computer manipulation of fuzzy variables. Fuzzy logic makes it possible to create mathematical formulations that allow computer-based handling of humanlike interpretations of the world, and as such, provides a fundamental tool for building control systems that simulate human behavior at the physical (or control) level.

To better understand how fuzzy logic can be used to simulate human behavior, we must first provide some basic principles associated with fuzzy set theory. A core concept associated with fuzzy set theory is partial membership. In traditional set theory, an element either belongs to a set or it does not. In fuzzy sets, it is possible to assign partial membership to a set. The *membership* variable is a variable whose range is inclusively between 0 and 1, and it represents the degree to which an element belongs to a set. Let us look at an example that demonstrates how we can to apply partial membership to the velocity of a vehicle. Let us assume that set $F$ contains velocities that are considered *fast*. In traditional set theory, we would pick a threshold, let us say 75 mph, and any speed greater than or equal to that threshold would be considered fast, thus belonging to the set, whereas any speed below that would not be considered fast; that is, it would not belong to set $F$. Figure 9.1 depicts this membership variable for this threshold when using traditional set theory logic. The membership variable (i.e., the variable representing if a speed belongs to the set) is binary, taking the value of 0 or 1 exclusively.

Notice the conceptual discontinuity associated with this definition. A person driving at 74.9 mph is *not* traveling fast, but a person driving 75 mph is traveling fast. This definition is unnatural and inconsistent with how we view things. A

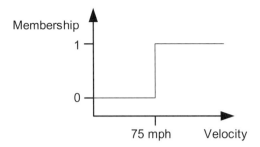

**Figure 9.1** Binary set membership.

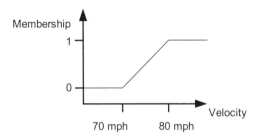

**Figure 9.2** Fuzzy set membership.

more natural way to classify speed is to provide a transition range, let us say starting at 70 mph and ending at 80 mph. A person driving below 70 is surely not going fast, and a person driving above 80 is definitely going fast. Any speed between 70 and 80 would be considered borderline, with an ever-increasing bias toward fast as we approach the upper end of the range; in fuzzy set nomenclature, as the speed approaches 80 mph, the membership to the set of driving fast will also increase. Figure 9.2 depicts the fuzzy set membership based on this definition.

Note that the membership to the "fast" set now varies between 0 and 1; for example, when driving 73 mph, the membership variable would be 0.3.

The basic operations applied to traditional sets can also be applied to fuzzy sets. Consider two sets whose membership profiles are shown in Figure 9.3; set A represents velocities that are "about the speed limit" and set B represents velocities that are "safe," defined as more than about 60 and less than about 75.

The intersection of these sets is any velocity that complies with membership to set A and set B. The union of these sets is any velocity that complies with membership to set A or set B. The membership function for the intersection and union of these sets is shown in dashes in Figure 9.4. Similarly, it is possible to define set subtraction and set negation.

Having established the basic tools associated with the fuzzy set theory, let us now see how we can model a controller using fuzzy logic. Our goal is to

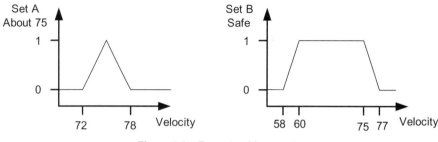

**Figure 9.3** Example of fuzzy sets.

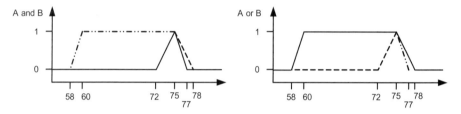

**Figure 9.4** Intersection and union of fuzzy sets A and B.

simulate how a human would control the environment in order to achieve a goal. We represent this effort as a set of rules, written as if-then statements and using qualitative descriptions both for the observations of the environment but also for the actions that need to be taken to maintain the desired control. In fuzzy logic nomenclature, these rules are referred to as the *linguistic rules*. Let us use a simple example to demonstrate the process. The example is of a driver controlling the speed of a vehicle to maintain a safe and legal speed, provided the speed limit is 65 mph. In order to apply fuzzy logic, we must first identify the variables that will be observed and the qualitative assessment of these variables. Then, we need to define the control actions, again using fuzzy terms, and finally we must provide the linguistic rules.

The variable to be controlled is the speed of our vehicle, and we create three rough assessments (fuzzy sets) of the speed, as follows:

(1) *legal speed*—from about 45 mph to about 65 mph
(2) *too slow*—any speed below about 45 mph
(3) *risky*—any speed above about 70 mph

Note that even though these definitions mention specific values for the velocity, terms such as "about" create a qualitative or fuzzy definition. Figure 9.5 depicts one possible definition of the three fuzzy sets mentioned above. The controller designer has a choice on how to operationalize the definition of "about" by appropriately shaping the membership functions for each of

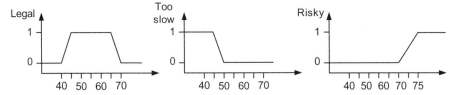

**Figure 9.5** Intersection and union of fuzzy sets A and B.

these fuzzy sets. The steeper the boundaries of the membership function, the less fuzzy the definition. For example, the risky speed could be defined as starting at 63 mph and reaching full membership at 65 mph; this would provide an earlier but more gradual transition into the risky zone. Also, it is not mandatory that the fuzzy set definitions utilize linear segments. Other curves could also be used; for example, a bell curve could be used to represent "about." To keep matters simple in the example, we use linear segments as shown in Figure 9.5.

After defining the input sets, the next step involves defining fuzzy sets that capture possible actions. Car speed is controlled through the use of either the throttle or the brake pedal. Gentle deceleration can often be achieved by releasing the throttle and letting friction reduce the speed, and the brake pedal can always be used to provide authoritative deceleration. In this example, we define a single variable that ranges from −100 to 100, with −100 referring to maximum braking and 100 referring to maximum acceleration; the advantage of this assignment is that a value of 0 indicates that neither pedal is pressed and the vehicle is coasting. We now build three fuzzy sets, each representing possible actions, as follows:

(1) *gentle speed up*—gradually increase speed with no urgency
(2) *maintain speed*—current speed is acceptable
(3) *slow down*—must reduce speed immediately

The definition of these fuzzy sets is depicted in Figure 9.6.

A few notes are warranted with regard to the specific definitions shown in Figure 9.6. The gentle acceleration action is centered at 50 percent of throttle; this implies some system-specific knowledge regarding the throttle setting that will keep the car traveling at near the speed limit. When a human drives, such assessments are made instinctively and automatically. When building a fuzzy controller, such calibration can be done experimentally, but the overall design is not dependent on precise selection on this value. A second issue to note is that these actions have areas of overlapping control. For example, a setting of 0 percent can belong to the "maintain speed" or "slow down" action. Such overlaps are perfectly acceptable and, in fact, are very common as the controller will eventually blend the affects of these actions according to the membership of the corresponding input variables. Finally, note that the slow down

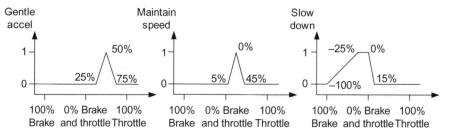

**Figure 9.6** Fuzzy sets for possible actions.

action uses a more complicated shape than the triangles used for the other two actions. As mentioned earlier, the selection of the particular shape should depend on the associated meaning of the action and is not constrained in any particular shape. In this case, the sharp rise on the right side implies that even though small throttle inputs can be considered part of the "slow down" set, the proportional membership value is very small compared with applying the brake pedal.

The final step in the definition of the fuzzy control algorithm involves the linguistic rules. We will use three rules as follows:

- *Rule 1:* If speed is legal, then maintain speed.
- *Rule 2:* If speed is too slow, then gently speed up.
- *Rule 3:* If speed is risky, then slow down.

Few simple rules are used here in order to keep the example manageable. Additional variables could be defined (e.g., another variable may be the current acceleration) and rules involving arbitrary predicates can be used; for example, a more complicated rule could be: "If speed is risky and acceleration is positive, then decelerate briskly." Once the implementation has been automated, adding new rules and variables can be done with relative ease as their treatment is uniform.

At a conceptual level, simulating human control behavior using fuzzy logic is done by using a measured value of each fuzzy variable, determining the fuzzy sets to which the actual value belongs, and then determining which actions are affected by these "active" fuzzy sets. This process was outlined in Mamdani and Assilian [19]. Actions that contain active sets in their predicate are referred to as "firing." Once the firing actions have been identified, their contribution to the eventual control input is calculated according to the degree of membership that each rule provides. One way to visually explain this process is by creating a "rules chart," which on the horizontal axis contains the input fuzzy sets and on the vertical axis contains the action fuzzy sets. The rules chart for our example system is shown in Figure 9.7, which uses different line styles to separate among the input and action fuzzy sets. Also, vertical lines have been drawn to show the ranges of each input fuzzy set, and hori-

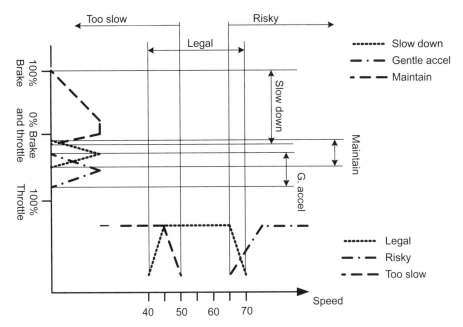

**Figure 9.7** Rules chart for fuzzy controller example.

zontal lines have been drawn to show the application area of each action. Note that each input fuzzy set creates a horizontal region; in this case, the "too slow" input covers any speed below 50 mph, the "legal" input covers the region from 40 to 70 mph, and finally the "risky" input covers any speed over 65 mph.

Given the rules chart, the mapping between variables and rules can be interpreted as the rectangular region that overlaps the input range and the corresponding rule range. Figure 9.8 illustrates these three regions, one per rule. The upper-right region represents rule 3, the middle region represents rule 2, and the lower-left region represents rule 1. Note that rules overlap, indicating that there are situations where a given speed may belong to more than one input set. Correspondingly, more than one of the rules will fire in order to produce the final action, which will consist of a blend of each of the actions associated with the firing rules.

Let us now look more specifically at how the controller works by assuming that the current speed of the car is 73 mph. This speed can only belong to the "risky" input set, so the only rule that can fire is rule 3. We determine this by observing that this is the only rule that references the "risky" input in its predicate. If additional rules referenced the risky group, they would also fire. Next, we determine the degree of membership of this value to the "risky" set. Given the visual definition of the fuzzy sets, calculating the membership value can be done by inspecting the rightmost shape of Figure 9.5, but of course, it can also be programmed on a digital computer. In this case, by inspection, we determine that the value of membership of the 73-mph speed to the risky

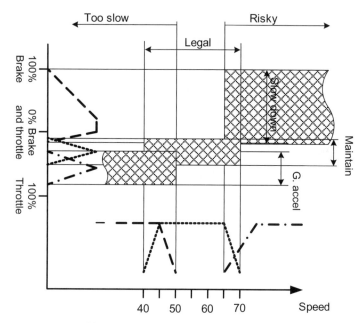

**Figure 9.8** Rules overlaid in rules chart.

fuzzy set is 0.8. This membership value is the link between the observation and the action. We use that same value to slice the action set and produce a new set that represents all possible actions with a maximum membership of 0.8. Figure 9.9 depicts both of these operations. First, a vertical line is placed at 73 mph to determine the membership; note that the rule 3 region is intersected by this line, something that we have already determined. Then, a vertical line is drawn parallel to the action axis at a membership value of 0.8 and the "slow down" set is reduced by this value. Visually, this yields the cross-hatched pattern as the applicable subset of the "slow down" action.

At this point, the fuzzy controller is suggesting that the appropriate action is reflected by the fuzzy set associated with the "slow down" action as reduced by the membership value of 0.8; however, this is still a fuzzy set, and in order to control the vehicle, it is necessary to perform a last step that de-fuzzyfies the set and produces a specific control input. The traditional approach for this step is to utilize the center of gravity of the fuzzy set, as shown in Figure 9.10. The intercept between the centroid and the action axis yields the final control value; in this case, this value is approximately 24 percent, braking.

A more complicated (and much more common) situation occurs when the input variable intersects more than one of the input sets, and in turn, fires more than a single action. Let us look at such an example when the input speed is 67 mph. Drawing a vertical line at 67 mph, we observe that it intersects both the legal and risky speed regions; so driving at 67 mph is both "legal" and "risky." Specific membership values can be calculated by computing the inter-

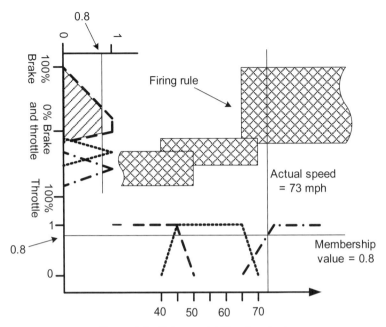

**Figure 9.9** Rules overlaid in rules chart.

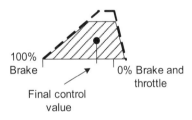

**Figure 9.10** Final control value.

cept between the 67-mph line and the membership profile functions. Figure 9.11 depicts the intercepts, which occur at 0.2 for the risky set and 0.4 for the legal set. To determine which rules fire, we can either inspect the rules and identify the ones that include the risky or legal sets in their predicate or graphically identify which region the 67-mph line intersects in the rules chart shown in Figure 9.9. Both approaches yield rule 1 and rule 3 as the rules that fire in this situation. To determine the value applied to the slicing of each of the actions, we select the membership value of legal for the maintain speed action (in this case a value of 0.4) and the membership value of risky for the slow down action (in this case a value of 0.2).

Figure 9.12 illustrates the appropriate regions and the application of the membership values to the two firing actions. The resultant action sets are shown on the right side of the figure. As in the previous example, the final

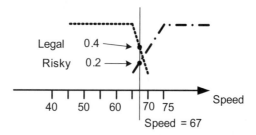

**Figure 9.11** Input intercepts for the second example.

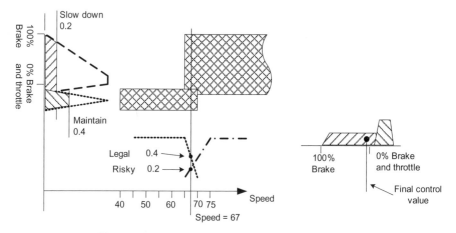

**Figure 9.12** Rules chart illustrating controller operation.

control output can be derived by selecting the centroid of the union of the fuzzy action sets. In this case, this value indicates a slight application of the brake, which is consistent with the earlier solution that demanded larger brake application due to the higher speed.

As demonstrated by the above examples, fuzzy logic can be used to build controllers that mimic the operation of human beings including the interpretation of input data, specification of linguistic rules that select control actions, and correlation of action intensity with root cause of the disturbance. This last feature is a critical component of fuzzy logic control. Whereas classical control approaches utilize differential equations to develop control commands that depend on specific assumptions about the system under control, fuzzy logic controllers utilize an arbitrary number of linguistic rules, each of which can handle a portion of the controlled system's performance envelope. This resembles low-level human behavior; we can easily switch between control strategies and manage to adapt along with a system, hence, the attractiveness of fuzzy logic for building humanlike control algorithms. Finally, note that even though the examples shown here focus on the low-level control, it is possible to apply fuzzy logic in higher-level decision-making processes; in fact, any area that can be defined in terms of fuzzy sets and formulated as a closed-loop control problem described by linguisitc rules can benefit from the direct or indirect application of fuzzy logic theory.

## Finite-State Machines (FSMs)

Another very popular technique used for building human behavior models is FSMs. A state machine consists of a set of states, a set of transitions among these states, and a set of actions. Actions are associated with states and along with transitions govern the behavior of the modeled entity. Each state and associated actions represent a distinct behavior phase that the entity can exhibit. The transitions control when the mode of operation changes based on external factors. Sometimes, a set of initialization and termination functions are associated with each state; the initialization function executes once when a state first takes control, and the termination function executes once when a state losses control.

The use of FSMs for modeling human behavior is better explained by constructing an example. In this case, we will use the example of the behavior of a driver that is attempting to pass the vehicle ahead while driving along a two-lane road and each lane having opposite direction. For state machines with relatively few states, a directed graph can be used to depict the structure of the state machine, so we will use this technique here. Figure 9.13 illustrates a directed graph representing the behavior.

There are seven states and eight transitions. Table 9.3 contains the meaning of each state along with the actions associated with each state.

Transitions control when the entity will change from one state to the other. Depending on the exact formalism used, transitions can be associated with the occurrence of singular external events or can contain complex logic for determining when states change. In the model described, transitions are generalized Boolean functions that can contain arbitrarily complex logic that eventually evaluates to true or false. Table 9.4 contains a sample of transitions and their associated logic as would apply to this example.

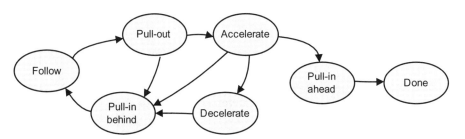

**Figure 9.13** Directed graph representing passing state machine.

**Table 9.3  Finite-state machine for passing maneuver**

| State | Meaning | Action |
|---|---|---|
| Follow | Maintain appropriate distance behind the lead vehicle. | Control vehicle to maintain desired following distance. Track the lane. |
| Pull-out | Veer to the opposite lane traffic in order to prepare for passing. | Accelerate and at the same time steer to place vehicle on opposite lane; remain behind the lead vehicle. |
| Accelerate | Now at the opposite lane; the goal is to drive ahead of the vehicle being passed. | Accelerate rapidly to pass; track the opposite lane. |
| Pull-in ahead | Remain ahead of the vehicle being passed and drive back on the lane. | Maintain desired speed and steer back onto own lane. |
| Done | Maneuver is completed. | Drive as usual. |
| Decelerate | Maneuver is aborted— slow down to get back behind the lead vehicle. | Decelerate slower than other vehicle; track the opposite lane. |
| Pull-in behind | Maneuver is aborted—get back behind the lead vehicle. | Maintain appropriate following distance. Steer to place vehicle back onto own lane. |

**Table 9.4  Logic for selected transitions in passing maneuver**

| Transition | Logic |
|---|---|
| Follow → pull-out | Following distance less than a threshold and no oncoming traffic exists. |
| Pull-out → accelerate | Vehicle has shifted to oncoming lane and no oncoming traffic exists. |
| Pull-out → pull-in behind | Oncoming traffic exists. |
| Accelerate → pull-in ahead | Rear bumper of own vehicle has moved ahead of front bumper of vehicle being passed. |
| Accelerate → decelerate | Oncoming traffic exists and front bumper of own vehicle is ahead of rear bumper of vehicle being passed. |
| Accelerate ( pull-in behind | Oncoming traffic exists and front bumper of own vehicle is behind rear bumper of vehicle being passed. |

There are few noteworthy items in the above example:

(1) The entity implementing this behavior can only be at one state at any given time, so states can be traversed in a finite number of sequences within the directed graph. Because of the finite number of states, there are a finite number of nonoverlapping paths through the system.

(2) The model as described so far is not associated with any particular execution style; this model could execute with discrete-event or continuous time simulation semantics. FSMs are very effective tools that can be used independent of the execution semantics—of course, the implementation details will vary depending on the actual execution semantics.

(3) The actions described in Table 9.3 resemble to some degree the description of the state itself. This is common as the essence of each state is captured in the actions or activities implemented by the entity while in that state. The state names serve as shorthand descriptions of the activities that the entity will engage while in the state.

(4) It is generally expected that among the transitions emanating from a state, no more than one will evaluate to true at any given time. If none of the transitions are true, the current state remains current.

Even though not explicitly specified, most state machines have a designated start and end state, which provides for a clean start and end points in simulating the behavior.

Details on the approach used for action implementation have been intentionally ignored so far, primarily to avoid unnecessary details. However, the discussion would not be complete without addressing various techniques for incorporating actions into a human behavior model implemented using state machines. First, let us assume that this state machine will execute by using continuous time semantics. Furthermore, let us assume that the state machine is represented by the two data structures listed in Figure 9.14:

The first data structure represents transitions. It contains a Boolean function that evaluates if the transition should take place and also contains an integer variable that refers to the target state. Note that depending on the programming language semantics, the target variable could be a pointer, reference, or similar construct that can be used to refer to another data structure.

```
Transition
  Function evaluate();
  Integer target;
End Data Structure Transition

StateEntry
  Procedure activation();
  Procedure termination();
  Procedure actions();
  Set transitions of type Transition
End Data Structure StateEntry
```

**Figure 9.14** *Transition* and *state* data structures.

```
state = initState;
call FSM[state].activation();

for time = 0 to endTime
    call FSM[state].actions();
    for each transition i emanating from FSM[state]
        B = FSM[state].transition[i].evaluate();
        If ( b = true )
            call FSM[state].termination();
            State = FSM[state].transition[i].target
            call FMS[state].activation();
        endif
    endfor
endfor
```

**Figure 9.15** Typical state machine execution loop.

The second data structure represents states. It contains three procedures, one that is called when the specific state first takes over, another that is called when the state loses control, and finally one that is called while the state is active. Finally, the state contains a set of transitions, completing the FSM definition.

Figure 9.15 lists a typical execution loop that can be used to execute an arbitrary state machine, provided that the variable FSM represents an array of StateEntry data structures that has been properly initialized to reflect the structure of the state machine.

Implicit in the above pseudocode segment is the assumption that all procedures and functions have access to the global clock as well as all the data representing the virtual environment within which the agent exhibiting this behavior resides.

Let us get back to addressing the issue of action implementation. In this particular example, we are dealing with a driving behavior so we need to account for one additional issue that is specific to driving, namely using a realistic model of the movement of the vehicle. It turns out that the complexity of precisely controlling a vehicle increases as the fidelity of the vehicle model increases. This is because vehicle dynamics are nonlinear and thus less amenable to traditional closed-loop control techniques. Whereas we can select the fidelity of the movement model for a simulated vehicle, when the simulated human behavior must control an actual vehicle we have no such control [20,21]. This would be a perfect application of fuzzy control as described earlier. To simplify matters here, let us assume that the vehicle model is purely kinematic and thus can be controlled by directly providing the desired acceleration and direction of travel. A set of Newtonian equations can then handle the actual time-based motion of the vehicle that will comply exactly to the commands of the behavior model.

Given the aforementioned assumptions, we will review two approaches for action implementation, one in which control is embedded in the tactical level directly, and one in which control is implemented by a different but coexecuting model. Even though the differences may appear subtle when viewed at the implementation level, they reflect the difference between the physical and tactical level of behavior described earlier.

Using embedded control, the action procedure associated with each state is responsible for generating the desired acceleration and direction of travel. That means that the logic embedded in the behavioral model must perform calculations to determine the desired acceleration and travel. For example, while in the pull-out state, the acceleration should ensure that the own vehicle does not collide with the lead vehicle and that the track of the vehicle is performing a gradual shift in lane position. In effect, the action procedure must not only define but also implement the low-level behavior that the entity exhibits while in the specific state.

When using external control implementation, the action procedure need only specify what it needs done, not how. The low-level implementation of the control would be done elsewhere; in this case, a separate vehicle control model would be tasked with the implementation. For example, a possible vocabulary of commands may include "Follow Vehicle X," "Change Lane to L0," and so on. This approach decouples the state machine implementation from the specific issues associated with the physical control of the underlying vehicle.

Both approaches have advantages and disadvantages, although using an external control implementation is recommended for all but the simplest behavior models. Whereas using embedded control provides for a more centralized representation of the state machine and tends to reduce the initial effort of developing the system, it also mixes numerous details associated with the low-level physical control into the upper behavior levels. That is contrary to general behavior traits that tend to remain constant across different physical interactions. Except for extreme situations, the general behavioral traits of a human driver do not change when driving different automobiles; our writing style does not change when we use a different pen and a capable shooter maintains a high hit ratio independent of the weapon in use. This is because motor skills are generally independent of our decision making, and thus an implementation that mimics this separation of higher-level behaviors from lower-level motor skills is better able to match the overall human behavior.

**State Machine Limitations, Extensions, and Alternatives**

Despite their usefulness and widespread use, state machines have significant limitations. At a conceptual level, because the set of states is finite, the possible set of behaviors is also finite; an agent build using an FSM never learns! The field of machine learning focuses on developing approaches that aim to address this limitation, and some of this material will be covered in the pattern recognition section [22].

## 292  MODELING HUMAN BEHAVIOR

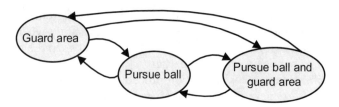

Figure 9.16  Soccer player state machine.

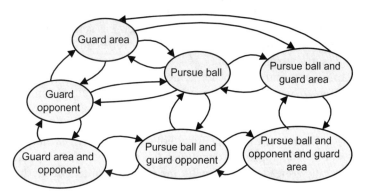

Figure 9.17  Extended soccer player state machine.

Another limitation of FSMs is state-space explosion. As the number of non-mutually exclusive behaviors increases, the number of states that must be used to capture all permutations of managing such behavior increases exponentially. For example, consider an FSM that implements a defensive soccer player who must guard a specific portion of the playing field and also pursue the ball beyond the specific area when other teammates are beyond range. Because these behaviors are not mutually exclusive, we must implement three states, one for handling area coverage, one for handling ball pursuit, and one for handling both. A state diagram representing this FSM is shown in Figure 9.16.

Now consider adding a third behavior, that of guarding a specific opponent. The number of states will now be more than double, because all basic permutation of behaviors must be implemented. The revised diagram is shown in Figure 9.17; note that only a subset of the possible transitions is included in the diagram.

Simple combinatorial analysis shows that the number of states required to implement $N$ nonmutually exclusive behaviors is $2^N$, something that can quickly become unmanageable. Even when behaviors are mutually exclusive, the number of states and associated transitions required to model complex behaviors quickly escalates.

There are several ways to address this limitation, primarily by extending the basic formalism of an FSM. Work in domain-specific modeling has extended

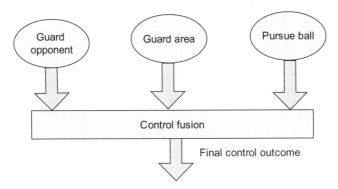

**Figure 9.18** Fusing concurrent state control.

basic state machines to include hierarchy and concurrency as well as domain-specific extension for enhancing the ability of controlling behavior [23]. Hierarchical state machines include the notion of state *containment*, whereas one or more states are nested inside a superstate. The ability to provide hierarchical specifications significantly reduces the number of states and decomposes the problem in manageable pieces. *Concurrency* is another useful extension that allows more than one states to be active at the same time. The code associated with each active state executes during each time step. This allows isolated implementation of individual behaviors since they can all be active at the same time. The only added complexity of this approach is that when it comes to the control of the associated agent, it is possible to have contradictory commands. Looking back in the soccer player example, an implementation that utilizes concurrent states would contain only three states as shown in Figure 9.18. Because the states are now concurrent, there is no need to code transitions; the associated code executes at each time step. However, there can only be a single point of control for the physical layer, and thus a layer is inserted between the output of concurrent state machines and the final physical layer; the task of this layer is to fuse the multiple control inputs into a single and operationally logical control outcome.

As shown in Cremer et al., it is possible to implement the fusion logic by using a conservative approach that simply selects the least aggressive control input [23]. That, however, necessitates that control inputs are subject to an ordering relationship, something that is not always possible. Specifically, in the soccer example shown in Figure 9.18, if the pursue ball behavior requests that the player seeks to point A but the guard area behavior requests that the player runs to point B, there is no automatic way to compare aggressiveness. In such cases, domain-specific handling must be implemented, something that tends to create ad hoc and nongeneralizable solutions.

An alternative to concurrent states is the use of context-based reasoning as described in Gonzalez et al. [24,25]. Contexts are "heavyweight" states that

**294** MODELING HUMAN BEHAVIOR

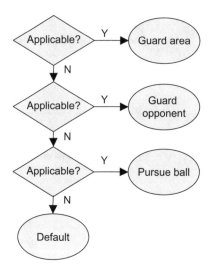

**Figure 9.19** Context-based reasoning implementation of a soccer player.

are effectively prioritized through a series of sentinel functions that determine if a particular context should be engaged. A sentinel function associated with each context is sequentially engaged in context priority order. The sentinel function assesses the situation and if it determines that its context can handle it, it engages the context; otherwise, the next sentinel function is called. Once engaged, a context gives up control voluntarily when it considers its task completed, or it can be preempted by a higher priority context, since the sentinel functions are consulted on a continuous basis. A possible implementation of the soccer behavior using context-based reasoning is depicted in Figure 9.19. Note that a default behavior is always present and will engage if none of the other contexts engage in order to provide some reasonable, although purposeless, behavior. An example of combining the formalism of hierarchical state machines and context-based reasoning for guiding an autonomous vehicle in an urban environment is described in Patz et al. [26].

Finally, a formalism that incorporates hierarchy, concurrency, communication, and a flexible visual specification language is *statecharts* [27,28]. As initially presented, statecharts target discrete-event systems but can easily be extended to support behavior modeling in discrete time simulations. Because they support concurrent execution, referred to as AND states in statechart nomenclature, they address state-space explosion; in addition, statecharts support hierarchy, where states can belong to other states and are only active when the parent state is active. Statecharts also formalize the notion of activity functions associated with each state and the general form of transitions as Boolean functions that are driven by either external events or internal logic. Statecharts also introduce communication in the form of broadcast events, which are events that are received by more than one concurrent state machine and result in transitions of the same name in all superstates. Statecharts were initially developed to support practical specification of real-time systems, but are used extensively for numerous applications including human behavior modeling.

## RBSs

The RBSs paradigm is an alternative approach to developing human behavior models. At its core, an RBS is rather simple, consisting of a set of rules, working memory, and an inference engine. The rules are pairs of if-then statements that each encodes a distinct condition and rational action both of whose domain is application specific. All associated data reside on the working memory, which initially presents the input of the external world, effectively a snapshot of the situation that an agent must handle. A rule fires when the *if* portion is true. When a rule fires, the action portion of the rule can have an arbitrary effect on the working memory. The inference engine engages when multiple rules are eligible to fire and implements a conflict resolution scheme to select the one rule that fires. This process continues until a terminal rule fires or until no rule is eligible to fire. At that point, it is typical that the contents of the working memory represent the best information available about the system. An alternative interpretation of the function of the RBS for behavior modeling is the association with actions with each rule, allowing the inference engine to act as a means by which to select appropriate actions based on current conditions.

Originally, the RBS paradigm was used in *expert systems* whose goal was to mimic the decision making of an expert when faced with facts relevant to the area of expertise. Automated medical diagnosis is the most common example of the use of RBS in expert systems. Beyond expert systems, however, an RBS can be used to implement behavior (1) by associating actions with each rules or (2) by associating actions with the terminal state of the working memory. At each time step, an RBS is initialized by encoding the situation facing the agent; the rules are evaluated with the inference engine used to select which rules take priority in case of conflicts. In the former variation, each rule firing corresponds to an action that is part of the behavior that the agent exhibits. In the latter variation, the state of the working memory at the end provides the best possible assessment of the situation at hand, and the appropriate activity is engaged to manage the physical aspect of the behavior. Note that this is similar to how the physical model obeys the commands of the currently active state in an FSM. Figure 9.20 illustrates the structure of an RBS used for behavior modeling, structured according to the latter variation.

RBS have several benefits when used to represent human behavior. Because there needs to be no linkage between rules, it is possible to add an arbitrary number of rules to a system without having to worry about explicit dependencies among them. Each rule represents a small amount of knowledge that can

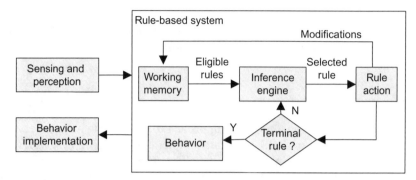

**Figure 9.20** Structure of an RBS used for behavior modeling.

be used to improve the understanding of the situation, and any number of such rules can be added during development. This scalability is a significant advantage for behaviors that must handle a wide range of situations and for behaviors whose definition is evolving during development. Another advantage of RBS for representing behavior is the fact that behavior need not always be deterministic. In fact, the basic paradigm does not specify specific constraints for the order in which rules are evaluated or the order that they execute. Modifications of the inference engine can be used to create variations, either stochastic or situation-based that lead to an agent taking a slightly different course of action when presented with what seems to be identical situations. Both of these reasons have made RBS a popular architecture for building the logic used for nonplayer characters in games.

By far the most comprehensive implementation of RBSs is *Soar*, a cognitive architecture for developing intelligent behaviors that is based on RBS [29]. Some notable extensions to the basic RBS paradigm include the use of multiple level memory and explicit treatment of perception, cognition, and motor skills as part of the overall architecture. In effect, the Soar architecture not only supports the physical, tactical, and strategic behavior levels, but also includes the input portion of an agent's behavior, which controls the flow of information from the environment to the cognitive engine. Finally, Soar includes a limited learning mechanism that can increase the available rules dynamically, thus simulating learning and adaptation. The Soar architecture implementation is in the public domain, and extensive user communities provide materials supporting its use [30,31]. Soar has been used extensively in constructing behaviors for military simulations and computer-generated forces and games [32].

### Pattern Recognition

We will now discuss pattern recognition as it applies to simulation of human behavior. Broadly speaking, *pattern recognition* is defined as the conversion of raw data into meaningful and contextualized data structures that can be used for further processing. Pattern recognition relates to human behavior modeling in two ways: first as a means to simulate or replace sensory inputs and second as a means of actual decision making.

The former relation is focused on understanding the world through sensing stimuli, using pattern matching algorithms that focus on visual imagery, sounds, and so on. Despite popularized visions of humanoid robots that can "see" and "hear," the ability to recognize speech or reliably react to visual stimuli is not as related to human behavior modeling as may initially appear. For example, data fed into a decision-making algorithm used in a game or military simulation does not need to be generated by a visual recognition task; similarly, recognizing speech may be convenient for a human operator but does not significantly affect the process of developing human behavior models that depend on audio stimuli. In effect, behavior modeling focuses on the decision-making and cognitive issues, whereas this class of pattern recognition technologies imitates human sensory inputs.

The latter relationship, which is focused on recognizing emerging patterns as a means to decide on the course of action, is more interesting as it relates to human behavior modeling. As mentioned earlier, it has been shown that a significant portion of human behavior is rooted in consciously or unconsciously recognizing patterns and tapping in stored memories of prior actions when faced with such patterns as a means for deciding the beset course of action. It thus makes sense to pursue human behavior modeling approaches that utilize pattern recognition as the primary driver for decision making. In this context, pattern recognition is a much higher-level process than sensory-type pattern matching, and it is better equated with situational awareness and intuition.

As a way to demonstrate the differences between sensory pattern recognition and higher-level situational awareness recognition, consider modeling the behavior of a member of a sports soccer team. The low-level pattern recognition tasks would be engaged in identifying the position and future trajectory of the ball and remaining players. The higher-level pattern recognition task, on the other hand, would try to identify possible opponent offensive maneuvers or weakness in the current formation of the opponent as a means to decide on appropriate future actions. In effect, the player continuously assesses the situation by trying to identify known patterns formed by assessing the current situation. Any action taken in response to the current situation yields an outcome, which in turn affects the situation, thus closing the loop. A simplified depiction of this process is shown in Figure 9.21. Note that this process resembles the architecture of RBSs, with one rule stored per pattern and the associated action tied to recognition of the particular pattern.

In addition to supporting a meaningful approach to modeling human behavior, pattern matching also provides the opportunity of encoding knowledge. The information used to represent known patterns, typically referred to as exemplars or templates, can be augmented with new patterns and

**298** MODELING HUMAN BEHAVIOR

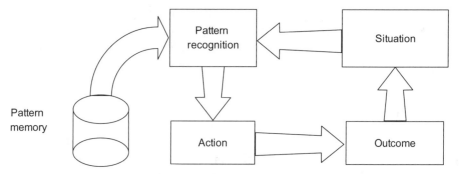

**Figure 9.21** Simplified model of pattern-recognition-driven actions.

associated rules. Depending on the techniques used for pattern recognition, it is also possible to reiterate or reject templates based on the success of the recognition tasks, in effect simulating learning by experience, or by example. In addition to mimicking how human beings learn, pattern recognition has several advantages over other techniques used for behavior modeling. Because learning occurs by example, it is not necessary to develop explicit rules that describe the patterns. In fact, it is not even necessary to be explicitly aware of the specific features that comprise a situation, it is only enough to have enough examples to train the algorithm to recognize the pattern.

Let us now summarize two related but distinct techniques used for pattern recognition. The first is artificial neural networks (ANNs) and the second is HMMs. They both share traits that support learning and ruleless specification of patterns, but take a different mathematical approach.

**ANNs** ANNs are an information processing paradigm inspired by human brain physiology [33,34]. ANNs consist of a large number of interconnected elements, referred to as neurons, a term borrowed from biology and refers to a single cell that can process and transmit information using electrical signals. In living organisms, neurons work together to perform recognitions tasks. Input stimuli is provided as input to neurons which in turn produce output signals that are routed to other neurons, eventually reaching a terminal output that controls higher cognitive functions. Neurons are interconnected through synapses; these connections govern the eventual output, given a specific set of inputs. In a similar manner, ANNs are constructs that mimic the interconnection and operation of neurons. The ability of ANNs to recognize patterns is stored in the structure and strength of the interconnections among the individual neurons.

A key aspect of neural networks is that the connections and connection strengths can evolve based on received signals; in biologic neurons, electrochemical processes create new connections and adjust the strength of such connections, and in ANNs, a variety of algorithms are used to adjust the

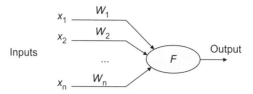

**Figure 9.22** Operational illustration of an artificial neuron.

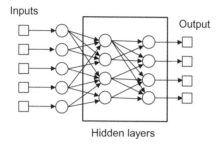

**Figure 9.23** Feed-forward ANN with hidden layers.

weights and connections. The important point is that in both biologic and artificial neural networks, this process of adjusting connections and weights represents learning. Depending on the type of learning, the network either automatically reconfigures itself to identify patterns that appear frequently or can be trained to identify specific patterns of interest. The former is referred to as supervised learning, while the latter is referred to as unsupervised learning. The ability to train by example combined with the ability of adaptive learning makes ANNs an attractive technique for model human behavior.

Figure 9.22 depicts the structure of an artificial neuron, where $X_i$ refers to inputs, $W_i$ refers to weights associated with each input, and $F$ is the summation function $F = \sum_{i=1}^{n} W_i X_i$ that generates the output. A neuron *fires* when the output value exceeds a prespecified threshold. ANNs consist of multiple neurons by connecting the output of each neuron to the input of one or more other neurons creating a network. There are two primary ANN architectures: the feed-forward and the feedback.

The *feed-forward network* supports only forward connections, whereas the feedback network allows the creation of loops within the network. Feed-forward networks are often organized in one or more layers as shown in Figure 9.23. Any number of hidden layers may exist, but as long as there is no loop in the directed graph formed by the network, the network remains a feed-forward network.

**300** MODELING HUMAN BEHAVIOR

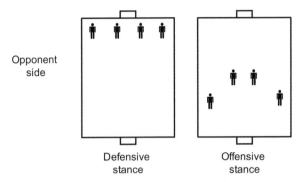

**Figure 9.24** Example of coding.

Feed-forward networks are static, in the sense that once an input is presented, they produce an output in the form of a binary signal. *Feedback networks* on the other hand exhibit dynamic behavior since the looped connections create a time-evolving system that changes as the output of the network is fed back into the input.

For each problem domain, effective use of an ANN involves two key steps:

(1) Identification of the mapping between problem domain and ANN input
(2) Training of the ANN

In the first step, it is necessary to convert the domain-specific knowledge and awareness into specific inputs that can be provided to the ANN. This requires the development of a mapping between world observations and inputs. Let us take a simple example of a generic team sport where the opposing team's position affects our short-term strategy. Figure 9.24 depicts two formations that are important to identify: a defensive stance (on the left) and an offensive stance (on the right).

A straightforward way for encoding such pattern is to overlay a grid over the playing field and mark the respective grid point as 1 when an opponent is present or 0 when an opponent is absent. The grid can then be represented by an occupancy matrix as shown in Figure 9.25; the occupancy matrix can then serve as the input to the ANN.

Once the problem domain has been mapped into ANN inputs, the ANN must be trained. Training an ANN is performed by using a set of inputs called the training set. The idea is that once trained with the training set, the ANN will then be able to recognize patterns that are identical or similar to the patterns in the training set.

As indicated earlier, there are two approaches to training an ANN. In supervised training, the ANN is provided with specific patterns and the appropriate answer. Note that depending on the ANN topology, each pattern may

TECHNIQUES FOR HUMAN BEHAVIOR MODELING 301

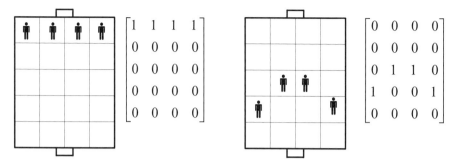

**Figure 9.25** Illustration of mapping environment into ANN input.

require its own ANN for recognition, or a composite ANN can be trained for multiple patterns. In either case, training involves adjusting the weights to ensure that the output of the network matches, as much as possible, the correct answer. A variation of this approach is reinforcement learning, in which the ANN is given guidance as to the degree of correctness of each answer it produces as opposed to a strict comparison with the known correct answers. The overall approach is called supervised because it resembles a teacher supervising a student during learning. In the example shown in Figures 9.24 and 9.25, the matrices reflecting the defensive and offensive stances would be used to train the ANN so that when either formation (or any formation that resembles them) appears, appropriate action can be taken.

Extending the example even further, another possible way to utilize an ANN is by trying to identify formations that lead to goal scoring. For example, by using games recorded a priori, the formation of the opposing team can be sampled at regular intervals along with the outcome of the formation in terms of scoring or failing to score a goal during a short period after the formation. The ANN can then be trained with a training set that consists of observed formations with the correct answer being the goal scoring outcome. Once the ANN has been trained, its output can be used to identify formations that lead to the opposite team scoring a goal, which in turn can be used to initiate specific defensive maneuvers or other action whose aim is to counteract the opposing team.

In unsupervised learning, the ANN is provided with streams of inputs, and the weights are adjusted so that similar patterns yield similar answers. Unlike supervised learning, unsupervised learning does not require a correct answer to be associated with each input pattern in the training set. Instead, the ANN is sequentially exposed to all inputs in the training set and the weights are adjusted in an iterative manner so that the output will be correlated highly with similar patterns in the training set. Unsupervised learning is useful because it inherently explores similarities in the structure of the input data and through the association of similar output scores clusters inputs into similar groups, in effect organizing the inputs into categories. Unsupervised learning is often used in data mining applications, whose goal is to make sense of large sequences of data for which no known or obvious pattern exists. Unsupervised learning provides less utility for pragmatic human behavior modeling because it is difficult to associate actions without a priori knowledge of possible outcomes. An alternative to pure unsupervised learning is to utilize a hybrid approach, where the initial training is done in a supervised manner, but unsupervised learning is used to refine the ANN. The advantage of this approach is that the supervised portion provides the necessary associations between recognized patterns and actions, while the unsupervised learning acts as a self-calibrating mechanism that adapts to slight variations to known patterns that occur after the initial training has taken place.

There are several approaches used for implementing ANN training. The most straightforward approach is to use an error-correcting rule that utilizes the error signal, that is, the difference between the correct answer and the actual output of the ANN, as a means for changing the weights. Possibly the oldest approach for training ANNs was published in Hebb, and it is referred to as the Bebbian rule [35]. It uses firings of neurons (caused by the training set) to perform local changes to the ANN weights. Other approaches include Boltzmann learning, which is derived from thermodynamic principles [36] A summary of various training approaches in tutorial form, along with further examples of ANN usage, is given in Hertz et al. and Haykin [33,34].

**HMMs** An alternative paradigm for pattern recognition as it relates to human behavior modeling is the HMM. Similar to an ANN, an HMM can be used to recognize patterns following a period of successful training from a training set. What is attractive about the HMM, however, is that the model itself often has a structural resemblance to the system being modeled. Also, the HMM can be used as a generator of behavior, not just a recognizer of patterns, and as such is worth some independent coverage.

The formal definition of an HMM involves several constructs. First, an HMM consists of a set $S$ of $N$ states, much like a state machine. The system changes states based on probabilities that dictate the chance of the system transitioning from one state to another. These probabilities are specified in the state transition probability matrix $A$, which is an $N \times N$ matrix whose element $(i, j)$ contains the probability that the state will transition from state $i$ to state $j$. Unlike state machines though, the actual states and transitions of an HMM cannot be directly observed. Instead, we can only observe symbols emitted by the system while at a given state. For a given HMM, there is a set $V$ containing $M$ observable symbols; typically, this set of symbols is referred to as the model alphabet. The symbols emitted by each state is governed by the observation symbol probability distribution function, defined as a set $B$ containing the probability that a given symbol will be emitted by the system while on a given state. The specification of an HMM becomes complete after providing the initial state of an HMM, symbolized by $\pi$. In summary, an HMM is specified by the following elements:

$$S = \{S_1, S_2, \ldots, S_N\};$$
$$V = \{v_1, v_2, \ldots, v_M\};$$
$$A = \{a_{ij}\}, a_{ij} = P[q_{t+1} = S_j | q_t = S_i];$$
$$B = \{b_j(k)\}, b_j(k) = P[v_k @ t | q_t = S_j];$$
$$\pi = \{\pi_i\}, \pi_i = P[q_1 = S_i].$$

In the above, $q_t$ represents the state of the system at time $t$. Given an HMM, we observe a series of symbols emitted by the system being modeled; this sequence is represented by the set O: $O = \{O_1, O_2, \ldots, O_T\}$, where $T$ is the number of observations. This notation is established in Rabiner, which also provides an excellent tutorial on HMMs along with the core algorithms associated with their use [37].

It is interesting to note the resemblance of an HMM to a human performing a specific mechanical task. We know that a person's actions are driven by what is in his or her mind loosely operating according to a sequence that can be encoded in a state machine. At the same time, we cannot directly observe the person's mind; this is reflected by the hidden portion of the model. Instead, we can only observe the person's actions, a fact which is reflected in the model by the emitted symbols. Finally, the probabilistic nature of the state transitions and emitted symbols reflects the inherent variation of human task performance and provides a natural mechanism for specifying varying levels of accuracy while driven by a single motivation.

HMMs have been used for a wide range of applications that benefit from their similarity to human actions. In speech recognition, HMMs are used to detect individual words by detecting sequences of phonemes that together form words; by training the model, it is possible to obtain recognition that is independent of accents or other phonetic variations [38]. Similar approaches are used in various other pattern recognition fields, such as character recognition and gesture recognition [39,40]. When it comes to human behavior modeling, HMMs are a useful tool for recognizing high-level patterns, much like ANNs, but also for generating realistic behavior. Before explaining the specific approach, let us summarize the basic issues associated with an HMM formulation. As described in Rabiner, there are three basic procedures that need to be addressed while working with HMMs [37]. The first is evaluating or recognizing a sequence. The second procedure involves training the HMM so it recognizes a given class of observation sequences. Finally, the last procedure involves producing a sequence of observations that is the best match for the model. The mathematical treatment of each of these procedures is explained in Rabiner and will not be repeated here [37]. However, we will discuss how each of them relates to human behavior modeling.

The evaluation or recognition procedure assumes that a model is given and seeks to compute the probability that the observed sequence was generated by the model. The forward–backward procedure can be used to efficiently compute this probability, providing a recognizer with the same utility as an

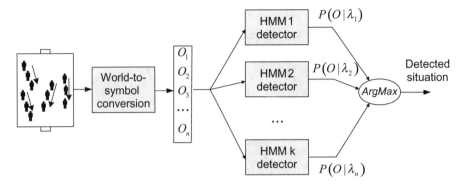

**Figure 9.26** Use of HMMs for situational awareness.

ANN [41]. When using HMMs, one model would be used to detect each required pattern. Figure 9.26 demonstrates how to use multiple HMM detectors, loosely related to the earlier example on team sports formation detection. The first step involves the conversion of world observations in a sequence of symbols consistent with the HMM alphabet. The issues associated with this step are similar to when using ANN and will not be repeated here. The observed sequence is then run through the recognition algorithm for multiple HMMs, each calibrated to recognize a particular series of formations. The resultant probabilities can then be compared and the maximum probability can be used to arbitrate among the outputs, thus yielding the best estimate for the current situation.

One key difference between ANN and HMMs is that HMM utilize sequence of observations versus a single observation used in ANNs. In the aforementioned example, the observation sequence reflects a series of formations that is identified not only by detecting isolated player placement but also by the specific sequence that these formations appear. Conceivably, the series of subformations would reflect sequential states from the opposing team's playbook; consistent with the HMM paradigm, states are not visible directly, instead only observable through the actions of the opposite team.

The conceptual issues associated with the training process are similar to the issues encountered when training ANNs. Given a set of observations, we seek to adjust the model so that it recognizes these observations. In mathematics terms, we seek to adjust the model in order to maximize the probability of the observation sequence, given the model. The classical solution to this problem is to use the *Baum–Welch iterative* procedure, which yields a model that locally maximizes the probability of detection [42]. Here, the term "locally" indicates that as an iterative algorithm, the Baum–Welch approach cannot guarantee that the final model will be globally optimal. In practice, it is necessary to have input observations that provide enough variation but do not contain large distinct clusters of data, as the model may be trained for a specific cluster.

Unlike ANN, when training an HMM we can only provide observation sequences that we wish to recognized; there is no equivalent concept of supervised or unsupervised learning. Once trained, the resultant HMM can be used for recognition as indicated in Figure 9.26.

The last procedure associated with HMMs is the generation of a state sequence based on observations. As in the other problems, straightforward application of the mathematics is computationally expensive, and alternative algorithms have been developed to produce equivalent results. In this case, the *Viterbi algorithm* is used to generate a sequence of state transitions that best match an observation [43,44]. Stated simply, this procedure seeks to unhide the model operation based on an observation. Assuming the structure of the model is known, and given an observation, this procedure will provide a series of state transitions that maximize the probability of generating the same observation sequence as the one observed. Possible applications of this technique include the reverse engineering of observed behavior into a probabilistic sequential model.

The ability of HMMs to adjust based on observations can be used in ways beyond straightforward recognition of patterns. In fact, the trainability of HMMs can be leveraged for purposes of encoding knowledge and creating behavioral models that have an inherent ability of learning from experience. For example, the work described in Shatkay and Kaelbling extends the basic HMM model with odometry information as a means to facilitate an autonomous robot learn about its environment [45]. In this case, each state loosely represents the state of the robot near a position in the map, and transitions represent traveling between these configurations. The HMM model is augmented by an odometric relation matrix that represents the spatial relationship between such states. In this context, observations reflect the movement of a robot; the probabilistic nature of observations compensates for the noisy measurements provided by sensors. By extending the HMM model with domain-specific information, this example demonstrates how the ability of an HMM to learn based on observations in this case can be leveraged for environment discovery and mapping.

## HUMAN FACTORS

In recent years, researchers have begun to use several computational modeling techniques to describe, explain, predict, and demonstrate human behavior and performance, particularly in environments embodying time pressure, stress and close coupling between humans and machines. Such attempts can be classified loosely into the field of *human factors*. Modeling efforts in this field are usually based on data collected in traditional human factors behavioral experiments in laboratory-based simulators and in the actual task environments. These models aim to extend the value of empirical results by making engineering predictions and attempting to explain phenomena related to basic human performance variables in contexts ranging from human–computer interaction (HCI), workplace design, driving, health care, military and defense, and aviation. One example of such modeling techniques are discrete-event simulation packages such as Arena, ProModel, and Microsaint that are used to generate stochastic predictions of human performance unfolding in real time. The ultimate goal is the integration of cognitive modeling techniques with physical and mathematical modeling techniques and task analysis to generate comprehensive models of human performance in the real world.

Research in the newly emerged field of *augmented cognition* (AugCog) has demonstrated great potential to develop more intelligent computational systems capable of monitoring and adapting the systems to the changing cognitive state of human operators in order to minimize cognitive bottlenecks and improve task performance. Limitations in human cognition are due to intrinsic restrictions in the number of mental tasks that a person can execute at one time, and this capacity itself may fluctuate from moment to moment depending on a host of factors including mental fatigue, novelty, boredom, and stress. As computational interfaces have become more prevalent in society and increasingly complex with regard to the volume and type of information presented, it is important to investigate novel ways to detect these bottlenecks and devise strategies to understand and improve human performance by effectively accommodating capabilities and limitations in human information processing and decision making.

The primary goal of incorporating modeling and simulation techniques in studying human behavior, therefore, is to research and develop technologies capable of extending, by an order of magnitude or more, the information management capacity of individuals working with modern day computing technologies. This includes the study of methods for addressing cognitive bottlenecks (e.g., limitations in attention, memory, learning, comprehension, visualization abilities, and decision making) via technologies that assess the user's cognitive status in real time. A computational interaction employing such novel system concepts monitors the state of the user, through behavioral, psychophysiological, and/or neurophysiological data acquired from the user in real time, and then adapts or augments the computational interface to significantly improve its performance on the task at hand based on a series of complex computations.

At the most general level, human behavior modeling has the explicit goal of utilizing methods and designs that harness computation and explicit knowledge about human limitations to open bottlenecks and address the biases and deficits in human cognition. This can be done through continual background sensing, learning, and inferences to understand trends, patterns, and situations relevant to a user's context and goals. At its most basic level, at least four variables are critical—a method to determine and quantify current user state, a technique to evaluate change of state in cognitive terms, a mechanism to translate the change in cognition into computational terms, and an underlying computational architecture to integrate these components. In order to understand better the cognitive variables that drive the modeling process, it is first important to begin with a clear understanding of the basic process of human information processing.

## Model of Human Information Processing

Human information processing theory addresses how people receive, store, integrate, retrieve, and use information. There are a few basic principles of information processing that most cognitive psychologists agree with:

(1) The mental system has limited capacities; that is, bottlenecks in the flow and processing of information occur at very specific points.
(2) A control mechanism is required to oversee the encoding, transformation, processing, storage, retrieval, and utilization of information. This control mechanism requires processing power and varies as a function of the difficulty of the task.
(3) Sensory input is combined with information stored in memory in order to construct meaning. Stimuli that have been associated with some meaning lead to more detailed processing eventually leading to action execution.
(4) Information flow is typically two way—environmental stimuli influence information processing; processed information in turn leads to action execution, which alters environmental stimuli.

A comprehensive model of human information processing recently proposed provides a useful framework for analyzing the different psychological processes used in interacting with systems as described below [46].

*Stage 1: Sensory Processing.* At this stage, environmental events gain initial access to the brain. Any incoming information is temporarily prolonged and stored in the short-term sensory store (STSS). The STSS typically prolongs the experience of a stimulus for durations as short as 0.5 s (visual stimuli) to as long as 2–4 s (auditory stimuli). If the human is distracted during this brief period, the information is permanently lost from the system.

*Stage 2: Perception.* Raw sensory relayed from the STSS must next be interpreted, or given meaning, through the stage of perception. At this stage, information processing proceeds automatically and rapidly, and requires minimal attention. Also, perception is driven both by raw sensory input (called "bottom-up" processing) as well as inputs and expectations derived from long-term memory (called "top-down" processing). The speed and relative automaticity of perception is what distinguishes it from the next stage of "cognition," which is discussed below.

*Stage 3: Cognition and Memory.* Cognitive operations generally require more time, mental effort, and attention. This is because cognition comprises reasoning, thinking, rehearsal, and several mental transformations, all of which are carried out in a temporary memory store known as working memory. Processes in working memory are highly vulnerable to disruption when attention is diverted to other activities. If information is processed uninterrupted at this stage, it is permanently stored in long-term memory and drives response selection and action execution.

*Stage 4: Response Selection and Execution.* The understanding of the situation, achieved through perception and augmented by cognitive transformations, often triggers an action transformation—the selection of a response. It is important to note that the mere selection of a response depicts the stage of decision making. Actual execution of the response requires coordination of muscles for controlled movement to assure that the goal is correctly obtained.

Across all stages of perception to cognition to response selection and execution, the presence of *continuous feedback* and *attention* is an important determinant of accuracy of information processing. In the section below, we will now discuss how these human information processing variable can be translated into improvements in human interaction with technology.

## HUMAN–COMPUTER INTERACTION

An increasing portion of computer use is dedicated to supporting intellectual and creative enterprise. The media is so filled with stories about computers that raising public consciousness of these tools seems unnecessary. However, humans in large part are still uncomfortable interacting with computers. In addition to perceived system complexities, humans are subject to the underlying fear of making mistakes, or might even feel threatened by a computer that is "smarter than" the user. These negative perceptions are generated, in part, by poor designs, hostile and vague error messages, unfamiliar sequences of actions, or deceptive anthropomorphic style [47].

Human–computer interaction (HCI) is the detailed study of the interaction between people (users) and computers [48,49]. It is often regarded as the intersection of computer science, behavioral sciences, design, and several other fields of study such a communication, business, and language. Interaction between users and computers occurs at the user interface, which includes both software and hardware. Because HCI studies a human and a machine in conjunction, it draws from supporting knowledge on both the machine and the human side. On the machine side, techniques in computer graphics, operating systems, programming languages, and development environments are relevant. On the human side, communication theory, graphic and industrial design disciplines, linguistics, cognitive psychology, and human factors are relevant.

HCI differs from human factors, a closely related field, in that in the former there is more of a focus on users working with *computers* rather than other kinds of machines or designed artifacts, and an additional focus on how to implement the (software and hardware) mechanisms behind computers to support HCI. HCI also differs with ergonomics in that in HCI there is less of a focus on repetitive work-oriented tasks and procedures, and much less emphasis on physical stress and the physical form or industrial design of physical aspects of the user interface, such as the physical form of keyboards and mice.

The ultimate goal of HCI is to improve the interaction between users and computers by making computers more usable and receptive to the user's needs. Specifically, HCI is concerned with:

(1) Methodologies and processes for designing interfaces (i.e., given a task and a class of users, design the best possible interface within given constraints [47], optimizing for a desired property such as learning ability or efficiency of use)
(2) Methods for implementing interfaces (e.g., software toolkits, efficient algorithms)
(3) Techniques for evaluating and comparing interfaces
(4) Developing new interfaces and interaction techniques
(5) Developing descriptive and predictive models and theories of interaction

Two areas of study, however, have substantial overlap with HCI even as the focus of inquiry shifts [47]. In *computer-supported cooperative work* (CSCW), emphasis is placed on the use of computing systems in support of the collaborative work of a group of people. In the study of *personal information management* (PIM), human interactions with the computer are placed in a larger informational context. People may work with many forms of information, some computer-based, others not (e.g., whiteboards, notebooks, sticky notes, refrigerator magnets) in order to understand and effect desired changes in their world. In the next section, we examine different methods for demystifying HCI and the reasons for its inception.

## User Interface

In the 1950s, the approach to system design was primarily one of comparing human and machine attributes. Function allocation was accorded based on the strengths of each [50]. For example, humans are better at complex decision-making tasks that involve using information from past experience; on the other hand, machines are better at performing large tasks of a monotonous nature such as multiplying six-digit numbers! This allocation of function that attempts to capitalize on the strengths of each entity is known as "compensatory" principle. Fitts' list that delineated strengths of each (humans vs. machines) was initially known as men are best at-machines are best at (MABA-MABA). In the modern day, this has been changed to the more politically correct humans are best at-machines are best at (HABA-MABA) [51].

***Human Strengths*** Based on the HABA-MABA model, human strengths have been identified as:

(1) Speed—Humans can work fast for short bursts when feeling energetic. However, human reaction times are typically much slower than machines.
(2) Accuracy—Humans are capable of accurate performance when attention to detail is a requirement. However, it is difficult to sustain high levels of accuracy for long periods of time.
(3) Repetitive actions—Humans are better than machines at applying past experience and learning when a task is to be performed repeatedly; however, humans are easily prone to boredom and fatigue.
(4) Perceptual skills—Humans possess excellent perceptual skills, especially applicable to tasks that require pattern detection and interpretation.
(5) Memory—Human memory is very situation dependent. Information that is meaningful to the individual is remembered and recalled with alarming clarity; however, human memory is often selective and not always reliable.
(6) Decision making—Humans are particularly good at using rules of thumb or "heuristics" under time pressure and resource deprivation; additionally, humans are very good at extrapolating from past experience.
(7) Intelligence—It is no secret that human intelligence has been identified as superior to any other species.

***Computer Strengths*** Based on the HABA-MABA model, machine strengths have been identified as:

(1) Speed—Machines can work fast for long periods of time. Machine reaction times are typically much faster than humans.
(2) Accuracy—Machines are capable of maintaining high standards of accuracy when programmed accurately. However, they are prone to catastrophic breakdowns in accuracy when faced with new situations or design glitches.
(3) Repetitive actions—Machines are excellent at repeating actions in a consistent manner.
(4) Memory—Machine memory is very reliable in that all items stored in memory can be accessed quickly; machine memory is limited but typically has large capacity.
(5) Decision making—Machines typically function on a rule-based approach; they are good at deductive reasoning, but perform poorly in new situations.
(6) Intelligence—Machine intelligence is limited to what it is hardwired to do by the designer.

***Desired Interaction between Humans and Computers*** Although the above list provides very clear distinctions between humans and computers, the distinctions are oversimplified and somewhat sterile. Generally, machines evolve much faster than suggested by the list above; at the same time, human performance can be progressively improved to overcome the aforementioned shortcomings via training and experience. Furthermore, the list above is an oversimplification in that although some tasks are more efficiently performed by a human, there are often compelling reasons for using a machine to perform them (e.g., moving hazardous material). On the other hand, humans are unique in that they have an affective (emotional, motivational) component in most of their activities, which may well provide an additional reason for allocating work to humans versus machine. In summary, allocation of functions based purely on the strengths and weaknesses of humans and machines is a very limited approach, which rarely works on its own [52]. It is more important to create a degree of synergistic activity or collaboration between systems and humans such that the two can work as "teammates."

"Computer consciousness" in the general public is a possible solution to creating the abovementioned synergy between humans and computers. Exploratory applications of HCI where such consciousness can be created include World Wide Web browsing, search engines, and different types of scientific simulation [47,48]. Creative environments include writing workbenches, artist or programmer workstations, and architectural design systems. *Collaborative interfaces* enable two or more people to work together, even if the users are separated by time and space, through the use of electronic text, voice, and video mail; through electronic meeting systems that facilitate face-to-face meetings; or through groupware that enables remote collaborators to work concurrently on a document, map, spreadsheet, or image. In these tasks, users may be knowledgeable in the work domain but novices in underlying computer concepts. One of the most effective solutions to humans' lack of knowledge has been the creation of "point-and-click" direct manipulation (DM) representations of the world of action, supplemented by keyboard shortcuts. In the next section, we discuss two popular metaphors for implementing effective HCI—DM and dialog boxes.

## Metaphors and Designing for HCI

DM is a style of human machine interaction design, which features a natural representation of task objects and actions promoting the notion of people performing a task themselves (directly) not through an intermediary like a computer. DM is based on the *principle of virtuality*—a representation of reality that can be physically manipulated. A similar concept was conveyed by the *principle of transparency*—the user should be able to apply intellect directly to the task; that is, the user should feel involved directly with the objects to be manipulated rather than communicate with an intermediary [53]. DM essentially involves continuous representation of objects of interest, and rapid, reversible, incremental actions and feedback. The intention is to allow a user to directly manipulate objects presented to them, using actions that correspond at least loosely to the physical world. Having real-world metaphors for objects and actions make it easier for a user to learn and use an interface (thereby making the interface more natural or intuitive). Rapid, incremental feedback allows a user to make fewer errors and complete tasks in less time, because they can see the results of an action before completing the action. An example of DM is resizing a graphic shape, such as a rectangle, by dragging its corners or edges with a mouse [48].

DM is a topic to which computer science, psychology, linguistics, graphic design, and art all contribute substantially. The computer science foundation areas of computer architecture and operating systems provide us with an understanding of the machines upon which human–computer interface styles such as DM are implemented. This understanding allows us to determine capabilities and limitations of computer platforms, thereby providing boundaries for realistic HCI designs. As with any computer application development, DM interface development benefits greatly from the computer science foundation areas of algorithms and programming languages. The specialized field of computer graphics can also make a key contribution.

A DM interface possesses several key characteristics. As mentioned earlier, a visual representation of objects and actions is presented to a person in contrast to traditional command line languages. The visual representation usually takes the form of a metaphor related to the actual task being performed. For example, computer files and directories are represented as documents and file cabinets in a desktop publishing system. The module used to permanently delete documents from a computer (the "recycle bin") physically resembles an actual trash dumpster into which "waste" materials can be dragged and dropped. The use of metaphors allows a person to tap their analogical reasoning power when determining what actions to take when executing a task on the computer [54]. For example, this property is drawn upon heavily by desktop metaphors in their handling of windows like sheets of paper on a desk. This enhances the impression that the person is performing the task and is in control at all times.

Given a thoughtful design and strong implementation, an interactive system employing DM principles can realize many benefits. Psychology literature cites the strengths of visual representations in terms of learning speed and retention. DM harnesses these strengths resulting in systems whose operation is easy to learn and use and difficult to forget [55]. Because complex syntax does not have to be remembered and analogical reasoning can be used, fewer errors are made than in traditional interfaces. In situations where errors are unavoidable, they are easily corrected through reversible actions. Reversible actions also foster exploration because the fear of permanently damaging something has been diminished. This increases user confidence and mastery and improves the overall quality of HCI.

***Dialog Box Manipulation*** In graphic user interfaces, a *dialog box* is a special window used in user interfaces to display information to the user, or to get a response if needed [56]. They are so called because they form a dialog between the computer and the user—either informing the user of something or requesting input from the user, or both. It provides controls that allow the user to specify how to carry out an action. Dialog boxes are classified as modal or modeless, depending on whether they block interaction on the application that triggered the dialog or not. Different types of dialog boxes are used for different sorts of user interaction. The simplest type of dialog box is the alert that displays a message and requires only an acknowledgment (by clicking "OK") that the message has been read. Alerts are used to provide simple confirmation of an action, or include program termination notices or intentional closing by the user.

When used appropriately, dialog boxes are a great way to give power and flexibility to the program. When misused, dialog boxes are an easy way to annoy users, interrupt their flow, and make the program feel indirect and tedious to use. Dialog boxes typically demand attention, especially when they are modal; furthermore, since they pop out unexpectedly on some portion of the screen, there is a danger that they will obscure relevant information. Therefore, they are most effective when design characteristics match usage requirements.

A dialog box's design is largely determined by its purpose (to offer options, ask questions, provide information or feedback), type (modal or modeless), and user interaction (required, optional response, or acknowledgment). Its usage is largely determined by the context (user or program initiated), probability of user action, and frequency of display [57].

*Characteristics of "Effective" Dialog Boxes* Modal dialog boxes that are typically used to signal critical system events should have the following characteristics:

(1) Displayed in a window that is separate from the user's current activity.
(2) Require interaction—can break the user's flow such that users must close the dialog box before continuing work on the primary task.
(3) Use a delayed commit model—changes are not made until explicitly committed.
(4) Have command buttons that commit to a specific action (e.g., shutdown program, report error).

Modeless dialog boxes that are used to signal repetitive and less critical incidents should have these characteristics:

(1) Displayed in context using a task pane or with an independent window.

(2) Do not require interaction—users can switch between the dialog box and the primary window as desired and can delay response.
(3) Can use an immediate commit model—changes are made immediately.
(4) Have standard command buttons (e.g., OK, yes, close) that close the window.

Overall, the primary criterion for dialog box design must ensure that they are distinct enough for users to distinguish them from the background. However, at the same time, they should be minimally disruptive visually. The key among these requirements is that they should be accompanied by a meaningful title, be consistent in style and layout (margins, font, white space, justification), and must be controlled entirely by DM (or point-and-click mechanisms).

**Documenting Interface Design** *Storyboards* are graphic organizers such as a series of illustrations or images displayed in sequence for the purpose of previsualizing a motion picture, animation, motion graphic, or interactive media sequence, including Web site interactivity. Although traditionally used in the motion picture industry, the term *storyboard* has more recently come to be used in the fields of Web development, software development, and instructional design to present and describe user interfaces and electronic pages in written format as well as via audio and motion [58].

Storyboards are typically created in a multiple step process. They can be created by hand drawing or digitally on the computer. If drawing by hand, the first step is to create or download a storyboard template. These appear like a blank comic strip, with space for comments and dialog. Then a thumbnail storyboard is sketched. "Thumbnails" are rough sketches, not much bigger than a thumbnail. This is followed by the actual storyboarding process wherein more detailed and elaborate storyboard images are created.

The actual storyboard can be created by professional storyboard artists on paper or digitally by using 2D storyboarding programs. Some software applications even supply a stable of storyboard-specific images making it possible to quickly create shots that express the designer's intent for the story. These boards tend to contain more detailed information than thumbnail storyboards and convey components of mood, emotions, personality, and other psychological variables [59]. If needed, 3D storyboards can be created (called "technical previsualizations"). The advantage of 3D storyboards is that they show exactly what the final product will look like. The disadvantage of 3D is the amount of time it takes to build and construct the storyboard.

The primary advantage of using storyboards to HCI is that it allows the user to experiment with changes in the storyline to evoke stronger reaction or interest. Flashbacks, for instance, are often the result of sorting storyboards out of chronological order to help build suspense and interest. The process of visual thinking and planning allows a group of people to brainstorm together, placing their ideas on storyboards and then arranging the storyboards on the wall [59]. This fosters more ideas and generates consensus inside the group leading to more effective HCI.

*Prototypes* are common. During the process of design, there is great uncertainty as to whether a new design will actually do what is desired to do. A prototype is often used as part of the product design process to allow engineers and designers the ability to explore design alternatives, test theories, and confirm performance prior to starting production of a new product. The creation of prototypes is based on the counterintuitive philosophy that the easiest way to build something is to first build something else! In general, an iterative series of prototypes will be designed, constructed, and tested as the final design emerges and is prepared for production [60]. With rare exceptions, multiple iterations of prototypes are used to progressively refine the design. A common strategy is to design, test, evaluate, and then modify the design based on the analysis of the prototype.

In many products, it is common to assign the prototype iterations in Greek letters. For example, a first iteration prototype may be called an "Alpha" prototype. Often, this iteration is not expected to perform as intended and some amount of failures or issues are anticipated. Subsequent prototyping iterations (Beta, Gamma, etc.) will be expected to resolve issues and perform closer to the final production intent.

In many product development organizations, prototyping specialists are employed—individuals with specialized skills and training in general fabrication techniques that can help bridge between theoretical designs and the fabrication of prototypes. There is no general agreement on what constitutes a "good prototype," and the word is often used interchangeably with the word "model," which can cause confusion. In general, "prototypes" fall into four basic categories:

(1) *Proof-of-principle prototypes* (also called "breadboards") are used to test some aspect of the intended design without attempting to exactly simulate the visual appearance, choice of materials, or intended manufacturing process. Such prototypes can be used to demonstrate "behavioral" characteristics of a particular design such as range of motion, mechanics, sensors, architecture, and so on. These types of prototypes are generally used to identify which design options will not work, or where further development and testing are necessary.

(2) *Form study prototypes* allow designers to explore the basic size, look, and feel of a product without simulating the actual function, behavior, or exact physical appearance of the product. They can help assess ergonomic factors and provide insight into gross visual aspects of the product's final form. Form study prototypes are often hand-carved or machined models from easily sculpted, inexpensive materials (e.g., urethane foam), without representing the intended color, finish, or texture. Due to the materials used, these models are intended for internal

decision making and are generally not durable enough or suitable for use by representative users or consumers.

(3) *Visual prototypes* capture the intended design aesthetic and simulate the exact appearance, color, and surface textures of the intended product but will not actually embody the functions or "behaviors" of the final product. These prototypes are suitable for use in market research, executive reviews and approval, packaging mock-ups, and photo shoots for sales literature.

(4) *Functional prototypes* (also called "working prototypes") are, to the greatest extent practical, attempt to simulate the final design, aesthetics, materials, and functionality of the intended design all in one package. The functional prototype may be reduced in size (scaled down) in order to reduce costs. The construction of a fully working full-scale prototype is the engineer's final check for design flaws and allows last minute improvements to be made before larger production runs are ordered.

***Characteristics and Limitations of Prototypes*** Engineers and prototyping specialists seek to understand the limitations of prototypes to exactly simulate the characteristics of their intended design. First, a degree of skill and experience is necessary to effectively use prototyping as a design verification tool [61]. Second, it is important to realize that by their very definition, prototypes will represent some compromise from the final production design. Due to differences in materials, processes, and design fidelity, it is possible that a prototype may fail to perform acceptably, whereas the production design may have been sound. A counterintuitive idea is that prototypes may actually perform acceptably, whereas the final production design may be flawed since prototyping materials and processes may occasionally outperform their production counterparts.

Although iterative prototype testing can possibly reduce the risk that the final design may not perform acceptably, prototypes generally cannot eliminate all risk. There are pragmatic and practical limitations to the ability of a prototype to match the intended final performance of the product and some allowances and engineering judgments are often required before moving forward with a production design. Building the full design is often expensive and can be time-consuming, especially when repeated several times. As an alternative, "rapid prototyping" or "rapid application development" techniques are used for the initial prototypes, which implement part, but not all, of the complete design. This allows designers and manufacturers to rapidly and inexpensively test the parts of the design that are most likely to have problems, solve those problems, and then build the full design.

***Unified Modeling Diagrams*** Unified Modeling Language (UML) is a standardized general-purpose modeling language in the field of software engineering. UML includes a set of graphic notation techniques to create abstract models of specific systems. UML is an open method used to specify, visualize, construct, and document the artifacts of an object-oriented software-intensive system under development. UML offers a standard way to write a system's blueprints including conceptual components such as business processes. In addition, UML can also help document concrete variables such as programming language statements, database schemas, and reusable software components. UML combines the best practice from data modeling concepts and component modeling. It can be used with all processes, throughout the software development cycle, and across different implementation technologies. UML is extensible and models may be automatically transformed to other representations (e.g., Java) by means of transformation languages.

Typically UML diagrams represent two different views of a system model [62]:

(1) *Static (or structural) view* emphasizes the static structure of the system using objects, attributes, operations, and relationships. Structure diagrams emphasize what components must be incorporated in the system being modeled. Since structure diagrams represent the structure of a system, they are used extensively in documenting the architecture of software systems:
  - *Class diagram* describes the structure of a system by showing the system's classes, their attributes, and the relationships among the classes.
  - *Component diagram* depicts how a software system is split into components and shows the dependencies among these components.
  - *Composite structure diagram* describes the internal structure of a class and the collaborations that this structure makes possible.
  - *Deployment diagram* serves to model the hardware used in system implementations and the execution environments and artifacts deployed on the hardware.
  - *Object diagram* shows a complete or partial view of the structure of a modeled system at a specific time.
  - *Package diagram* depicts how a system is split up into logical groupings by showing the dependencies among these groupings.

(2) *Dynamic (or behavioral) view* emphasizes the dynamic behavior of the system by showing collaborations among objects and changes to the internal states of objects. Behavior diagrams emphasize what must happen in the system being modeled. Since behavior diagrams illustrate the behavior of system, they are used extensively to describe the functionality of software systems:
  - *Activity diagram* represents the business and operational step-by-step workflows of components in a system. An activity diagram shows the overall flow of control.

- *State machine diagram* is a standardized notation to describe many systems, from computer programs to business processes.
- *Use case diagram* shows the functionality provided by a system in terms of actors, their goals represented as use cases, and any dependencies among those use cases.

Although UML is a widely recognized and used modeling standard, it is frequently criticized for being gratuitously large and complex. In addition, weak visualizations that employ line styles that are graphically very similar frequently disrupt learning, especially when required of engineers lacking the prerequisite skills. As with any notational system, UML is able to represent some systems more concisely or efficiently than others. Thus, a developer gravitates toward solutions that reside at the intersection of the capabilities of UML and the implementation language. This problem is particularly pronounced if the implementation language does not adhere to orthodox object-oriented doctrine, as the intersection set between UML and implementation language may be that much smaller. Lastly, UML has been proven to be aesthetically inconsistent due to the arbitrary mixing of abstract notation (2D ovals, boxes, etc.) that make UML appear jarring and visually disruptive to the user.

**Web Design** A Web site is a collection of information about a particular topic or subject. Designing a Web site is defined as the arrangement and creation of Web pages, each page presenting some information of relevance to the primary theme of the Web site. There are many aspects (primarily design concerns) in the process of Web design, and due to the rapid development of the Internet new aspects constantly emerge [63]. For typical commercial Web sites, the basic aspects of design are the following:

- *Content*: Information on the site should be relevant to the site and should target the area of the public that the Web site is concerned with.
- *Usability*: The site should be user-friendly, with the interface and navigation simple and reliable.
- *Appearance*: The graphics and text should include a single style that flows throughout, to show consistency. The style should be professional, appealing, and relevant.
- *Visibility*: The site must be easy to find by typing in common keywords via most major search engines and advertisement media.

Web site design crosses multiple disciplines of information systems, HCI, and communication design. Typically, the observable content (e.g., page layout, graphics, text, audio, etc.) is known as the "front end" of the Web site. The "back end" comprises the source code, invisible scripted functions, and the server-side components that process the output from the front end.

Depending on the size of a Web development project, it may be carried out by a multiskilled individual or may represent collaborative efforts of several individuals with specialized skills. As in any collaborative design, there are conflicts between differing goals and methods of Web site designs. Some of the ongoing challenges in Web design are discussed below.

*Form versus Function*  Frequently, Web designers may pay more attention to how a page "looks" while neglecting other important functions such as the readability of text, the ease of navigating the site, or ease of locating the site [64]. As a result, the user is often bombarded with decorative graphics at the expense of keyword-rich text and relevant text links. Assuming a false dichotomy that form and function are mutually exclusive overlooks the possibility of integrating multiple disciplines for a collaborative and synergistic solution. In many cases, form follows function. Because some graphics serve communication purposes in addition to aesthetics, how well a site works may depend on the graphic design ideas in addition to the professional writing considerations [64].

To be optimally accessible, Web pages and sites must conform to certain accessibility principles. Some of the key accessibility principles most relevant to HCI can be grouped into the following main points:

(1) Use semantic markup that provides a meaningful structure to the document (i.e., Web page). Semantic markup also refers to semantically organizing the Web page structure and publishing Web services description accordingly so that they can be recognized by other Web services on different Web pages. Standards for semantic Web are set by the Institute of Electrical and Electronics Engineers (IEEE).

(2) Provide text equivalents for any nontext components (e.g., images, multimedia).

(3) Use hyperlinks that make sense when read out of context (e.g., avoid "Click Here"; instead, have the user click on a meaningful word that is relevant to the context).

(4) Author the page so that when the source code is read line-by-line by user agents (such as screen readers), it remains intelligible.

*Liquid versus Fixed Layouts*  On the Web, the designer generally has little to no control over several factors, including the size of the browser window, the Web browser used, the input devices used (mouse, touch screen, keyboard, number pad, etc.), and the size, design, and other characteristics of the fonts users have available (installed) on their own computers [65]. Some designers choose to control the appearance of the elements on the screen by using specific width designations. When the text, images, and layout do not vary among browsers, this is referred to as *fixed-width design*. Advocates of fixed-width design argue for the designers' precise control over the layout of a site and the placement of objects within pages.

Other designers choose a more liquid approach, wherein content can be arranged flexibly on users' screens, responding to the size of their browsers' windows. Proponents of *liquid design* prefer greater compatibility with users' choice of presentation and more efficient use of the screen space available. Liquid design can be achieved by setting the width of text blocks and page modules to a percentage of the page, or by avoiding specifying the width for these elements altogether, allowing them to expand or contract naturally in accordance with the width of the browser. It is more in keeping with the original concept of HTML, that it should specify, not the appearance of text, but its contextual function, leaving the rendition to be decided by users' various display devices.

Web page designers (of both types) must consider how their pages will appear on various screen resolutions [66]. Sometimes, the most pragmatic choice is to allow text width to vary between minimum and maximum values. This allows designers to avoid considering rare users' equipment while still taking advantage of available screen space.

*Adobe Flash* (formerly Macromedia Flash) is a proprietary, robust graphics animation or application development program used to create and deliver dynamic content, media (such as sound and video), and interactive applications over the Web via the browser. Many graphic artists use Flash because it gives them exact control over every part of the design, and anything can be animated as required. Flash can use embedded fonts instead of the standard fonts installed on most computers.

Criticisms of Flash include the use of confusing and nonstandard user interfaces, the inability to scale according to the size of the Web browser, and its incompatibility with common browser features (such as the back button). An additional criticism is that the vast majority of Flash Web sites are not accessible to users with disabilities (for vision-impaired users). A possible HCI-relevant solution is to specify alternate content to be displayed for browsers that do not support Flash. Using alternate content will help search engines to understand the page and can result in much better visibility for the page.

## CONCLUSION

Based on human behavior taxonomy, several levels of human behavior were identified, and a variety of tools and techniques supporting modeling of this behavior were presented in this section. These techniques include fuzzy logic, for low-level humanlike control of external processes, and state machines, RBSs, and pattern recognition methods for higher-level behaviors. It is important to note that these tools and techniques are not meant to work in isolation; in fact, comprehensive solutions for behavior modeling often utilize multiple techniques at the same time. As an extreme example that utilizes all of the aforementioned approaches, consider the simulation of an intelligent agent that involves movement within a vehicle, maybe a car or boat. Any of the state machine paradigms can be used to contextualize the behavior of the agent according to some goal. Pattern recognition techniques can be used to assess the state of the outside environment, which, combined with an RBS, can produce decisions that are in turn forwarded to a low-level controller that utilizes fuzzy logic to control a realistic simulation of the vehicle. At the end, it is important to be aware of the relative merits of each approach and utilize each in a way that maximizes its usefulness.

Several computational modeling techniques are engaged to describe, explain, predict, and characterize human behavior—this effort finds its place in the field of human factors. A number of modeling methods are used and with that comes a variety of human information processing. Numerous software programs exist to enhance and document interface and Web design.

## REFERENCES

[1] Skinner BF. *Science and Human Behavior*. New York: B.F. Skinner Foundation; 1965.

[2] Nina KV. Measurement of human muscle fatigue. *Journal of Neuroscience Methods*, 74:219–227; 1997.

[3] Posner MI, Petersen SE. The attention system of the human brain. *Annual Review of Neuroscience*, 13:25–42; 1990.

[4] Gawron V. *Human Performance Measures Handbook*. New York: Lawrence Erlbaum Associates; 2000.

[5] Zadeh LA. Fuzzy sets. *Information & Control*, 8(3):338–353; 1965.

[6] Zadeh LA. The role of fuzzy logic in modeling, identification and control. In *Fuzzy Sets, Fuzzy Logic, and Fuzzy Systems: Selected Papers*, World Scientific Series in Advances in Fuzzy Systems. Zadeh LA, Klir GJ, Yuan B (Eds.). Vol. 6. River Edge, NJ: World Scientific Publishing Co.; 1996, pp. 783–795.

[7] Gabay E, Merhav SJ. Identification of a parametric model of the human operator in closed-loop control tasks. *IEEE Transactions on Systems, Man, and Cybernetics*, SMC-7:284–292; 1997.

[8] Lee D. Neural basis of quasi-rational decision making. *Current Opinion in Neurobiology*, 16:1–8; 2006.

[9] James CA. Irrationality in philosophy and psychology: The moral implications of self-defeating behavior. *Journal of Consciousness Studies*, 5(2):224–234; 1998.

[10] Klein G. *The Power of Intuition*. New York: Bantam Dell; 2004.

[11] Mellers B, Schwartz A, Ritov I. Emotion-based choice. *Journal of Experimental Psychology General*, 128:332–345; 1999.

[12] Coricelli G, Critchley HD, Joffily M, O'Doherty JP, Sirigu A, Dolan RJ. Regret and its avoidance: A neuroimaging study of choice behavior. *Nature Neuroscience*, 8:1255–1262; 2005.

[13] Mehrabian A. Pleasure-arousal-dominance: A general framework for describing and measuring individual differences in temperament. *Current Psychology: Developmental, Learning, Personality, Social*, 14:261–292; 1996.

[14] Preston SH, Heuveline P, Guillot M. *Demography: Measuring and Modeling Population Processes.* New York: Blackwell Publishers; 2001.

[15] Allen LJS. *An Introduction to Stochastic Processes with Applications to Biology.* Upper Saddle River, NJ: Prentice Hall; 2003.

[16] Verhulst PF. Notice sur la loi que la population pursuit dans son accroissement. *Correspondance Mathématique et Physique*, 10:113–121; 1838.

[17] Nagel E. *The Structure of Science: Problems in the Logic of Scientific Explanation.* New York: Harcourt Brace; 1961.

[18] Simon HA. Artificial intelligence: An empirical science. *Artificial Intelligence*, 77:95–127; 1995.

[19] Mamdani EH, Assilian S. An Experiment in linguistic synthesis with a fuzzy logic controller. *International Journal of Man-Machine Studies*, 7:1–13; 1974.

[20] DARPA Urban Challenge Web site. Available at http://www.darpa.mil/grandchallenge/index.asp. Accessed May 10, 2009.

[21] *Journal of Field Robotics*, Special issue on the 2007 DARPA Urban Challenge, I, II, and III, 25(8–10); 2008.

[22] Anderson JR, Michalski RS, Carbonell JG, Mitchell TM (Eds.). *Machine Learning: An Artificial Intelligence Approach.* New York: Morgan Kaufmann; 1983.

[23] Cremer J, Kearney J, Papelis YE. HCSM: A framework for behavior and scenario control in virtual environments. *ACM Transactions on Modeling and Computer Simulation*, 5(3):242–267; 1995.

[24] Gonzalez AJ, Ahlers R. Context-based representation of intelligent behavior in training simulations. *Transactions of the Society for Computer Simulation International*, 15(4):153–166; 1998.

[25] Gonzalez AJ, Stensrud BS, Barrett G. Formalizing context-based reasoning: A modeling paradigm for representing tactical human behavior. *International Journal of Intelligent Systems*, 23(7):822–847; 2008.

[26] Patz BJ, Papelis YE, Pillat R, Stein G, Harper D. A practical approach to robotic design for the DARPA Urban Challenge. *Journal of Field Robotics*, 25(8):528–566; 2008.

[27] Harel D, Pnueli A, Lachover H, Naamad A, Politi M, Sherman R, Shtull-Trauring A, Trakhenbrot M. STATEMATE: A working environment for the development of complex reactive systems. *IEEE Transactions on Software Engineering*, 16(4):403–414; 1990.

[28] Harel D. STATECHARTS: A visual formalism for complex systems. *Science of Computer Programming*, 8:231–274; 1987.

[29] Laird JE, Newell A, Resenbloom PS. Soar: An architecture for general intelligence. *Artificial Intelligence*, 7:289–325; 1991.

[30] University of Michigan Soar Web site. Available at http://sitemaker.umich.edu/soar/home. Accessed May 10, 2009.

[31] University of Southern California Soar Web site. Available at http://www.isi.edu/soar/soar-homepage.html. Accessed May 10, 2009.

[32] Taylor G, Wray R. Behavior design patterns: Engineering human behavior models. *Behavior Representation in Modeling and Simulation.* Alexandria, VA: SISO; 2004.

# REFERENCES

[33] Hertz J, Krogh A, Palmer RG. *Introduction to the Theory of Neural Computation.* Reading, MA: Addison-Wesley; 1991.

[34] Haykin S. *Neural Networks: A Comprehensive Foundation.* New York: Macmillan College Publishing; 1994.

[35] Hebb DO. *The Organization of Behavior.* New York: John Willey & Sons; 1949.

[36] Anderson AA, Rosenfeld E. *Neurocomputing: Foundations of Research.* Cambridge, MA: MIT Press; 1988.

[37] Rabiner LR. A tutorial on hidden Markov models and selected applications in speech recognition. *Proceedings of the IEEE*, 77(2):257–286; 1989.

[38] Jiang H, Li X, Liu C. Large margin hidden Markov models for speech recognition. *IEEE Transaction on Audio, Speech and Language Processing*, 14(5):1587–1595; 2006.

[39] Arica N, Yarman-Vural F. An overview of character recognition focused on off-line handwriting. *IEEE Transaction on Systems, Man, and Cybernetics. Part C: Applications and Reviews*, 31(2):216–233; 2001.

[40] Lee HK, Kim JH. An HMM-based threshold model approach for gesture recognition. *IEEE Transactions on Pattern Analysis and Machine Intelligence*, 21(10):961–973; 1999.

[41] Baum LE, Petrie T. Statistical inference for probabilistic functions of finite state Markov chains. *Annals of Mathematical Statistics*, 37:1554–1563; 1966.

[42] Baum LE, Petrie T, Soules A, Weiss N. A maximization technique occurring in the statistical analysis of probabilistic functions of Markov chains. *Annals of Mathematical Statistics*, 41(1):164–171; 1970.

[43] Viterbi AJ. Error bounds for convolutional codes and an asymptotically optimal decoding algorithm. *IEEE Transactions on Information Theory*, IT-13:260–269; 1967.

[44] Forney GD. The Viterbi algorithm. *Proceedings of the IEEE*, 61:268–278; 1973.

[45] Shatkay H, Kaelbling LP. Learning geometrically-constrained hidden Markov models for robot navigation: Bridging the topological-geometrical gap. *Journal of Artificial Intelligence Research*, 16:167–207; 2002.

[46] Wickens CD, Hollands JG (Eds.). *Engineering Psychology and Human Performance.* Upper Saddle River, NJ: Prentice Hall; 2000.

[47] Shneiderman B. Universal usability: Pushing human-computer interaction research to empower every citizen. *Communications of the ACM*, 43(5):84–91; 2000.

[48] Shneiderman B, Plaisant C (Eds.). *Designing the User Interface: Strategies for Effective Human-Computer Interaction.* College Park, MD: Pearson Education; 2005.

[49] Myers BA. A brief history of human-computer interaction technology. *Interactions*, 5(2):44–54; 1998.

[50] Fitts PM. Engineering psychology and equipment design. In *Handbook of Experimental Psychology.* Stevens SS (Ed.). New York: Wiley; 1951.

[51] Noyes J (Ed.). *Designing for Humans.* New York: Psychology Press; 2001.

[52] Marchionini G, Sibert J. An agenda for human-computer interaction: Science and engineering serving human needs. *SIGCHI Bulletin*, 23(4):17–31; 1991.

[53] Rutlowski C. An introduction to the human applications standard computer interface, part I: Theory and principles. *Byte*, 7(11):291–310; 1982.

[54] Shneiderman B. Direct manipulation: A step beyond programming languages. *IEEE Computer*, 16(8):57–69; 1983.

[55] Irani P, Ware C. Diagramming information structures using 3D perceptual primitives. *ACM Transactions on Computer-Human Interaction*, 10(1):1–19; 2003.

[56] Raskin J (Ed.). *The Humane Interface*. Reading, MA: Addison-Wesley; 2000.

[57] Galitz WO (Ed.). *The Essential Guide to User Interface Design: An Introduction to GUI Design Principles and Techniques*. 2nd ed. New York: John Wiley & Sons; 2002.

[58] Tribus M (Ed.). *Quality First: Selected Papers on Quality and Productivity Improvement*. 4th ed. National Society of Professional Engineers; 1992.

[59] Jones I. Storyboarding: A method for bootstrapping the design of computer-based educational tasks. *Computers & Education*, 51(3):1353–1364; 2008.

[60] Christensen BT, Schunn CD. The role and impact of mental simulation in design. *Applied Cognitive Psychology*, 23(3):327–344; 2009.

[61] Schafroth D, Bouabdullah S, Bermes C, Siegwart R. From the test benches to the first prototype of the muFLY micro helicopter. *Journal of Intelligent & Robotic Systems*, 54:245–260; 2009.

[62] Zimmermann A (Ed.). *Stochastic Discrete Event Systems: Modeling, Evaluation, Applications*. Berlin: Springer Verlag; 2007.

[63] Savoy A, Salvendy G. Foundations of content preparation for the web. *Theoretical Issues in Ergonomics Science*, 9(6):501–521; 2008.

[64] Chevalier A, Fouquereau N, Vanderonckt J. The influence of a knowledge-based system on designers' cognitive activities: A study involving professional web designers. *Behaviour & Information Technology*, 28(1):45–62; 2009.

[65] Chevalier A, Chevalier N. Influence of proficiency level and constraints on viewpoint switching: A study in web design. *Applied Cognitive Psychology*, 23(1):126–137; 2009.

[66] van Schaik P, Ling J. Modeling user experience with web sites: Usability, hedonic value, beauty and goodness. *Interacting with Computers*, 20(3):419–432; 2008.

# 10

# VERIFICATION, VALIDATION, AND ACCREDITATION

Mikel D. Petty

Verification and validation (V&V) are essential prerequisites to the credible and reliable use of a model and its results. As such, they are important aspects of any simulation project and most developers and users of simulations have at least a passing familiarity with the terms. But what are they exactly, and what methods and processes are available to perform them? Similarly, what is accreditation, and how does it relate to V&V? Those questions are addressed in this chapter.* Along the way, three central concepts of verification, validation, and accreditation (VV&A) will be identified used to unify the material.

This chapter is composed of five sections. This first section motivates the need for VV&A and provides definitions necessary to their understanding. The second section places VV&A in the context of simulation projects and discusses overarching issues in the practice of VV&A. The third section

---

*This chapter is an advanced tutorial on V&V. For a more introductory treatment of the same topic, see Petty [1]; much of this source's material is included here, though this treatment is enhanced with clarifications and expanded with additional methods, examples, and case studies. For additional discussion of VV&A issues and an extensive survey of V&V methods, see Balci [2]. For even more detail, a very large amount of information on VV&A, including concept documents, method taxonomies, glossaries, and management guidelines, has been assembled by the U.S. Department of Defense Modeling and Simulation Coordination Office [3].

---

*Modeling and Simulation Fundamentals: Theoretical Underpinnings and Practical Domains*,
Edited by John A. Sokolowski and Catherine M. Banks
Copyright © 2010 John Wiley & Sons, Inc.

categorizes and explains a representative set of specific V&V methods and provides brief examples of their use. The fourth section presents three more detailed case studies illustrating the conduct and consequences of VV&A. The final section identifies a set of VV&A challenges and offers some concluding comments.

## MOTIVATION

In the civil aviation industry in the United States and in other nations, a commercial airline pilot may be qualified to fly a new type of aircraft after training to fly that aircraft type solely in flight simulators (the simulators must meet official standards) [4,5]. Thus, it is entirely possible that the first time a pilot actually flies an aircraft of the type for which he or she has been newly qualified, there will be passengers on board, that is, people who, quite understandably, have a keen personal interest in the qualifications of that pilot. The practice of qualifying a pilot for a new aircraft type after training only with simulation is based on an assumption that seems rather bold: The flight simulator in which the training took place is sufficiently accurate with respect to its recreation of the flight dynamics, performance, and controls of the aircraft type in question that prior practice in an actual aircraft is not necessary.

Simulations are often used in situations that entail taking a rather large risk, personal or financial, on the assumption that the model used in the simulation is accurate. Clearly, the assumption of accuracy is not made solely on the basis of the good intentions of the model's developers. But how can it be established that the assumption is correct, that is, that the model is in fact sufficiently accurate for its use? In a properly conducted simulation project, the accuracy of the simulation, and the model upon which the simulation is based, is assessed and measured via V&V, and the sufficiency of the model's accuracy is certified via accreditation. V&V are processes, performed using methods suited to the model and to an extent appropriate for the application. Accreditation is a decision made based on the results of the V&V processes. The goals of VV&A are to produce a model that is sufficiently accurate to be useful for its intended applications and to give the model credibility with potential users and decision makers [6].

## BACKGROUND DEFINITIONS

Several background definitions are needed to support an effective explanation of VV&A.* These definitions are based on the assumption that there is some

---

*A few of the terms to be defined, such as *model* and *simulation*, are likely to have been defined earlier in this book. They are included here for two reasons: to emphasize those aspects of the definitions that are important to VV&A and to serve those readers who may have occasion to refer to this chapter without reading its predecessors.

real-world system, such as an aircraft, which is to be simulated for some known application, such as flight training.

A *simuland* is the real-world item of interest. It is the object, process, or phenomenon to be simulated. The simuland might be the aircraft in a flight simulator (an object), the assembly of automobiles in a factory assembly line simulation (a process), or underground water flow in a hydrology simulation (a phenomenon). The simuland may be understood to include not only the specific object of interest, but also any other aspects of the real world that affect the object of interest in a significant way. For example, for a flight simulator, the simuland could include not just the aircraft itself but weather phenomena that affect the aircraft's flight. Simulands need not actually exist in the real world; for example, in combat simulation, hypothetical nonexistent weapons systems are often modeled to analyze how a postulated capability would affect battlefield outcomes.*

A *referent* is the body of knowledge that the model developers have about the simuland. The referent may include everything from quantitative formal knowledge, such as engineering equations describing an aircraft engine's thrust at various throttle settings, to qualitative informal knowledge, such as an experienced pilot's intuitive expectation for the feeling of buffet that occurs just before a high-speed stall.

In general terms, a *model* is a representation of something else, for example, a fashion model representing how a garment might look on a prospective customer. In modeling and simulation (M&S), a model is a representation of a simuland.** Models are often developed with their intended application in mind, thereby emphasizing characteristics of the simuland considered important for the application and de-emphasizing or omitting others. Models may be in many forms, and the modeling process often involves developing several different representations of the same simuland or of different aspects of the same simuland. Here, the many different types of models will be broadly grouped into two categories: *conceptual* and *executable*. *Conceptual models* document those aspects of the simuland that are to be represented and those that are to be omitted.*** Information contained in a conceptual model may include the physics of the simuland, the objects and environmental phenomena to be modeled, and representative use cases. A conceptual model will also

---

*In such applications, the term of practice used to describe a hypothetical nonexistent simulands is *notional*.

**Because the referent is by definition everything the modeler knows about the simuland, a model is arguably a representation of the referent, not the simuland. However, this admittedly pedantic distinction is not particularly important here, and model accuracy will be discussed with respect to the simuland, not the referent.

***The term *conceptual model* is used in different ways in the literature. Some define a conceptual model as a specific type of diagram (e.g., UML class diagram) or documentation of a particular aspect of the simuland (e.g., the classes of objects in the environment of the simuland and their interactions), whereas others define it more broadly to encompass any nonexecutable documentation of the aspects of the simuland to be modeled. It is used here in the broad sense, a use of the term also found in Sargent [9].

document assumptions made about the components, interactions, and parameters of the simuland [6]. Several different forms or documentation, or combinations of them, may be used for conceptual models, including mathematical equations, flowcharts, Unified Modeling Language (UML) diagrams [7], data tables, or expository text. The *executable model*, as one might expect, is a model that can be executed.* The primary example of an executable model considered here is a computer program. Execution of the executable model is intended to simulate the simuland as detailed in the conceptual model, so the conceptual model is thereby a design specification for the executable model, and the executable model is an executable implementation of the conceptual model.

*Simulation* is the process of executing a model (an executable model, obviously) over time. Here, "time" may mean simulated time for those models that model the passage of time, such as a real-time flight simulator, or event sequence for those models that do not model the passage of time, such as the Monte Carlo simulation [6,8]. For example, the process of running a flight simulator is simulation. The term may also refer to a single execution of a model, as in "During the last simulation, the pilot was able to land the aircraft successfully."**

The *results* are the output produced by a model during a simulation. The results may be available to the user during the simulation, such as the out-the-window views generated in real time by a flight simulator, or at the end of the simulation, such as the queue length and wait time statistics produced by a discrete-event simulation model of a factory assembly line. Regardless of when they are available and what form they take, a model's results are very important as a key object of validation.

A model's *requirements* specify what must be modeled, and how accurately. When developing a model of a simuland, it is typically not necessary to represent all aspects of the simuland in the model, and for those that are represented, it is typically not necessary to represent all at the same level of detail and degree of accuracy.*** For example, in a combat flight simulator with computer-controlled hostile aircraft, it is generally not required to model whether the enemy pilots are hungry or not, and while ground vehicles may

---

*The executable model may also be referred to as the *operational model*, for example, in Banks et al. [6].

** In practice, the term *simulation* is also often used in a third sense. The term can refer to a large model, perhaps containing multiple models as subcomponents or submodels. For example, a large constructive battlefield model composed of various submodels, including vehicle dynamics, intervisibility, and direct fire, might be referred to as a simulation. In this chapter, this third sense of simulation is avoided, with the term model used regardless of its size and number of component submodels.

***The omission or reduction of detail not considered necessary in a model is referred to as *abstraction*. The term *fidelity* is also often used to refer to a model's accuracy with respect to the represented simuland.

be present in such a simulator (perhaps to serve as targets), their driving movement across the ground surface will likely be modeled with less detail and accuracy than the flight dynamics of the aircraft. With respect to V&V, the requirements specify which aspects of the simuland must be modeled, and for those to be included, how accurate the model must be. The requirements are driven by the intended application.

To illustrate these definitions and to provide a simple example, which will be returned to later, a simple model is introduced. In this example, the simuland is a phenomenon, specifically gravity. The model is intended to represent the height over time of an object freely falling to the earth. A mathematical (and nonexecutable) model is given by the following equation:

$$h(t) = -16t^2 + vt + s,$$

where

$t$ = time elapsed since the initial moment, when the object began falling (seconds);

$v$ = initial velocity of the falling object (feet/second), with positive values indicating upward velocity;

$s$ = initial height of the object (feet);

$-16$ = change in height of the object due to gravity (feet), with the negative value indicating downward movement;

$h(t)$ = height of the object at time $t$ (feet).

This simple model is clearly not fully accurate, even for the relatively straightforward simuland it is intended to represent. Both the slowing effect of air resistance and the reduction of the force of gravity at greater distances from the earth are absent from the model. It also omits the surface of the earth, so applying the model with a value of $t$ greater than the time required for the object to hit the ground will give a nonsensical result.

An executable version of this model of gravity follows, given as code in the Java programming language. This code is assuredly not intended as an example of good software engineering practices, as initial velocity $v$ and starting height $s$ are hard coded. Note that the code does consider the surface of the earth, stopping once before a nonpositive height is reached; so in that particular way, it has slightly more accuracy than the earlier mathematical model.

```
// Height of a falling object
public class Gravity
{
  public static void main (String args[])
  {
    double h, s = 1000.0, v = 100.0;
    int    t = 0;
    h = s;
```

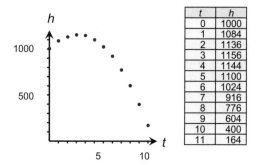

**Figure 10.1** Results of the simple gravity model.

```
    while (h >= 0.0)
    {
      System.out.println("Height at time " + t + "=" + h);
      t++;
      h = (-16 * t * t) + (v * t) + s;
    }
  }
}
```

Executing this model, that is, running this program is simulation. The results produced by the simulation are shown in both graphic and tabular form in Figure 10.1. The simulation commences with the initial velocity and height hard coded into the program, for example, the object was initially propelled upward at 100 ft/s from a vantage point 1000 ft above the ground. After that initial impetus, the object falls freely. The parabolic curve in the figure is not meant to suggest that the object is following a curved path; in this simple model, the object may travel only straight up and straight down along a vertical trajectory. The horizontal axis in the graph in the figure is time, and the curve shows how the height changes over time.

## VV&A DEFINITIONS

Of primary concern in this chapter are *verification* and *validation*. These terms have meanings in a general quality management context as well as in the specific M&S context, and in both cases, the latter meaning can be understood an M&S special case of the more general meaning. These definitions, as well as that of the related term *accreditation*, follow.

In general quality management, *verification* refers to a testing process that determines whether a product is consistent with its specifications or compliant with applicable regulations. In M&S, verification is typically defined analogously, as the process of determining if an implemented model is consistent with its specification [10]. Verification is also concerned with whether the model as designed will satisfy the requirements of the intended application.

Verification examines transformational accuracy, that is, the accuracy of transforming the model's requirements into a conceptual model and the conceptual model into an executable model. The verification process is frequently quite similar to that employed in general software engineering, with the modeling aspects of the software entering verification by virtue of their inclusion in the model's design specification. Typical questions to be answered during verification include:

(1) Does the program code of the executable model correctly implement the conceptual model?
(2) Does the conceptual model satisfy the intended uses of the model?
(3) Does the executable model produce results when needed and in the required format?

In general quality management, *validation* refers to a testing process that determines whether a product satisfies the requirements of its intended customer or user. In M&S, validation is the process of determining the degree to which the model is an accurate representation of the simuland [10]. Validation examines representational accuracy, that is, the accuracy of representing the simuland in the conceptual model and in the results produced by the executable model. The process of validation assesses the accuracy of the models.* The accuracy needed should be considered with respect to its intended uses, and differing degrees of required accuracy may be reflected in the methods used for validation. Typical questions to be answered during validation include:

(1) Is the conceptual model a correct representation of the simuland?
(2) How close are the results produced by the executable model to the behavior of the simuland?
(3) Under what range of inputs are the model's results credible and useful?

*Accreditation*, although often grouped with V&V in the M&S context in the common phrase "verification, validation, and accreditation," is an entirely different sort of process from the others. V&V are fundamental testing processes and are technical in nature. Accreditation, on the other hand, is a decision process and is nontechnical in nature, though it may be informed by technical data. Accreditation is the official certification by a responsible authority that a model is acceptable for use for a specific purpose [10]. Accreditation is concerned with official usability, that is, the determination that the model may be used. Accreditation is always for a specific purpose, such as a particular training exercise or analysis experiment, or a particular class of applications.

---

*Validation is used to mean assessing a model's utility with respect to a purpose, rather than its accuracy with respect to a simuland, in Cohn [11, pp. 200–201]. That meaning, which has merit in a training context, is not used here.

## 332  VERIFICATION, VALIDATION, AND ACCREDITATION

Models should not be accredited for "any purpose," because an overly broad accreditation could result in a use of a model for an application for which it has not been validated or is not suited. The accrediting authority typically makes the accreditation decision based on the findings of the V&V processes. Typical questions to be answered during accreditation include:

(1) Are the capabilities of the model and requirements of the planned application consistent?
(2) Do the V&V results show that the model will produce usefully accurate results if used for the planned application?
(3) What are the consequences if an insufficiently accurate model is used for the planned application?

To summarize these definitions, note that V&V are both testing processes, but they have different purposes.* The difference between them is often summarized in this way: Verification asks "Was the model made right," whereas validation asks "Was the right model made?" [2,12]. Continuing this theme, accreditation asks "Is the model right for the application?"

### V&V AS COMPARISONS

In essence, V&V are processes that compare things. As will be seen later, in any verification or validation process, it is possible to identify the objects of comparison and to understand the specific verification or validation process based on the comparison. This is the first of the central concepts of this chapter. The defining difference between V&V is what is being compared. Figure 10.2 illustrates and summarizes the comparisons. In the figure, the boxes represent the objects or artifacts involved in a simulation project.** The solid arrows connecting them represent processes that produce one object or artifact by transforming or using another. The results, for example, are produced by executing the executable model. The dashed arrows represent comparisons between the artifacts.

Verification refers to either of two types of comparison to the conceptual model. The first comparison is between the requirements and the conceptual model. In this comparison, verification seeks to determine if the requirements

---

*V&V are concerned with accuracy (transformational and representational, respectively), which is only one of several aspects of quality in a simulation project; others include execution efficiency, maintainability, portability, reusability, and usability (user-friendliness) [2].
**Everything in the boxes in Figure 10.2 is an *artifact* in the sense used in Royce [13], that is, an intermediate or final product produced during the project, except the simuland itself, hence the phrase *objects or artifacts*. Hereinafter, the term *artifacts* may be used alone with the understanding that it includes the simuland. Of course, the simuland could itself be an artifact of some earlier project; for example, an aircraft is an artifact, but in the context of the simulation project, it is not a product of that project.

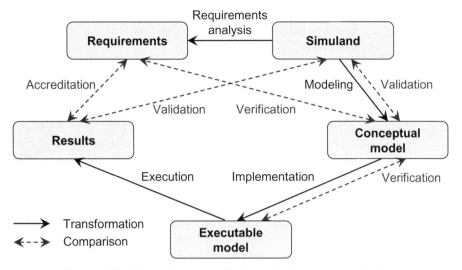

**Figure 10.2** Comparisons in verification, validation, and accreditation.

of the intended application will be met by the model described in the conceptual model. The second comparison is between the conceptual model and the executable model. The goal of verification in this comparison is to determine if the executable model, typically implemented as software, is consistent and complete with respect to the conceptual model.

Validation likewise refers to either of two types of comparisons to the simuland. The first comparison is between the simuland and the conceptual model. In this comparison, validation seeks to determine if the simuland and, in particular, those aspects of the simuland to be modeled, have been accurately and completely described in the conceptual model. The second comparison is between the simuland and the results.* The goal of validation in this comparison is to determine if the results, which are the output of a simulation using the executable model, are sufficiently accurate with the actual behavior of the simuland, as defined by data documenting its behavior. Thus, validation compares the output of the executable model with observations of the simuland.

## PERFORMING VV&A

V&V have been defined as both testing processes and comparisons between objects or artifacts in a simulation project, and accreditation as a decision

---

*To be precise, the results are not compared with the simuland itself but to observations of the simuland. For example, it is not practical to compare the height values produced by the gravity model with the height of an object as it falls; rather, the results are compared with data documenting measured times and heights recorded while observing the simuland.

regarding the suitability of a model for an application. This section places VV&A in the context of simulation projects and discusses overarching issues in the practice of VV&A.

## VV&A within a Simulation Project

V&V do not take place in isolation; rather, they are done as part of a *simulation project*. To answer the question of when to do them within the project, it is common in the research literature to find specific V&V activities assigned specific phases of a project (e.g., Balci [12]).* However, these recommendations are rarely consistent in detail, for two reasons. First, there are different types of simulation projects (simulation studies, simulation software development, and simulation events are examples) that have different phases and different V&V processes.** Even when considering a single type of simulation project, there are different project phase breakdowns and artifact lists in the literature. For examples, compare the differing project phases for simulation studies (given in References 2, 3, 9, and 14), all of which make sense in the context of the individual source. Consequently, the association of verification and verification activities with the different phases and comparison with the different artifacts inevitably produces different recommended processes.***

---

*Simulation project phase breakdowns are often called *simulation life cycles*, for example, in Balci [2].
**A *simulation study* is a simulation project where the primary objective is to use simulation to obtain insight into the simuland being studied. The model code itself is not a primary deliverable, and so it may be developed using software engineering practices or an implementation language that reflect the fact that it may not be used again. A *simulation software development* project is one where the implemented executable model, which may be a very large software system, is the primary deliverable. There is an expectation that the implemented executable model will be used repeatedly, and most likely modified and enhanced, by a community of users over a long period of time. In this type of project, proper software engineering practices and software project management techniques (e.g., [13]) move to the forefront. A *simulation event* is a particular use of an existing model, or some combination of models, to support a specific objective, for example, the use of an operational-level command staff training model to conduct a training exercise for a particular military headquarters. In simulation events, software development is often not an issue (or at least not a major one), but event logistics, scenario development, database preparation, and security can be important considerations.
***Further complicating the issue is that there are two motivations for guidelines about performing V&V found in the literature. The first motivation is technical effectiveness; technically motivated discussions are concerned with when to perform V&V activities and which methods to use based on artifact availability and methodological characteristics, with an ultimate goal of ensuring model accuracy. The second motivation is official policies; policy-motivated discussions relate official guidelines for when to perform V&V activities and how much effort to apply to them based on organizational policies and procedures, with an ultimate objective of receiving model accreditation. Of course, ideally, the two motivations are closely coupled, but the distinction should be kept in mind when reviewing V&V guidelines. In this chapter, only technical effectiveness is considered.

Nevertheless, it is possible to generalize about the different project phase breakdowns. Figure 10.2 can be understood as a simplified example of such a breakdown, as it suggests a sequence of activities (the transformations) that produce intermediate and final products (the artifacts) over the course of the simulation project. The more detailed phase breakdowns cited earlier differ from Figure 10.2 in that they typically define more project phases, more objects and artifacts, and more V&V comparisons between the artifacts than those shown in the figure.* Going further into an explanation and reconciliation of the various simulation project types and project phase breakdowns detailed enough to locate specific V&V activities within them is beyond the scope of this chapter. In any case, although the details and level of granularity differ, the overall concept and sequence of the breakdowns are essentially the same.

Despite the differences between the various available simulation project phase breakdowns found in the literature, and the recommendations for when to perform V&V within the project, there is an underlying guideline that is common across all of them: V&V in general, and specific V&V comparisons in particular, should be conducted as soon as possible. (Two of the 15 "principles of VV&A" in Balci [2] get at this idea.) But when is "as soon as possible?" The answer is straightforward: As soon as the artifacts to be compared in the specific comparison are available. For some verification activities, both the conceptual model (in the form of design documents) and the executable model (in the form of programming language source code) are needed. For some validation activities, the results of the simulation, perhaps including event logs and numerical output, as well as the data representing observations of the simuland, are required. The detailed breakdowns of V&V activity by project phase found in the literature are consistent with the idea that V&V should proceed as soon as the artifacts to be compared are available, assigning to the different project phases specific V&V activities and methods (methods will be discussed later) that are appropriate to the artifacts available in that phase.

In contrast to V&V, accreditation is not necessarily done as soon as possible. Indeed, it can be argued that accreditation should be done as late as possible, so that the maximum amount of information is available about the accuracy of the model and its suitability for the intended application. However, just how late is "as late is possible" may depend on programmatic considerations as much as on technical ones; for example, a decision about the suitability of a model for an application may be needed in order to proceed with

---

*For example, Balci [2] identifies the *conceptual model* and the *communicative model*, both of which are forms or components of the *conceptual model* in Figure 10.2, and includes a verification comparison between them. He also identifies the *programmed model* and the *experimental model*, both of which are forms or components of the *executable model* in Figure 10.2, and includes another verification comparison between them.

|  | Model<br>valid | Model<br>not valid | Model<br>not relevant |
|---|---|---|---|
| Results<br>accepted,<br>model used | Correct | **Type II error**<br>Use of<br>invalid model;<br>incorrect V&V;<br>model user's risk;<br>**more** serious error | **Type III error**<br>Use of<br>irrelevant model;<br>accreditation mistake;<br>accreditor's risk;<br>**more** serious error |
| Results<br>not accepted,<br>model not used | **Type I error**<br>Nonuse of<br>valid model;<br>insufficient V&V;<br>model builder's risk;<br>**less** serious error | Correct | Correct |

Figure 10.3  Verification and validation errors.

the next phase of model implementation. The accrediting authority should weigh the risks of accrediting an unsuitable model against those of delaying the accreditation or not accrediting a suitable one. Those risks are discussed later.

## Risks, Bounds of Validity, and Model Credibility

V&V are nontrivial processes, and there is the possibility that they may not be done correctly in every situation. What types of errors may occur during V&V, and what risks follow from those errors? Figure 10.3 summarizes the types of V&V errors and risks.*

In the figure, three possibilities regarding the model's accuracy are considered; it may be accurate enough to be used for the intended application ("valid"), it may not be accurate enough ("not valid"), or it may not be relevant to the intended application. Two possibilities regarding the model's use are considered; the model's results may be accepted and used for the intended application, or they may not. The correct decisions are, of course, when a valid model is used or when an invalid or irrelevant model is not used.

A *type I error* occurs when a valid model is not used. For example, a valid flight simulator is not used to train and qualify a pilot. This may be due to

---

*The figure is adapted from a flowchart that shows how the errors might arise found in Balci [2]. A similar table appears in Banks et al. [6].

insufficient validation to persuade the accrediting authority to certify the model. A type I error can result in model development costs that are entirely wasted if the model is never used or needlessly increased if model development continues [2]. Additionally, whatever potential benefits that using the model might have conferred, such as reduced training costs or improved decision analyses, are delayed or lost. The likelihood of a type I error is termed *model builder's risk* [15].

A *type II error* occurs when an invalid model is used. For example, an invalid flight simulator is used to train and qualify a pilot. This may occur when validation is done incorrectly but convincingly, erroneously persuading the accrediting authority to certify the model for use. A type II error can result in disastrous consequences, such as an aircraft crash because of an improperly trained pilot or a bridge collapsing because of faulty analyses of structural loads and stresses. The likelihood of a type II error is termed *model user's risk* [15].

A *type III error* occurs when an irrelevant model, that is, one not appropriate for the intended application, is used. This differs from a type II error, where the model is relevant but invalid; in a type III error, the model is in fact valid for some purpose or simuland, but it is not suitable for the intended application. For example, a pilot may be trained and qualified for an aircraft type in a flight simulator valid for some other type. Type III errors are distressingly common; models that are successfully used for their original applications often acquire an unjustified reputation for broad validity, tempting project managers eager to reduce costs by leveraging past investments to use the models inappropriately. Unfortunately, the potential consequences of a type III error are similar, and thus similarly serious, to those of a type II error. The likelihood of a type III error is termed *model accreditor's risk*.

Reducing validation risk can be accomplished, in part, by establishing a model's bounds of validity. The goal of V&V is not to simply declare "the model is valid," because for all but the simplest models, such a simple and broad declaration is inappropriate. Rather, the goal is to determine when (i.e., for what inputs) the model is usefully accurate, and when it is not, a notion sometimes referred to as the model's bounds of validity. The notion of bounds of validity will be illustrated using the example gravity model. Consider these three versions of a gravity model:

(1) $h(t) = 776$;
(2) $h(t) = (-420/9)t + 1864$;
(3) $h(t) = -16t^2 + vt + s$.

In these models, let $v = 100$ and $s = 1000$.

The results (i.e., the heights) produced by these models for time values from 0 to 11 are shown in Figure 10.4. Model (1) is an extremely simple and low-fidelity model; it always returns the same height regardless of time. It corresponds to the horizontal line in the figure. Essentially by coincidence, it

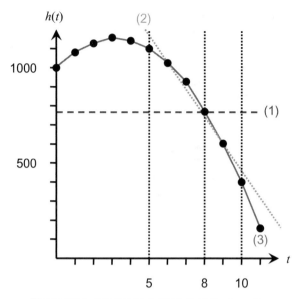

**Figure 10.4** Results from three models of gravity.

is accurate for one time value ($t = 8$). Model (2) is a slightly better linear model, corresponding to the downward sloping line in the figure. As can be seen there, model (2) returns height values that are reasonably close to correct over a range of time values (from $t = 5$ to $t = 10$). Model (3) is the original example gravity model, which is quite accurate within its assumptions of negligible air resistance and proximity to the surface of the earth.

Assume that the accuracy of each of these three models was being determined by a validation process that compared the models' results with observations of the simuland, that is, measurements of the height of objects moving under gravity. If the observations and validation were performed only for a single time value, namely $t = 8$, model (1) would appear to be accurate. If the observations were performed within the right range of values, namely $5 \leq t \leq 10$, then the results of model (2) will match the observations of the simuland fairly well. Only validation over a sufficient range of time values, namely $0 \leq t \leq 11$, would reveal model (3) as the most accurate.

These three models are all rather simple, with only a small range of possible inputs, and the accuracy of each is already known. Given that, performing validation in a way that would suggest that either model (1) or model (2) was accurate might seem to be unlikely. But suppose the models were 1000 times more complex, a level of complexity more typical of practical models, with a commensurately expanded range of input values. Under these conditions, it is more plausible that a validation effort constrained by limited resources or data availability could consider models (1) or (2) to be accurate.

Two related conclusions should be drawn from this example. The first is that validation should be done over the full range of input values expected in the intended use. Only by doing so would the superior accuracy of model (3) be distinguished from model (2) in the example. One objective of validation is to determine the range of inputs over which the model is accurate enough to use, that is, to determine the bounds of validity. This is the second central concept of this chapter. In the example, there were three models to choose from. More often, there is only one model and the question is thus not which model is most accurate, but rather when (that is, for what inputs) the one available model is accurate enough. In this example, if model (2) is the only one available, an outcome of the validation process would be a statement that it is accurate within a certain range of time values.

The second conclusion from the example is that finding during validation that a model is accurate only within a certain range of inputs is not necessarily a disqualification of the model. It is entirely possible that the intended use of that model will only produce inputs within that range, thus making the model acceptable. In short, the validity of a model depends on its application. This is the third central concept of this chapter. However, even when the range of acceptable inputs found during validation is within the intended use, that range should be documented. Otherwise, later reuse of the model with input values outside the bounds of validity could unknowingly produce inaccurate results.

*Model credibility* can be understood as a measure of how likely a model's results are to be considered acceptable for an application. VV&A all relate to credibility, V&V are processes that contribute to model credibility, and accreditation is an official recognition that a model has sufficient credibility to be used for a specific purpose. Developing model credibility requires an investment of resources in model development, verification, and validation; in other words, credibility comes at a cost.

Figure 10.5 suggests the relationship between model cost, credibility, and utility.* In the figure, model credibility increases along the horizontal axis, where it notionally varies from 0 percent (no credibility whatsoever) to 100 percent (fully credible). The two curves show how model credibility as the independent variable relates to model utility (how valuable the model is to its user) and model cost (how much it costs to develop, verify, and validate the model) as dependent variables. The model utility curve shows that model utility increases with model credibility, that is, a more credible model is thus a more useful one, but that as credibility increases additional fixed increments of credibility produce diminishing increments of utility. In other words, there is a point at which the model is sufficiently credible for the application, and adding additional credibility through the expenditure of additional resources on development, verification, and validation is not justified in terms of utility gained.

---

*The figure is adapted from Balci [2] (which in turn cites References 16 and 17; it also appears in Sargent [9]).

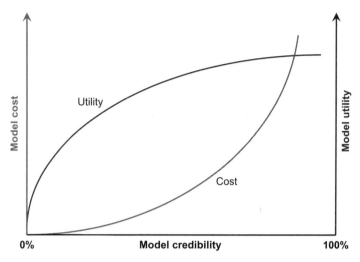

**Figure 10.5** Relationship between model cost, credibility, and utility.

The model cost curve shows that additional credibility results in increased cost, and moreover, additional fixed increments of credibility come at progressively increasing cost. In other words, there is a point at which the model has reached the level of credibility inherent in its design, and beyond that point, adding additional credibility can become prohibitively expensive. Simulation project managers might prefer to treat cost as the independent variable and credibility as the dependent variable; Figure 10.5 shows that relationship as well through a reflection of the credibility and cost axes and the cost curve. The reflected curve, showing credibility as a function of cost, increases quickly at first and then flattens out, suggesting that the return in credibility reaches a point of diminishing returns for additional increments of cost.

It is up to the simulation project manager to balance the projects requirements for credibility (higher for some applications than for others) against the resources available to achieve it, and to judge that utility that will result from a given level of credibility.

## V&V METHODS

The previous section discussed when to do V&V, and how much effort to expend on them. The model developer must also know how to do them. A surprisingly large variety of techniques, or methods, for V&V exist. The methodological diversity is due to the range of simulation project types, artifacts produced during the course of simulation projects, subjects (simulands) of those projects, and types of data available for those subjects. Some of the methods (especially verification methods) come from software engineering, because the executable models in simulation projects are almost always realized as software, while others (especially validation methods) are specific to M&S, and typically involve data describing the simuland. However, all of the methods involve comparisons of one form or another.

Over 90 different V&V methods, grouped into four categories (informal, static, dynamic, and formal), are listed and individually described in Balci [2] (and that list, while extensive, is not complete).* Repeating each of the individual method descriptions here would be pointlessly duplicative. Instead, the four categories from that source will be defined, and representative methods from each category will be defined. For some of those methods, examples of their use will be given.

## Informal Methods

*Informal V&V methods* are more qualitative than quantitative and generally rely heavily on subjective human evaluation, rather than detailed mathematical analysis. Experts examine an artifact of the simulation project, for example, a conceptual model expressed as UML diagrams, or the simulation results, for example, variation in service time in a manufacturing simulation, and assess the model based on that examination and their reasoning and expertise. Informal methods *inspection*, *face validation*, and the *Turing test* are defined here; other informal methods include *desk checking* and *walkthroughs* [2].

**Inspection**   *Inspection* is a verification method that compares project artifacts to each other. In inspection, organized teams of developers and testers inspect model artifacts, such as design documents, algorithms, physics equations, and programming language code. Based on their own expertise, the inspectors manually compare the artifacts being inspected with the appropriate object of comparison, for example, programming language code (the executable model) might be compared with algorithms and equations (the conceptual model). The persons doing the inspection may or may not be the developers of the model being inspected, depending on the resources of the project and the developing organization. Inspections may be ad hoc or highly structured, with members of an inspection team assigned specific roles, such as moderator, reader, and recorder, and specific procedure steps used in the inspection [2]. The inspectors identify, assess, and prioritize potential faults in the model.

**Face Validation**   *Face validation* is a validation method that compares simuland behavior to model results. In face validation, observers who may be potential users of the model and/or subject matter experts with respect to the

---

*See Balci [18,19] for earlier versions of the categorization with six categories instead of four.

simuland review or observe the results of a simulation (an execution of the executable model). Based on their knowledge of the simuland, the observers subjectively compare the behavior of the simuland as reflected in the simulation results with their knowledge of the behavior of the actual simuland under the same conditions, and judge whether the former is acceptably accurate. Differences between the simulation results and the experts' expectations may indicate model accuracy issues. Face validation is frequently used in interactive real-time virtual simulations where the experience of a user interacting with the simulation is an important part of its application. For example, the accuracy of a flight simulator's response to control inputs can be evaluated by having an experienced pilot fly the simulator through a range of maneuvers.*

While face validation is arguably most appropriate for such interactive simulations, it is often used as a validation method of last resort, when a shortage of time or a lack of reliable data describing simuland behavior precludes the use of more objective and quantitative methods. While moving beyond face validation to more objective and quantitative methods should always be a goal, face validation is clearly preferable to no validation at all.

As an example, face validation was used to validate the *Joint Operations Feasibility Tool* (JOFT), a model of military deployment and sustainment feasibility developed by the U.S. Joint Forces Command Joint Logistics Transformation Center [20]. JOFT was intended to be used to quickly assess the feasibility of deployment transportation for military forces to an area of operations and logistical sustainment for those forces once they have been transported. The process of using JOFT had three basic stages. First, based on a user-input list of military capabilities required for the mission, JOFT identifies units with those capabilities, and the user selects specific units. Second, the user provides deployment transportation details, such as points of embarkation and debarkation, type of transportation lift, and time available. JOFT then determines if the selected force can be deployed within the constraints and provides specific information regarding the transportation schedule. Third, given the details of the initial supplies accompanying the units, the supplies available in the area of operations, a rating of the expected operations tempo and difficulty of the mission, and a rating of the expected rate of resupply, JOFT calculates the sustainment feasibility of the force and identifies supply classes for which sustainment could be problematic.

JOFT was assessed using a highly structured face validation by a group of logistics subject matter experts.** Several validation sessions, with different groups of experts participating in each session were conducted, all with this procedure:

---

*However, the utility of face validation in such applications is called into question in Grant and Galanis [21], where it is asserted that subject matter experts often perform a task in a manner different from the way they verbalize it.
**Some sources might classify this validation method as a *Delphi test* [22] (a method that does not appear in the list of Balci [2]).

(1) The procedure and intent for the assessment session was explained to the experts.
(2) The experts were given a tutorial briefing and a live demonstration of the JOFT software.
(3) The experts used JOFT hands-on for two previously developed planning scenarios.
(4) The experts provided written feedback on the JOFT concepts and software.

A total of 20 experts participated in the assessment in four different sessions. Collectively, they brought a significant breadth and depth of military logistics expertise to the validation. Of the 20 experts, 17 were currently or had previously been involved in military logistics as planners, educators, or trainers. The remaining three were current or former military operators, that is, users of military logistics.

The experts' assessments were secured using questionnaires. Categories of questions asked the experts to validate JOFT's accuracy and utility in several ways:

(1) Suitability for its intended uses (e.g., plan feasibility "quick look" analysis).
(2) Accuracy of specific features of the JOFT model (e.g., resource consumption rate).
(3) Utility within the logistical and operational planning processes.

The face validation of JOFT compared the model's estimates of deployment transportation and logistical sustainment feasibility with the expectations of experts. The face validation was quite effective at identifying both strengths and weaknesses in the model. This was due both to the high degree of structure and preparation used for the validation process and the expertise of the participating subject matter experts. The test scenarios were carefully designed to exercise the full range of the model's functionality, and the questionnaires contained questions that served to secure expert assessment of its validity in considerable detail. The effectiveness of face validation as a validation method is often significantly enhanced by such structure.

**The Turing Test** The *Turing test* is an informal validation method well suited to validating models of human behavior, a category of models that can be difficult to validate [23,24]. The Turing test compares human behavior generated by a model to the expectations of human observers for such behavior. First proposed by English mathematician Alan Turing as a means to evaluate the intelligence of a computer system [25], it can be seen as a specialized form of face validation. In the Turing test as conventionally formulated, a computer system is said to be intelligent if an observer cannot reliably distinguish between system-generated and human-generated behavior at a rate better than chance.* When applied to the validation of human behavior models, the model is said to pass the Turing test and thus to be valid if expert observers cannot reliably distinguish between model-generated and human-generated behavior. Because the characteristic of the system-generated behavior being assessed is the degree to which it is indistinguishable from human-generated behavior, this test is clearly directly relevant to the assessment of the realism of algorithmically generated behavior, perhaps even more so than to intelligence as Turing originally proposed.

The Turing test was used to experimentally validate the semiautomated force (SAF) component of the SIMNET (Simulator Networking) distributed simulation system, a networked simulation used for training tank crews in team tactics by immersing them in a virtual battlefield [26]. In general, *SAF systems* (also known as computer-generated force, or CGF, systems) use algorithms that model human behavior and tactical doctrine supported by a human operator to automatically generate and control autonomous battlefield entities, such as tanks and helicopters [24]. In the SIMNET SAF validation, two platoons of soldiers fought a series of tank battles in the SIMNET virtual battlefield. In each battle, one of the platoons defended a position against attacking tanks controlled by the other platoon of soldiers, the automated SAF system, or a combination of the two. Each of the two platoons of soldiers defended in two different battles against each of the three possible attacking forces, for a total of 12 battles. The two platoons of soldiers had no contact with each other before or during the experiment other than their encounters in the virtual battlefield. Before the experiment, the soldiers were told that the object of the test was not to evaluate their combat skills but rather to determine how accurately they could distinguish between the human and SAF attackers. When asked to identify their attackers after each battle, they were not able to do so at a rate significantly better than random chance. Thus, the SIMNET SAF system was deemed to have passed the Turing test and thus to be validated [26].

Although the Turing test is widely advocated and used for validating models of human behavior, its utility for that application is critically examined in Petty [27], where it is argued that in spite of a number of claims of its efficacy by experts, the Turing test cannot be relied upon as the sole means of validating a human behavior generation algorithm. Examples are given that demonstrate that the Turing test alone is neither necessary nor sufficient to ensure the validity of the algorithm. However, if attention is given to the

---

*In Turing's original form of the test, which he called the *Imitation Game*, a human interrogator conducts a question and answer session with two hidden respondents, one of whom may be either a human or a computer system. The interrogator's goal is to determine which of the two respondents is the man and which is the woman. The computer system is said to have passed the test if the interrogator is no more likely to give the correct answer when the computer system is one of the respondents than when both are humans.

questions of who the appropriate observers are and what information about the generated behavior is available to them, a well-designed Turing test can significantly increase confidence in the validity, especially in terms of realism, of a behavior generation algorithm that passes the test. Such an application of the Turing test, with its results analyzed by an appropriate statistical hypothesis test, was used as a complement to another validation method in evaluating a computer model of decision making by military commanders [28].

## Static Methods

*Static V&V methods* involve assessment of the model's accuracy on the basis of characteristics of the model and executable model that can be determined without the execution of a simulation. Static techniques often involve analysis of the programming language code of the implemented model, and may be supported by automated tools to perform the analysis or manual notations or diagrams to support it. Static methods are more often performed by developers and other technical experts, as compared with informal methods, which depend more on subject matter experts. Static methods *data analysis* and *cause–effect graphing* are defined here; other static methods include *interface analysis* and *traceability assessment* [2].

**Data Analysis** *Data analysis* is a verification method that compares data definitions and operations in the conceptual model to those in the executable model. Data analysis ensures that data are properly defined (correct data types, suitable allowable data ranges) and that proper operations are applied to the data structures in the executable model. Data analysis includes data dependency analysis (analyzing which data variables depend on which other variables) and data flow analysis (analyzing which variables are passed between modules in the executable model code).

**Cause–Effect Graphing** *Cause–effect graphing* is a validation method that compares cause-and-effect relationships in the simuland to those in the conceptual model. Causes are events or conditions that may occur in the simuland, and effects are the consequences or state changes that result from the causes. For example, lowering flaps in a flight simulator (a cause) will change the flight dynamics of the aircraft, increasing both drag and lift (the effects). Note that effects may themselves be causes of further effects; for example, the additional drag caused by lowering flaps will then cause a slowing of the aircraft. In cause–effect graphing, all causes and effects considered to be important in the intended application of the model are identified in the simuland and in the conceptual model and compared; missing and extraneous cause–effect relationships are corrected. Causes and effects are documented and analyzed through the use of cause–effect graphs, which are essentially directed graphs where causes and effects, represented by nodes in the graph, are connected by the effects that related them, represented by directed edges.

*Petri nets* are a graphic or diagrammatic notation widely used for a variety of modeling applications, including control systems, workflow management, logistics supply chains, and computer architectures. Cause–effect graphing was used as the basis for a tool that automatically generates test cases for the validation of Petri net models; in effect, cause–effect graphing is used by this tool to support the validation of any Petri net model [29].

**Dynamic Methods**  *Dynamic V&V methods* assess model accuracy by executing the executable model and evaluating the results. The evaluation may involve comparing the results with data describing the behavior or the simuland or the results of other models. Because the comparisons in dynamic methods are typically of numerical results and data, dynamic methods are generally objective and quantitative. Dynamic methods *sensitivity analysis*, *predictive validation*, and *comparison testing* are defined here; other dynamic methods include *graphic comparisons* and *assertion checking* [2]. An important subcategory of dynamic methods is statistical validation methods. Examples of statistical comparison methods applicable to validation include *regression analysis*, *hypothesis testing*, *goodness-of-fit testing*, *time series analysis*, and *confidence interval testing*; the former two are defined here.

**Sensitivity Analysis**  *Sensitivity analysis* is a validation method that compares magnitude and variability in simuland behavior to magnitude and variability in the model results. It is an analysis of the range and variability in model results. A test execution of the model is arranged so as to cause the inputs to the model to vary over their full allowable range. The magnitude and variability of the results produced are measured and compared with the magnitude and variability of the simuland's behavior over the same range of input values. Differences could suggest invalidity in the model; if there are significant differences for some input values but not for others, this could suggest invalidity for some specific ranges of inputs.

If sufficient data regarding the simuland is available, sensitivity analysis can be conducted by comparing the response surfaces of the model and the simuland for appropriately chosen independent variables (the input values) and dependent variables (the output results); the sign and magnitude of the difference between the two response surfaces can be calculated and analyzed [30]. Beyond validation, sensitivity analysis can also be used to evaluate model response to errors in the input and to establish which inputs have the greatest impact on the results, information which can focus efforts and establish accuracy requirements when preparing model input data [31].

**Predictive Validation**  *Predictive validation* is a validation method that compares specific outcomes in simuland behavior to corresponding outcomes in the model results. Predictive validation may be used when available information about the behavior of the simuland includes corresponding input and output values; that is, historical or experimental data are available that show how the simuland behaved under well-established conditions. Given such data, the model is executed with the same inputs, and its results are compared with the historical or experimental data.* For example, the airspeed of an aircraft at different altitudes and throttle settings in a flight simulator can be compared with actual flight test for the aircraft being modeled, if the latter is available. Similarity or dissimilarity between the simuland's behavior and the simulation's results suggest validity or invalidity. The actual comparison may be done in a variety of ways, some of which are considered validation methods in their own right (e.g., statistical methods); this is certainly acceptable and strengthens the validation power of the method.

Predictive validation is a valuable method, not only because it is based on a direct comparison between simuland behavior and model results, but also because it can be applied in some circumstances where other methods would be problematic. For some simulands, it is convenient to exercise the system and take measurements specifically for use during validation. For other actual systems, exercising the actual system is infeasible due to danger, expense, and impracticality. Combat, for example, clearly falls into the latter category; fighting a battle in order to collect data to validate a model of combat is not an option. However, combat models can be validated by using them to predict (or retrodict) the outcomes of historical battles and comparing the model's results with the historical outcomes. Once the outcome of a historical battle have been documented to a level of detail and accuracy sufficient for validation (often an unexpectedly difficult task) and the model results have been generated for the same scenario, the two sets of results can be compared. Several of the methods already discussed may be used to make the comparison.

Predictive validation was used in this way to validate three separate combat models in unrelated validation efforts that compared model results to historical outcomes. The Ironside model [32] and the alternative aggregate model [33] are both two-sided, stochastic, constructive models of combat with internal representations at the entity level (e.g., individual tanks are represented at some degree of detail). COMAND is theater-level representation of naval air campaigns, focused on the representation of command and control [34]. Ironside was validated by using it to retrodict the outcome of the Battle of Medenine (March 6, 1943, North Africa). The alternative aggregate model was validated by using it to retrodict the outcome of the Battle for Noville (December 19–20, 1944, Belgium). COMAND was validated by using it to retrodict the outcome of the Falkland Islands campaign (April 2–June 20, 1982).

---

*The method is called predictive validation because the model is executed to "predict" the simuland's behavior. However, because the events being predicted by the model are in the past, some prefer to call the process *retrodiction* and the method *retrodictive validation*.

In addition to their use in validating combat models, the common element of these battles is that they are well documented, an essential prerequisite for the predictive validation method. The validation of Ironside found that the model produced results that were reliably different from the historical outcome [35]. The validation of the alternative aggregate model found, after some adjustment to the model during testing, that the model produced results that were reasonably close to the historical outcome [33]. The validation of COMAND was mostly successful and revealed the strengths and the weaknesses of the model [34].

**Comparison Testing** *Comparison testing* is a dynamic verification method that can be used when multiple models of the same simuland are available. The models are executed with the same input, and their results are compared with each other. Even if neither of the two models can be assumed to be accurate, comparing their results is useful nonetheless, because differences between the two sets of results suggest possible accuracy problems with the models [2].

As an example, comparison testing was used to verify $C^2PAT$, a queuing theory-based, closed-form model of command and control systems [36]. $C^2PAT$ is intended to allow analysis and optimization of command and control system configurations. $C^2PAT$ models command and control systems as a network of nodes representing command and control nodes connected by edges that represent communications links. A set of cooperating agents, known as servers, located at the network nodes exchange information via the connecting links. When a unit of command and control information, known as a job, arrives at a node, that node's server processes it and passes information to connected nodes. Processing time at a node is determined by exponential and nonexponential distributions, and various priority queuing disciplines are used to sequence jobs waiting at nodes to be served. Preemption in the job queues, something likely to happen in command and control systems, is also modeled. $C^2PAT$ models the dynamic response of the command and control system in response to time-varying job arrival rates, determining response and delay times for both individual nodes and for threads of job execution. $C^2PAT$'s queuing theory-based model is analytic, computing values for the parameters of interest in the system using closed-form, queuing theory equations.

To verify $C^2PAT$, two additional distinct versions of the model were implemented: a time-stepped model written in a discrete-event programming environment and an event-driven model written in a general-purpose programming language. Unlike $C^2PAT$, the additional models were stochastic and numerical, simulating the flow of information in the command and control system over time using random draws against probability distributions describing service and delay times.

A series of progressively more complicated test networks were developed. The three models were executed for those test networks and their results compared. Differences between the results were identified and analyzed, and revisions were made to the appropriate model. The direct comparison of results was quite effective at discovering and focusing attention on potential model accuracy problems. The verification process proved to be iterative, as each comparison would reveal issues to resolve, often leading to revisions to one or another of the three models, necessitating an additional run and comparison. The verification effort was ultimately essential to successful modeling in $C^2PAT$ [36].

A case study of the model comparison method, comparing the theater missile defense capabilities of the EADSIM and Wargame 2000, reports some of the statistical issues involved [37].

**Regression Analysis** *Regression analysis* is a multipurpose statistical technique that can be used as a dynamic validation method. In general, regression analysis seeks to determine the degree of relatedness between variables, or to determine the extent that variation in one variable is caused by variation in another variable [38]. When directly related values can be identified, regression analysis compares specific model result values to simuland observation values.

As an example, regression analysis was used to validate a model of spacecraft mass [39]. The spacecraft propulsion system sizing tool (SPSST) model predicts the mass of the propulsion system of automated exploration spacecraft. The propulsion system can account for as much as 50 percent of a spacecraft's total mass before liftoff. The SPSST model is intended to support engineering trade studies and provide quick insight into the overall effect of propulsion system technology choices on spacecraft mass and payload. The mass prediction is calculated using physics-based equations and engineering mass estimation relationships. Inputs to the model include mission profile parameters, such as velocity change and thermal environment, and selected options for nine subsystems, including main propellant tanks, main propellant pressure system, and main engines. The model outputs predicted mass for the overall spacecraft propulsion system, both with and without propellant (called *wet mass* and *dry mass*, respectively), as well as for the spacecraft subsystems, for the given mission.

To validate the SPSST model, mass, mission, and subsystem option data were collected for 12 existing spacecraft, including Mars Odyssey, Galileo, and Cassini. The SPSST model was used to predict the propulsion system and subsystem masses for these spacecraft, given the characteristics of these spacecraft and their missions as input. The resulting values for wet mass, dry mass, and subsystem mass predicted by the model were compared with the actual spacecraft mass values using linear regression. When the 12 pairs of related predicted and actual wet mass values were plotted as points, they fell quite close to a line, suggesting accuracy in the model. The computed coefficient of regression statistic for wet mass was $R^2 = 0.998$; the statistic's value close to 1 confirmed the model's accuracy for this value. Figure 10.6 shows the actual

| Spacecraft | Propellant type | Actual mass | Predicted mass |
|---|---|---|---|
| Stardust | N2H4 | 104.4 | 201.4 |
| Mars Odyssey | NTO/N2H4 | 403.1 | 486.0 |
| MRO | N2H4 | 1164.5 | 1262.9 |
| MESSENGER | NTO/N2H4 | 682.4 | 761.7 |
| MGS | NTO/N2H4 | 451.5 | 529.4 |
| Genesis | N2H4 | 180.7 | 248.5 |
| AXAF | NTO/N2H4 | 1148.4 | 1270.7 |
| Galileo | NTO/MMH | 1145.0 | 1131.8 |
| Cassini | NTO/MMH | 3627.9 | 3514.2 |
| NEAR | NTO/N2H4 | 437.7 | 447.0 |
| Jason-1 | N2H4 | 34.1 | 119.9 |
| Phoenix | N2H4 | 121.6 | 206.0 |

**Figure 10.6** Actual and predicted spacecraft wet mass values for the SPSST model validation.

and predicted wet mass values for the 12 spacecrafts. Similar results were found for dry mass, though here the statistic's values was somewhat lower, $R^2 = 0.923$. On the other hand, the subsystem predictions were not nearly as consistent; the model did well on some subsystems, such as propellant tanks, and not as well on others, such as components.

Regression analysis provided a straightforward and powerful validation method for the SPSST model. When model results provide values that can be directly paired with corresponding simuland values, regression analysis may be applicable as a validation method.

**Hypothesis Testing** In general, a *statistical hypothesis* is a statistical statement about a population, which is evaluated on the basis of information obtained from a sample of that population [38]. Different types of hypothesis tests exist (e.g., see Box et al. [40]). They can be used to determine if a sample is consistent with certain assumptions about a population or if two populations have different distributions based on samples from them. Typically, when a hypothesis test is used for validating a model, the results of multiple simulations using the model are treated as a sample from the population of all possible simulations using that model for relevant input conditions. Then an appropriate test is selected and used to compare the distribution of the model's possible results to the distribution of valid results; the latter may be represented by parameters or a sample from a distribution given as valid.

As an example, hypothesis testing was use to validate a behavior generation algorithm in a SAF system [41,42]. The algorithm generated reconnaissance routes for ground vehicles given assigned regions of terrain to reconnoiter. The intent of the algorithm was that a ground vehicle moving along the generated route would sight hostile vehicles positioned in the terrain region and would do so as early as possible. Sighting might be blocked by terrain features, such as ridges or tree lines. The algorithm considered those obstacles to sighting in the terrain and planned routes to overcome them.

The algorithm was validated by comparing routes planned by the algorithm for a variety of terrain regions with routes planned by human subject matter experts (military officers) for the same terrain regions.* To quantify the comparison, a metric of a route's effectiveness was needed. A separate group of human subject matter experts were asked to position hostile vehicles on each of the test terrain regions. Each of the routes was executed, and the time at which the reconnaissance vehicle moving along the route sighted each hostile vehicle was recorded. The sighting times for corresponding ordinal sightings were compared (i.e., the $k$th sighting for one route was compared with the $k$th sighting for the other route, regardless of which specific vehicles were sighted).

A Wilcoxon signed-rank test was used to compare the distributions of the sighting times for the algorithm's and the human's routes. This particular statistical hypothesis test was chosen because it does not assume a normal distribution for the population (i.e., it is nonparametric), and it is appropriate for comparing internally homogenous sample data sets (corresponding sightings were compared) [38]. Using this test, the sighting times for the algorithm's routes were compared with the sighting times for each of the human subject matter experts' routes. Because the algorithm was intended to generate human behavior for the reconnaissance route planning task, the routes generated by the human subject matter experts were assumed to be valid.

The conventional structure of a hypothesis test comparing two distributions is to assume that the two distributions are the same, and to test for convincing statistical evidence that they are different. This conventional structure was used for the validation; the algorithm's routes and the humans' routes were assumed to be comparable (the null hypothesis), and the test searched for evidence that they were different (a two-sided alternative hypothesis). The Wilcoxon signed-rank test did not reject the null hypothesis for the sighting time data, and thus did not find evidence that the algorithm's routes and the humans' routes were different. Consequently, it was concluded that the routes were comparable, and the algorithm was valid for its purpose.

Subsequent consideration of the structure of the hypothesis test suggested the possibility that the test was formulated backward. After all, the goal of the validation was to determine if the algorithm's routes were comparable to the humans' routes, and that comparability was assumed to be true in the null hypothesis of the test. This formulation is a natural one, as it is consistent with the conventional structure of hypothesis tests.** However, such an

---

*The terrain regions chosen for the algorithm were selected to present the route planners (algorithm and human) with a range of different densities of sight-obstructing terrain elevation and features.
**Indeed, a textbook validation example using a $t$-test formulates the test in the same way, with the null hypothesis assuming that the model and simuland have the same behavior [6, p. 368].

assumption means that, with respect to the validation goal, the finding of comparability was weaker than it might have been.* In such tests, it should be understood that rejecting the null hypothesis is evidence that the two differ, but failing to reject is not necessarily evidence that they are the same [23, 37].

In retrospect, it might have been preferable to formulate the null hypothesis to be the opposite of the validation goal, that is, that the algorithm's and the humans' routes were not comparable, and the test should have been used to check for convincing statistical evidence that they were comparable. Hypothesis testing can be a powerful validation tool, but care must be used in structuring the tests.

As another example, a different hypothesis test was used to validate a model of human walking [43]. The algorithm to be validated, which used a precomputed heading chart data structure, generated both routes and movement along those routes for synthetic human characters walking within rooms and hallways in a virtual-world simulation. The goal was for the algorithm to produce routes that resembled those followed by humans in the same situations, so validity in this example can be understood as realism. The desired similarity included both the actual route traversed and the kinematics (acceleration and deceleration rates, maximum and average speed, and turn rate) of the character's movement along the route.

The validity of the routes generated by the algorithm was evaluated using quantitative metrics that measured different aspects of the differences between the algorithm routes and the human routes. Three numerical error metrics that measured route validity were defined: (1) distance error, the distance between the algorithm's route and a human's route at each time step, averaged over all time steps required to traverse the route; (2) speed error, the difference between the speed of the moving character (on the algorithm's route) and the moving human (on a human's route) at each time step, averaged over all time steps; and (3) area error, the area between the algorithm's route and a human's route on the plane of the floor.

To acquire data that the algorithm's routes and movements could be compared with, observations of human subjects walking routes were recorded. A total of 30 human routes were measured in two situations that varied by room configuration and destination. Because the algorithm's routes were intended to mimic human movements, the routes recorded for the human subjects were assumed to be realistic (i.e., valid). For each of these 30 route situations, routes were also generated by the algorithm to be validated as well as a recent version of the well-known and widely used A* route planning algorithm [44].

Each of the two algorithms' (the new algorithm and the A* algorithm) routes were compared with the human routes by calculating values for the three error metrics for each of the algorithms' routes with respect to the corresponding human routes. The error metric values for the two algorithm's

---

*The "backward" Wilcoxon signed-rank test was formulated by this chapter's author. This example, with its self-admitted shortcoming, is deliberately included for its instructional value.

routes that corresponded to a single human route were compared statistically using paired-sample *t*-tests [38]. For the distance error and area error metrics, the new algorithm was significantly more realistic than the A* algorithm. There was no significant difference for the speed error metric. The hypothesis tests showed that the new algorithm's routes was more similar to the human routes (i.e., more realistic, or more valid) than the conventional A* algorithm for the walking movement application. Note that, as in the previous hypothesis testing example, quantitative metrics that measured the pertinent behavior of the model had to be defined; this is often an important step in applying a statistical hypothesis test to validation.

In the context of validating a discrete-event simulation model of queue lengths in a notional bank, a *t*-test was also used to compare the average customer delay observed in the bank with the average delay in the model's results [6].

## Formal Methods

*Formal V&V methods* employ mathematical proofs of correctness to establish model characteristics. Statements about the model are developed using a formal language or notation and manipulated using logical rules; conclusions derived about the model are unassailable from a mathematical perspective. However, formal methods are quite difficult to apply in practice, as the complexity of most useful models is too great for current tools and methods to deal with practically [2]. Nevertheless, useful results can be achieved using formal methods in some highly constrained situations. Formal methods *inductive assertions* and *predicate calculus* are defined here; other formal methods include *induction* and *proof of correctness* [2].

**Inductive Assertions**  The *inductive assertions* verification method compares the programming language code for the executable model to the descriptions of the simuland in the conceptual model. It is closely related to techniques from program proving. Assertion statements, which are statements about the input-to-output relations for model variables that must be true for the executable model to be correct, are associated with the beginning and end of each of the possible execution paths in the executable model. If it can then be proven for each execution path that the truth of beginning assertion and the execution of the instructions along the path imply the truth of the ending assertion, then the model is considered to be correct. The proofs are done using mathematical induction.

**Predicate Calculus**  *Predicate calculus* is a validation method that compares the simuland to the conceptual model.* Predicate calculus is a formal logic system that allows the creation, manipulation, and proof of formal statements

---

*Predicate calculus is also known as *first-order predicate calculus* and *predicate logic*.

that describe the existence and properties of objects. Characteristics of both the simuland and the conceptual model can be described using predicate calculus. One procedure of predicate calculus is the proving of arguments, which can demonstrate that one set of properties of the object in question, if true, together implies additional properties. The goal of the method is that by describing properties of the simuland and conceptual model using predicate calculus, it can be possible to prove that the two are consistent.

### Selecting and Applying V&V Methods

The specific methods to be applied to a given model depend on many factors, including model type, availability of simuland data for comparison, and ease of conducting multiple executions of the model. The model developer should examine previous V&V efforts for similar models for possible applicable methods. It should not be assumed that the process of V&V for any given model will always use exactly one of these methods. More often, performing V&V on a large complex model, or even a simpler one, will involve multiple methods (e.g., separate methods will be used for verification and for validation). For example, V&V of the Joint Training Confederation, a large distributed system assembled from multiple constructive models that supports command and battle staff training, used at least four different methods (event validation, face validation, sensitivity analysis, and submodel testing) [45].*

## VV&A CASE STUDIES

This section presents three VV&A case studies of greater length than the examples in the previous section. Each is intended to illustrate some aspect of the proper (and improper) conduct of VV&A.

### Case Study: The U.S. Army's Virtual Targets Center (VTC) Validation Process

V&V activities are not executed in isolation, but rather in the context of a project or process of model development. The requirements of a specific model development process often drive the timing and methods used to perform V&V, and conversely, V&V results as they are produced can determine the sequence of events within the process. Both of those effects are present in the validation process described in this case study, which is intended

---

*Interestingly, the event validation method appears in early versions of Balci's taxonomy of V&V methods (e.g., [18]) but not in later versions (e.g., [2]). *Event validation* compares identifiable events or event patterns in the model and simuland behaviors [18].

**Figure 10.7** Example target: BM-21 Grad truck-mounted multiple rocket launcher.

to illustrate how validation should be integrated into the process of model development and use.

*Virtual Targets* New and enhanced weapon systems are often tested with M&S, using digital virtual models of the targets they are intended to engage. Those target models must be detailed, accurate, and authoritative if the testing that uses them is to be credible. The U.S. Army's VTC is developing a library of extremely detailed, highly accurate, and carefully validated digital virtual target models.* The VTC develops virtual target models of a wide variety of military targets, including armored vehicles, mounted weapon systems, and air vehicles including helicopters, aerial targets, and unmanned aircraft. Figure 10.7 is an example of such a target. The detailed virtual target models provide the base from which appropriately configured models can be prepared to provide inputs into physics-based simulations to produce high-fidelity predictions of the target's observable signatures in the infrared (IR), radio frequency (RF), or visual frequency ranges. The virtual target models, and the signature predictions generated from them, are subsequently used by weapon system developers, test and evaluation engineers, and trainers in various ways within the simulations employed in those communities [46]. The models are reusable and are freely available to qualified users via an online accessible repository.**

---

*The VTC is a collaborative activity of two U.S. Army organizations: the System Simulation and Development Directorate's End Game Analysis Functional Area of the Aviation and Missile Research, Development, and Engineering Center and the Program Manager–Instrumentation, Targets, and Threat Simulators Target Management Office of the Program Executive Office–Simulation, Training, and Instrumentation.

**The VTC model library currently has approximately 15,000 target models of approximately 400 distinct target items. (There may be multiple target models per target item to allow for variations in configuration, damage states, color, etc.) Approximately 4000–4500 model downloads have occurred per year since the VTC effort began in 1999.

A target model is a digital representation of the geometry of a target item (such as a truck or helicopter). The various surfaces, panels, protrusions, and parts of the target item are represented as facets with attributes that include their three-dimensional spatial coordinates relative to the overall target equipment item, material type, and reflectivity. The geometry data are stored using standard engineering geometry file formats. The development of a new target model may begin from a previously existing geometry file (such as a computer-aided design, or CAD, file for the target item) developed for some other application. If such a file is not available, it may be necessary to acquire the target item geometry directly by taking measurements from an actual physical example or from available photographs or design drawings of the target item. The resulting geometry data file might typically comprise 2 million facets, with some models comprised of up to 15 million facets. In any case, once acquired, the geometry file constitutes an unvalidated target model.

**Validation Process** The process employed by the VTC to validate an unvalidated target model has eight steps, illustrated in Figure 10.8. The steps of that process are described next; the description and the figure are adapted from the U.S. Army's VTC [47].

*Step (1) Request Validation.* The validation process begins with a request for validation support to validate a particular target model from the appropriate intelligence production center or project office. The request is accompanied by information needed in the validation process, including the following:

(1) Baseline description and technical details of the target item.
(2) Characteristics that distinguish the target item from similar items of its type (e.g., how does this truck differ from other trucks).
(3) Features that significantly affect the target item's RF, IR, and visual signatures.
(4) Authoritative signature data for the target item in the frequency ranges of interest; these referent data will serve as comparison data during the validation—the source of these data will be discussed later.
(5) Authorization for assistance from subject matter experts in the validation process.

*Step (2) Prepare Model.* Because the target model to be validated may come from a variety of sources, the validation process begins with a reengineering of the model. It is subdivided into its major components (e.g., turret and hull for a tank) and existing parent–child component relationships are checked. The geometry is examined for defects such as intersecting or misaligned facets. The components and attributes of the digital model are compared with the target descriptions provided in the request.

*Step (3) Reconstitute Model.* Once the underlying geometry model has been prepared, the virtual target model is "reconstituted," that is, converted, from

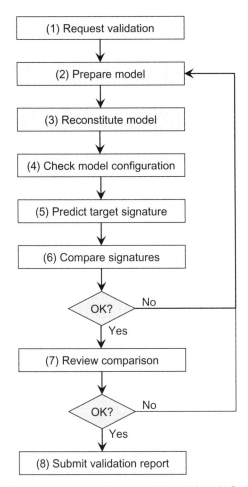

**Figure 10.8** VTC validation process flowchart (adapted from U.S. Army VTC, 2005).

the base file format to a file format appropriate to the intended application of the model. This step is necessary because the virtual target model may be input to RF signature prediction, IR signature prediction, or visualization simulations depending on application, each of which requires a different input file format.

*Step (4) Check Model Configuration.* Configuration refers to the positioning of articulated or movable portions of the target item, such as azimuth angle of a tank turret or the elevation status (up for launch or down for transport) of a truck-mounted rocket launcher. The configuration of the virtual target model must match the configuration of the actual target item used to collect the authoritative signature data. The target model's configuration is adjusted to match the configuration of the hardware item at the time signature data were collected.

**Figure 10.9** Example of a predicted RF signature (U.S. Army RDECOM, 2009).

*Step (5) Predict Target Signature.* The prepared, reconstituted, and checked target model produced in the preceding steps is input to signature prediction software, which generates output predicting the target's signature using the geometry and other target attribute information in the target model and its own physics-based models. The signature prediction software used and the nature of its output depend on the signature of interest (RF, IR, or visual).* For example, Figure 10.9 is a visualization of the RF scatter produced for the BM-21 [46].

*Step (6) Compare Signatures.* The predicted signature generated by the signature predication software is compared with the authoritative signature provided as part of the validation request. (The methods used to compare the signatures will be discussed later.) If the predicted and authoritative signatures are found to be sufficiently close, the process continues at step (7); if the signatures are not sufficiently close, the process reverts to step (2) to correct the model.

*Step (7) Review Comparison.* Subject matter experts review the results of comparing the predicted and authoritative signatures. They confirm that correct authorization signature data were used, that the comparison was conducted using an appropriate method, and that the differences between the signatures were within acceptable tolerances. If the experts judge the comparison to be acceptable, the process continues at step (8); if the signatures are not sufficiently close, the process reverts to step (2) to correct the model.

---

*RF signature prediction may be done using the Xpatch software; Xpatch is distributed by the U.S. Air Force Research Laboratory. IR signature prediction may be done using the MuSES software; MuSES is a proprietary software product of ThermoAnalytics, Inc.

*Step (8) Submit Validation Report.* A report documenting the virtual target model and the validation process is prepared. The report includes identification of the signature prediction software used, the results of the signature comparison, the intended uses for the target model assumed during validation, and any limitations to its use (bounds of validity) found during validation. The report is submitted to the validating authority and the target model is made available for authorized use.

**Validation Comparison** In step (6) of the validation process, the target item's reflectance or emissions signature in the frequency range of interest is compared with an authoritative signature in the same frequency range provided with the validation request. The authoritative signature consists of measurements taken under controlled conditions from an actual physical example of the modeled target item. For example, for an RF signature, measurements would typically be taken by irradiating the target item with a radar beam of an appropriate frequency and measuring the target item's radar reflection using sensors. The target item would be positioned on a large turntable, allowing measurements to be conveniently taken across the full range of azimuth angles, for a given elevation angle, by rotating the turntable between measurements.

The comparison of the predicted signature (model data) to the authoritative signature (referent data) is conducted using previously agreed upon methods, metrics, and tolerances. The specific comparison method depends on the frequency range. For example, for RF signatures, the plot of the predicted signature (radar cross section) must be within bands established either by multiple authoritative signature data sets or within a predetermined range of a single authoritative data set. Figure 10.10 illustrates an RF comparison.* The upper graph shows radar cross-section values for both the referent data (black line) and the model data (gray line) at each azimuth angle. The lower graph shows the difference between the two data sets, again at each azimuth angle. If the difference between the two exceeds a given magnitude at a given number of azimuth angles, the model is not validated; the specific error tolerances depend on the model's intended application and are set in advance by subject matter experts.

For IR signatures, temperature differences between the predicted signature and the authoritative signature at appropriate areas of interest must be within specified tolerances. For visual signatures, a human observer must be able to recognize target features on the predicted signature (which will be a visual image) that are also recognizable on the target item.

---

*The data values in the figure, while meant to be typical, are notional and do not correspond to any particular target item or target model signatures.

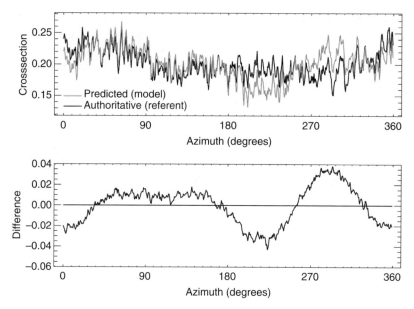

**Figure 10.10** Example of an RF signature comparison.

***Case Study Conclusions*** The VTC validation process exemplifies how a validation comparison must be set within the context of an overall process of model development and validation, and how the development process and the validation comparison can affect each other. Although the details will differ across projects and models, the essential idea is broadly applicable; effective validation depends on proper positioning of validation within the development process and response of the development process to validation findings.

Finally, it is worth addressing a question raised by this case study that applies more generally. One might wonder, if an authoritative signature is available (as it must be to perform the validation as described here), why it is necessary to bother with the predicted signature at all? In other words, why not simply use the authoritative signature for the target item and forgo the development and validation of the target model? The answer is that measuring a signature for an actual physical target item is time-consuming and expensive, and typically can only be done for a single, or limited number of, target item configurations and elevation angles. However, information about the target item's signature is very likely to be needed by model users for multiple configurations and elevation angles. Once the target model has been validated for a given configuration and elevation angle using the available referent data as described in this case study, the model can be used to predict signatures at other configurations and elevation angles with increased confidence and reliability.

## VV&A CASE STUDIES 361

**Case Study: The Crater Model and the Columbia Disaster**

As noted earlier, one objective of validation is to establish the range of input values for which a model's results can be considered usefully valid, that is, to determine the model's bounds of validity. This case study examines a well-known tragedy that included, as one aspect of a much larger chain of events, instances where models were used outside their validated bounds. The content of this case study is based largely on Gehman et al. [48].

***The Columbia Disaster*** National Aeronautics and Space Administration's (NASA) first space-worthy space shuttle *Columbia* flew for the first time in April 1981. Its final flight, designated STS-107, was a 17-day multidisciplinary earth science and microgravity research mission. On February 1, 2003, *Columbia* broke up over Texas during reentry, killing the entire seven-member crew.

The damage that ultimately caused the loss occurred at the beginning of the mission, approximately 82s after launch. A suitcase-sized piece of the foam insulation from the shuttle's external fuel tank broke off and struck the leading edge of *Columbia's* left wing. The foam fragment was later estimated to weigh approximately 1.2lb and have a volume of approximately 1200 in$^3$. It was estimated to have impacted the wing at a relative velocity of 625–840 ft/s and an impact angle of approximately 20 degrees. Ground experiments subsequent to the accident indicated that this impact could have created a hole in the wing's thermal insulation that was possibly as large as 10 in in diameter.

Examination of launch video the day after the launch, while *Columbia* was in orbit, showed the impact but not whether damage had resulted. Engineering analyses were conducted in an attempt to assess the extent of the damage and the resulting risk. The assessment did not find sufficient evidence to indicate an unsafe situation and *Columbia* was given permission for reentry. During the course of the reentry, the hole in the wing's thermal insulation allowed hot gases to enter and burn through the wing's structure, leading to catastrophic failure of the wing structure and the consequent loss of the orbiter and crew.

***The Crater Model and the Columbia Analysis*** The damage assessment process conducted during the mission examined possible damage to the orbiter's thermal protection system, which is essential to withstand the tremendous temperatures experience during reentry (at some points more than 2800°F). Because the exact impact location on the orbiter was not known, two components of the thermal protection system were considered: the thermal insulating tiles that covered the bottom of the orbiter, including the wings, and the reinforced carbon–carbon panels in the orbiter's wing's leading edges.

Possible damage to the thermal tiles was investigated using a model known as Crater. Crater is a mathematical model that predicts the depth of penetration to be expected from a foam, ice, or debris projectile impacting one of the thermal tiles. Figure 10.11 shows the Crater model. It is most often used to analyze possible damage caused by ice debris falling from the external tank

$$p = \frac{0.0195\,(L/d)0.45(d)(\rho_P)^{0.27}(V-V^*)^{2/3}}{(S_T)^{1/4}(\rho_T)^{1/6}}$$

where
- $p$ = penetration depth
- $L$ = length of foam projectile
- $d$ = diameter of foam projectile
- $\rho_P$ = density of foam
- $V$ = component of foam velocity at right angle to foam
- $V^*$ = velocity required to break through the tile coating
- $S_T$ = compressive strength of tile
- $\rho_T$ = density of tile
- 0.0195 = empirical constant

**Figure 10.11** The Crater model (Gehman et al. [48]).

during launch. Input parameters to the Crater model include length, diameter, and density of the projectile and the compressive strength and density of the tile. The Craft model also includes an empirical parameter that affects the model's predictions. That parameter, and the overall model, had been calibrated and validated by comparison to actual impact data several times, including ice droplet tests in 1978 and foam insulation projectile tests in 1979 and 1999. The projectiles used in these tests had a maximum volume of 3 in³. The tests showed that within the limits of the test data, Crater's damage predictions tended to be more severe than the actual damage observed during the test, a tendency referred to by engineers as "conservative."

For projectiles, the size of the *Columbia* foam fragment, approximately 400 times larger than those for which the model had been validated, the accuracy of the model was not known. For the *Columbia* fragment, Crater predicted a penetration depth that was greater than the tile thickness; that is, it predicted that the impact would have created a hole in the tile that would expose the orbiter's wing structure to the extreme temperatures of reentry. However, because Crater's predictions within its validated limits were known to be conservative, the model's penetration finding was discounted.

Another model, similar to Crater, was used to investigate possible damage to the reinforced carbon–carbon panels in the leading edge of the left wing. This model, which was designed to estimate the panel thickness needed to withstand impacts, was calibrated and validated in 1984 using impact data from ice projectiles with a volume of approximately 2.25 in³. Analysis using this model for the much larger *Columbia* foam fragment indicated that impact angles greater than 15 degrees would penetrate the panels. However, because the foam that impacted Columbia was less dense than the ice for which the model had been validated, the analysts concluded that the panel would not be penetrated by foam impacts at up to 21 degrees, and the model's penetration finding was discounted.

***Case Study Conclusions*** The size of foam fragment that damaged *Columbia* had an estimated volume well outside of the bounds for which the two models had been validated. (Several other parameters of the foam fragment were outside those values as well.) A simple conclusion would be to present this as an example of a model that, when used outside of its bounds of validity, gives an incorrect answer. However, the case of the *Columbia* damage analysis is not so simple. Although the two models were used for input values beyond what they had been validated for, they both indicated the possibility of significant and dangerous damage to the orbiter's thermal protection, which regrettably proved to be correct.

The actual conclusion is that it is possible that the fact that the models were used outside their bounds of validity misled the analysts into discounting the models' predictions and underestimating the risk. On this point, the Columbia Accident Investigation Board Report found that "the use of Crater in this new and very different situation compromised [the analysts'] ability to accurately predict debris damage in ways [they] did not fully comprehend" [48].

## Case Study: Validation Using Hypothesis Testing

As noted earlier, hypothesis testing is a broadly useful statistical technique that can be used as a dynamic validation method. This case study examines, in a bit more detail than the earlier examples, one application of a hypothesis test.

***Model and Validation Procedure*** The model to be validated was a model of human decision making [28]. The model's decision-making algorithm was based on recognition-primed decision making [49], a psychological model of decision making by experienced human decision makers. It was implemented using multiagent software techniques. The goal of the model, known as RPDAgent, was not to make theoretically optimum decisions, but rather to mimic the decisions of human decision makers. In particular, the model was challenged to model the decisions of senior military commanders at the operational level of warfare.

A test scenario based on an amphibious landing was devised to validate the model. The scenario included four specific decisions, or decision points: one of four landing locations, one of four landing times, one of three responses to changes in enemy deployment, and one of two responses to heavy friendly casualties had to be selected. The choices were numerically coded to allow the computation of mean decisions.

Relevant military information about the decision choices was included with the scenario and used by the decision makers. Thirty human military officers were asked to make selections for each of the four decision points within the test scenario. The RPDAgent model has stochastic elements, so 200 replications of 30 sets of the four decisions were generated using the model. For each of the four decision points, the distribution of the human decisions was compared with the distribution of RPDAgent's decisions choice by choice; the number of the humans selected a particular choice was compared with the mean number of times the choice was selected by the model over the 200 replications.

**Hypothesis Test Used in Validation** The statistical method employed for the comparison, equivalency testing, can be used to determine if the difference between two distributions is insignificant, as defined by a maximum difference parameter [50]. For the RPDAgent validation, the maximum difference was set to 20 percent, a value judged to be small enough for realism but large enough to allow for reasonable human variability.

The hypotheses and statistics used for the equivalency testing were

Test 1 hypotheses

$$h_0 : \bar{X} - \mu \leq \delta_1$$
$$h_a : \bar{X} - \mu > \delta_1$$

Test 2 hypotheses

$$h_0 : \bar{X} - \mu \geq \delta_2$$
$$h_a : \bar{X} - \mu < \delta_2$$

Test 1 test statistic

$$t_1 = \frac{(\bar{X} - \mu) \leq \delta_1}{S_{\bar{X} - \mu}}$$

Test 2 test statistic

$$t_2 = \frac{(\bar{X} - \mu) \leq \delta_2}{S_{\bar{X} - \mu}}$$

where

$\bar{X}$ = mean model decision response (times a choice was selected)
$\mu$ = human decision response (times a choice was selected)
$\delta_1$ = lower limit of the equivalency band
$\delta_2$ = upper limit of the equivalency band

Two statistical hypothesis tests are needed in equivalency testing. For this application, test 1 determines if the model mean is less than or equal to the human mean allowing for the equivalency band; rejecting the null hypothesis shows that it is not. Test 2 determines if the model mean is greater than or equal to the human mean, allowing for the equivalency band; rejecting the null hypothesis shows that it is not. Rejecting both null hypotheses shows that the distributions are equivalent within $\pm \delta$. For both tests, a one-tailed $t$-test was used with $\alpha = 0.05$, giving a critical value for the test statistic $t = 1.645$.

The model and human means were compared in this manner for each of the choices for each of the four decision points in the test scenario. In every comparison, the calculated test statistics exceeded the critical value for both tests, leading to the rejection of both null hypotheses, thereby supporting the conclusion that the model and human decision means were equivalent within the difference parameter.

***Case Study Conclusions*** Using a statistical validation technique is not always possible. To do so, the data needed to support it must be available, and the appropriate technique must be selected and properly employed. However, when these conditions are met, such methods can be quite powerful, as this case study shows. Note that the hypothesis test structure here avoided the "backward" structure of the example given earlier; here, the two null hypotheses were that the model behavior was different from the human behavior, and showing that the model behavior was comparable to the human behavior required rejecting both null hypotheses.

## CONCLUSION

This section first identifies a set of challenges that arise in the practice of VV&A. It then offers some concluding comments.

### VV&A Challenges

As with any complex process, the practice of VV&A involves challenges. In this section, five important VV&A challenges are presented.

***Managing VV&A*** V&V activities must be done at the proper times within a simulation project, and they must be allocated sufficient resources. As discussed earlier, the project artifacts to be compared in V&V become available at different times in the project; for example, the executable model will be available for comparison to the model before the results are available for comparison to the simuland. Because of this, the guideline that V&V should be done "as soon as possible" implies that V&V should be done over the course of the project.* Unfortunately, in real-world simulation projects, V&V are all too often left to the end of simulation projects. This has at least two negative consequences. First, problems are found later than they might have been, which almost always makes them more difficult and costly to correct. Second, by the end of the project, schedule and budget pressures can result in V&V being given insufficient time and attention [12]. All too often, V&V are curtailed in simulation projects facing impending deadlines,

---

*Indeed, in the list of 15 principles of V&V, this is the first [2].

producing a model that has not been adequately validated and results that are not reliable. The project manager must ensure that V&V are not skipped or shortchanged.

***Interpreting Validation Outcomes*** For all but the simplest models, it is rarely correct to simply claim that a model is "valid" or has "been validated."* Almost any model with practical utility is likely to be sufficiently complex that a more nuanced and qualified description of its validity is necessary. For such models, validation is likely to show that the model is valid (i.e., is accurate enough to be usable) under some conditions, but not under others. A model might be valid for inputs within a range of values, but not outside those values; for example, the simple gravity model presented earlier could be considered valid for small initial height values, but not height values so large that the distance implied by the height value would cause the acceleration due to gravity to be noticeably less. Or, a model might be valid while operating within certain parameters, but not outside those; for example, an aircraft flight dynamics model could be valid for subsonic velocities, but not for transonic or supersonic. Finally, the accuracy of a model could be sufficient for one application, but not for another; for example, a flight simulator might be accurate enough for entertainment or pilot procedures training, but not accurate enough for official aircraft type qualification.

In all of these cases, describing the model as "valid" without the associated conditions is misleading and potentially dangerous. The person performing the validation must take care to establish and document the bounds of validity for a model, and the person using the model must likewise take care to use it appropriately within those bounds of validity.

***Combining Models*** Models are often combined to form more comprehensive and capable models.** For example, a mathematical model of an aircraft's flight dynamics might be combined with a visual model of its instrument panel in a flight simulator. The means of combination are varied, including direct software integration, software architectures, interface standards, and networked distributed simulation; these details of these methods are beyond the scope of this chapter (see Davis and Anderson [51] and Weisel et al. [52] for details). However, the combination of models introduces a validation challenge. When models that have been separately validated are combined, what can be said about the validity of the combination? Quite frequently, in both the research literature and in the practice of M&S (especially the latter), it is assumed that the combination of validated models must be valid, and that validation of the combination is unnecessary or redundant. In fact, a combination of validated models is not necessarily valid; this has been recognized for some time [2] and more recently formally proven [53]. Consequently, when

---
*The second principle of V&V makes this point [2].
**In the literature, the terms *integrated* or *composed* [54] are used more often than *combined*.

models are combined, a complete validation approach will validate each submodel separately (analogous to unit testing in software engineering) and then validate the combined composite model as a whole (analogous to system testing).*

***Data Availability*** Validation compares the model's results to the simuland's behavior. To make the comparison, data documenting observations of the simuland is required. For some simulands such data may be available from test results, operational records, or historical sources, or the data may be easily obtainable via experimentation or observation of the simuland. However, not all simulands are well documented or conveniently observable in an experimental setting. A simuland may be large and complex, such as theater-level combat or national economies, for which history provides only a small number of observable instances and data sufficiently detailed for validation may not have been recorded. Or, a simuland may not actually exist, such as a proposed new aircraft design, for which observations of the specific simuland will not be available. The availability (or lack thereof) of reliable data documenting observations of the simuland will often determine which validation method is used. This is one reason for the relatively frequent application of face validation; subject matter experts are often available when data necessary to support a quantitative validation method, such as statistical analysis, are not.

***Identification of Assumptions*** Assumptions are made in virtually every model. The example gravity model assumes proximity to the earth's surface. A flight simulator might assume calm, windless air conditions. The assumptions themselves are not the challenge; rather, it is the proper consideration of them in V&V that is. Models that are valid when their underlying assumptions are met may not be outside when they are not, which implies that the assumptions, which are characteristics of the model, become conditions on the validity of the model. This is not a small issue; communicating all of the assumptions of a model to the person performing the validation is important, but the developer of a model can make many assumptions unconsciously and unrealized, and identifying the assumptions can be difficult. This point is made more formally in Spiegel et al. [55], where the assumptions upon which a model's validity depends are captured in the idea of validation constraints.

## Concluding Remarks

The three central concepts of VV&A asserted in this chapter were the following:

---

*In Balci [2], there are five levels of testing. The separate validation of submodels is level 2, *submodel (module) testing*, and the combined validation of the composite model is level 4, *model (product) testing*.

(1) V&V are processes that compare artifacts of model development to each other, or model results to observations or knowledge of the simuland. Specific V&V methods are structured means of making those comparisons.
(2) The outcome of V&V is most often not a simplistic determination that the model is entirely correct or incorrect [2,6]. Rather, V&V serve to determine the degree of accuracy a model has and the ranges or types of inputs within which that accuracy is present, that is, the model's bounds of validity.
(3) The accuracy of a model with respect to its simuland, or more precisely, the standard of accuracy a model must meet, depends on its intended application. The effort spent on V&V and the standard to be applied in accreditation depend on the need for model accuracy and the consequences of model inaccuracy within the model's application.

As already stated, V&V are essential prerequisites to the credible and reliable use of a model and its results. V&V reveal when and how a model should be. Shortchanging V&V is false economy, as the consequences of using an invalid model can in some circumstances be dire. Accreditation is the crucial decision regarding the suitability of a model for an application. The accreditation decision can only be made successfully when fully informed by the results of properly conducted V&V.

## ACKNOWLEDGMENTS

Stephanie E. Brown and Ann H. Kissell (U.S. Army VTC) provided information for the VTC case study. William V. Tucker (Boeing) suggested the topic for the Crater model case study. Wesley N. Colley (University of Alabama in Huntsville) helped to prepare several of the figures. Their support is gratefully acknowledged.

## REFERENCES

[1] Petty MD. Verification and validation. In *Principles of Modeling and Simulation: A Multidisciplinary Approach.* Sokolowski JA, Banks CM (Eds.). Hoboken, NJ: John Wiley & Sons; 2009, pp. 121–149.

[2] Balci O. Verification, validation, and testing. In *Handbook of Simulation: Principles, Methodology, Advances, Applications, and Practice.* Banks J (Ed.). New York: John Wiley & Sons; 1998, pp. 335–393.

[3] Modeling and Simulation Coordination Office. Verification, Validation, and Accreditation (VV&A) Recommended Practices Guide. September 15 2006. Available at http://vva.msco.mil. Accessed May 31, 2009.

[4]  Ford T. Helicopter simulation. *Aircraft Engineering and Aerospace Technology*, 69(5):423–427; 1997.

[5]  Kesserwan N. Flight Simulation. M.S. Thesis, McGill University, Montreal, Canada; 1999.

[6]  Banks J, et al. *Discrete-Event System Simulation*. 4th ed. Upper Saddle River, NJ: Prentice Hall; 2005.

[7]  Rumbaugh J, et al. *The Unified Modeling Language*. Reading, MA: Addison-Wesley; 1999.

[8]  Fontaine MD, et al. Modeling and simulation: Real-world examples. In *Principles of Modeling and Simulation: A Multidisciplinary Approach*. Sokolowski JA, Banks CM (Eds.). Hoboken, NJ: John Wiley & Sons; 2009, pp. 181–245.

[9]  Sargent RG. Verification, validation, and accreditation of simulation models. In *Proceedings of the 2000 Winter Simulation Conference*. Orlando, FL, December 10–13, 2000, pp. 50–59.

[10] U.S. Department of Defense. DoD Modeling and Simulation (M&S) Verification, Validation, and Accreditation (VV&A). *Department of Defense Instruction 5000.61*, May 13, 2003.

[11] Cohn J. Building virtual environment training systems for success. In *The PSI Handbook of Virtual Environment Training and Education: Developments for the Military and Beyond; Volume 1: Learning, Requirements, and Metrics*. Nicholson D, Schmorrow D, Cohn J (Eds.). Westport, CT: Praeger Security International; 2009, pp. 193–207.

[12] Balci O. Verification, validation, and accreditation. In *Proceedings of the 1998 Winter Simulation Conference*. Washington, DC, December 13–16, 1996, pp. 41–48.

[13] Royce W. *Software Project Management: A Unified Framework*. Reading, MA: Addison-Wesley; 1998.

[14] Jacoby SLS, Kowalik JS. *Mathematical Modeling with Computers*. Englewood Cliffs, NJ: Prentice Hall; 1980.

[15] Balci O, Sargent RG. A methodology for cost-risk analysis in the statistical validation of simulation models. *Communications of the ACM*, 24(4):190–197; 1981.

[16] Shannon RE. *Systems Simulation: The Art and Science*. Upper Saddle River, NJ: Prentice Hall; 1975.

[17] Sargent RG. Verifying and validating simulation models. In *Proceedings of the 1996 Winter Simulation Conference*. Coronado, CA, December 8–11, 1996, pp. 55–64.

[18] Balci O. Guidelines for successful simulation studies. In *Proceedings of the 1990 Winter Simulation Conference*. New Orleans, LA, December 9–12, 1990, pp. 25–32.

[19] Balci O. Validation, verification, and testing techniques throughout the life cycle of a simulation study. In *Proceedings of the 1994 Winter Simulation Conference*. Lake Buena Vista, FL, December 11–14, 1994, pp. 215–220.

[20] Belfore LA, et al. Capabilities and intended uses of the Joint Operations Feasibility Tool. *Proceedings of the Spring 2004 Simulation Interoperability Workshop*. Arlington, VA, April 18–23, 2004, pp. 596–604.

[21] Grant S, Galanis G. Assessment and prediction of effectiveness of virtual environments: Lessons learned from small arms simulation. In *The PSI Handbook of Virtual Environment Training and Education: Developments for the Military and Beyond; Volume 3: Integrated Systems, Training Evaluations, and Future Directions.* Nicholson D, Schmorrow D, Cohn J (Eds.). Westport, CT: Praeger Security International; 2009, pp. 206–216.

[22] Knepell PL, Arangno DC. *Simulation Validation: A Confidence Assessment Methodology.* New York: John Wiley & Sons; 1993.

[23] Moya LJ, McKenzie FD, Nguyen QH. Visualization and rule validation in human-behavior representation. *Simulation & Gaming*, 39(1):101–117; 2008.

[24] Petty MD. Behavior generation in semi-automated forces. In *The PSI Handbook of Virtual Environment Training and Education: Developments for the Military and Beyond; Volume 2: VE Components and Training Technologies.* Nicholson D, Schmorrow D, Cohn J (Eds.). Westport, CT: Praeger Security International; 2009, pp. 189–204.

[25] Turing AM. Computing machinery and the mind. *Mind*, 59(236):433–460; 1950.

[26] Wise BP, Miller D, Ceranowicz AZ. A framework for evaluating computer generated forces. In *Proceedings of the Second Behavioral Representation and Computer Generated Forces Symposium.* Orlando, FL, May 6–7, 1991, pp. H1–H7.

[27] Petty MD. The Turing test as an evaluation criterion for computer generated forces. In *Proceedings of the Fourth Conference on Computer Generated Forces and Behavioral Representation.* Orlando, FL, May 4–6, 1994, pp. 107–116.

[28] Sokolowski JA. Enhanced decision modeling using multiagent system simulation. *SIMULATION*, 79(4):232–242; 2003.

[29] Desel J, Oberweis A, Zimmer T. A test case generator for the validation of high-level petri nets. In *Proceedings of the International Conference on Emerging Technologies and Factory Automation.* Los Angeles, CA, September 9–12, 1997, pp. 327–332.

[30] Cohen ML, et al. *Statistics, Testing, and Defense Acquisition, New Approaches and Methodological Improvements.* Washington DC: National Research Council, National Academy Press; 1998.

[31] Miller DR. Sensitivity analysis and validation of simulation models. *Journal of Theoretical Biology*, 48(2):345–360; 1974.

[32] Harrison A, Winters J, Anthistle D. Ironside: A command and battle space simulation. In *Proceedings of the 1999 Summer Computer Simulation Conference.* Chicago, IL, July 11–15, 1999, pp. 550–554.

[33] Petty MD, Panagos J. A unit-level combat resolution algorithm based on entity-level data. In *Proceedings of the 2008 Interservice/Industry Training, Simulation and Education Conference.* Orlando, FL, December 1–4, 2008, pp. 267–277.

[34] Herington J, et al. Representation of historical events in a military campaign simulation model. In *Proceedings of the 2002 Winter Simulation Conference.* San Diego, CA, December 8–11, 2002, pp. 859–863.

[35] Poncelin de Raucourt VPM. The reconstruction of part of the Battle of Medenine. Unpublished M.Sc. Thesis, The Royal Military College of Science, Shrivenham, UK; 1997.

[36] Tournes C, Colley WN, Umansky M. C²PAT, a closed-form command and control modeling and simulation system. In *Proceedings of the 2007 Huntsville Simulation Conference*. Huntsville, AL, October 30–November 1, 2007.

[37] Simpkins SD, et al. Case study in modeling and simulation validation methodology. In *Proceedings of the 2001 Winter Simulation Conference*. Arlington, VA, December 9–12, 2001, pp. 758–766.

[38] Bhattacharyya GK, Johnson RA. *Statistical Concepts and Methods*. New York: John Wiley & Sons; 1977.

[39] Benfield MPJ. Advanced Chemical Propulsion System (ACPS) validation study. Unpublished presentation, University of Alabama in Huntsville, November 28, 2007.

[40] Box GEP, et al. *Statistics for Experimenters: An Introduction to Design, Data Analysis, and Model Building*. New York: John Wiley & Sons; 1978.

[41] Van Brackle DR, et al. Terrain reasoning for reconnaissance planning in polygonal terrain. In *Proceedings of the Third Conference on Computer Generated Forces and Behavioral Representation*. Orlando, FL, March 17–19, 1993, pp. 285–306.

[42] Petty MD, Van Brackle DR. Reconnaissance planning in polygonal terrain. In *Proceedings of the 5th International Training Equipment Conference*. The Hague, The Netherlands, April 26–28, 1994, pp. 314–327.

[43] Brogan DC, Johnson NL. Realistic human walking paths. In *Proceedings of International Computer Animation and Social Agents*. New Brunswick, NJ, May 7–9, 2003, pp. 94–101.

[44] Bandi S, Thalmann D. Path finding for human motion in virtual environments. *Computational Geometry: Theory and Applications*, 15:103–127; 2000.

[45] Page EH, et al. A case study of verification, validation, and accreditation for advanced distributed simulation. *ACM Transactions on Modeling and Computer Simulation*, 7(3):393–424; 1997.

[46] U.S. Army Research, Development, and Engineering Command. Virtual targets center modeling process. Unpublished presentation to the NATO RTO MSG-058 Group (Conceptual Modeling for M&S), January 16, 2009.

[47] U.S. Army Virtual Targets Center. Threat virtual target validation process. Unpublished document, July 26, 2005.

[48] Gehman HW, et al. *Columbia Accident Investigation Board Report Volume I*. National Aeronautics and Space Administration; August 2003.

[49] Klein G. Strategies of decision making. *Military Review*, 69(5):56–64; 1989.

[50] Rogers JL, Howard KI, Vessey JT. Using significance tests to evaluate equivalence between two experimental groups. *Psychological Bulletin*, 113(3):553–565; 1993.

[51] Davis PK, Anderson RH. *Improving the Composability of Department of Defense Models and Simulations*. Santa Monica, CA: RAND National Defense Research Institute; 2003.

[52] Weisel EW, Petty MD, Mielke RR. A survey of engineering approaches to composability. In *Proceedings of the Spring 2004 Simulation Interoperability Workshop*. Arlington, VA, April 18–23, 2004, pp. 722–731.

[53] Weisel EW, Mielke RR, Petty MD. Validity of models and classes of models in semantic composability. In *Proceedings of the Fall 2003 Simulation Interoperability Workshop.* Orlando, FL, September 14–19, 2003, pp. 526–536.

[54] Petty MG, Weisel EW. A composability lexicon. In *Proceedings of the Spring 2003 Simulation Interoperability Workshop.* Orlando, FL, March 30–April 4, 2003, pp. 181–187.

[55] Spiegel M, et al. A case study of model context for simulation composability and reusability. In *Proceedings of the 2005 Winter Simulation Conference.* Orlando, FL, December 4–7, 2005, pp. 436–444.

# 11

# AN INTRODUCTION TO DISTRIBUTED SIMULATION

Gabriel A. Wainer and Khaldoon Al-Zoubi

Distributed simulation technologies were created to execute simulations on distributed computer systems (i.e., on multiple processors connected via communication networks) [1]. Distributed simulation is a computer program that models real or imagined systems over time. On the other hand, distributed computer systems interconnect various computers (e.g., personal computers) across a communication network. Therefore, distributed simulation deals with executing simulation correctly over interconnected multiple processors. Correctness means that the simulation should produce the same results as if it was executed sequentially using a single processor. Fujimoto distinguished parallel from distributed simulation by their physical existence, used processors, communication network, and latency [1]. Parallel systems usually exist in a machine room, employing homogeneous processors, and communication latency is measured with less than $100\,\mu s$. In contrast, distributed computers can expand from a single building to global networks, often employing heterogeneous processors (and software), and communication latency is measured with hundreds of microseconds to seconds. The simulation is divided spatially (or temporally) and mapped to participating processors. Our focus here is on distributed simulation, which employs multiple distributed computers to execute the same simulation run over a wide geographic area.

---

*Modeling and Simulation Fundamentals: Theoretical Underpinnings and Practical Domains*, Edited by John A. Sokolowski and Catherine M. Banks
Copyright © 2010 John Wiley & Sons, Inc.

A focus of distributed simulation software has been on how to achieve model reuse via interoperation of heterogeneous simulation components. Other benefits include reducing execution time, connecting geographically distributed simulation components (without relocating people/equipment to other locations), interoperating different vendor simulation toolkits, providing fault tolerance, and information hiding—including the protection of intellectual property rights [1,2].

## TRENDS AND CHALLENGES OF DISTRIBUTED SIMULATION

The defense sector is currently one of the largest users of distributed simulation technology. On the other hand, the current adoption of distributed simulation in the industry is still limited. In recent years, there have been some studies (conducted in the form of surveys) to analyze these issues [2–4]. The surveys collected opinions, comments, and interviews of experts from different backgrounds in the form of questionnaires and showed that there is now an opportunity for distributed simulation in industry. It has been predicted that in the coming years, the sectors that will drive future advancement in distributed simulation are not only the defense sector, but also the gaming industry, the high-tech industry (e.g., auto, manufacturing, and working training), emergency, and security management [4].

The *high-level architecture* (HLA) is the preferred middleware standard in the defense sector [5]. However, its popularity in industry is limited. The HLA started as a large project mainly funded by the military, in order to provide the means for reusing legacy simulations in military training operations, so that this exercise could be conducted between remote parties in different fields, reusing existing simulation assets. On the other hand, the adoption of these technologies in the industry is based on return-of-investment policies. Therefore, most *commercial off-the-shelf* (COTS) simulation packages do not usually support distributed simulation due to a cost/benefit issue. In "A survey on distributed simulation in industry," the authors suggested that in order to make distributed simulation more attractive to the industrial community, we need a lightweight COTS-based architecture with higher cost/benefit ratio [2]. The middleware should be easy to understand (e.g., programming interface, fast development, and debugging), and interoperable with other vendor's simulation components. Distributed simulation might become a necessity when extending the product development beyond factory walls, particularly when such organizations prefer to hide detailed information [6]. New standards (for instance, COTS simulation package interoperability, core manufacturing simulation data, and Discrete-Event System Specification [DEVS]) can contribute to achieve these goals [7].

Another recent study, carried out by Strassburger et al., focused on surveying experts from the area of distributed simulation and distributed virtual environment [4]. This study found out that the highest rated applications in future distributed simulation efforts include the integration of heterogeneous resources, and joining computer resources for complex simulations and training sessions. The study also identified some research challenges:

(1) *Plug-and-play capability*: The middleware should be able to support coupling simulation models in such a way that the technical approach and standards gain acceptance in industry. In other words, interoperability should be achieved effortlessly.
(2) *Automated semantic interoperability between domains*: To achieve the plug-and-play challenge, interoperability must be achieved at the semantic level.

## A BRIEF HISTORY OF DISTRIBUTED SIMULATION

Simulations have been used for war games by the U.S. Department of Defense (DoD) since the 1950s. However, until the 1980s, simulators were developed as a stand-alone with single-task purpose (such as landing on the deck of an aircraft carrier). Those stand-alone simulators were extremely expensive compared with the systems that they are suppose to mimic. For example, the cost of a tank simulator in the 1970s was $18 million, while the cost of an advanced aircraft was around $18 million (and a tank was significantly less). By the 1980s, the need of performing cost-effective distributed simulation started to be used at the DoD to simulate war games [8].

The first large project in this area, the *SIMulator NETworking* (SIMNET) program, was initiated in 1983 by the Defense Advanced Research Projects Agency (DARPA) in order to provide virtual world environment for military training [8–11]. SIMNET was different from previous simulators in a sense that many objects played together in the same virtual war game. During a SIMNET exercise, a simulator sent/received messages to/from other simulators using a *local area network* (LAN). This distributed simulation environment enabled various simulation components to interact with each other over the communication network. Cost played as a major factor for developing SIMNET. However, the ability of having different types of simulations interacting with each other was another major factor. For example, warships, tanks, and aircraft simulators worked together, enhancing the individual systems' ability to interact with others in a real-world scenario. Further, the design of SIMNET was different from previous simulators. The goal became to derive the simulation requirements, and only then decide the hardware needed for the simulation environment. This caused many required hardware in the actual systems to be rolled out from simulation training.

The success of SIMNET led to developing standards for *distributed interactive simulation* (DIS) during the 1990s [12–14]. DIS is an open standard discussed in numerous articles of the Institute of Electrical and Electronics Engineers (IEEE) for conducting interactive and distributed simulations mainly within military organizations [15–18]. DIS evolved from SIMNET and applied many of SIMNET's basic concepts. Therefore, DIS can be viewed as a standardized version of SIMNET. The DIS standards introduced the concept of interoperability in distributed simulation, meaning that one can interface a simulator with other compliant DIS simulators, if they follow the DIS standards. Interoperability via simulation standards was a major step forward provided by SIMNET, but it only permitted distributed simulations in homogeneous environments. DIS was designed to provide consistency (from human observation and behavior) in an interactive simulation composed by different connected components. Consistency in these human-in-the-loop simulators was achieved via data exchange protocols and a common database. DIS exchanged data using standardized units called the *protocol data unit* (PDU), which allowed DIS simulations to be independent of the network protocol used to transmit those PDUs [13]. DIS was successful in providing distributed simulation in LANs, but it only supported interactive simulations restricted to military training [4]. These simulations did not scale well in wide area networks (WANs).

SIMNET and DIS use an approach in which a single virtual environment is created by a number of interacting simulations, each of which controls the local objects and communicates its state to other simulations. This approach led to new methods for integrating existing simulations into a single environment, and during the 1990s, the Aggregate Level Simulation Protocol (ALSP) was built. ALSP was designed to allow legacy military simulations to interact with each other over LANs and WANs. ALSP, for example, enabled Army, Air Force, and Navy war game simulations to be integrated in a single exercise [19–21].

The next major progress in the defense simulation community occurred in 1996 with the development of the HLA [5,22,23]. HLA was a major improvement because it combined both analytic simulations with virtual environment technologies in a single framework [1]. The HLA replaced SIMNET and DIS, and all simulations in DoD are required to be HLA compliant since 1999 [1].

The distributed simulation success in the defense community along with the popularity of the Internet in the early 1990s led to the emergence of nonmilitary distributed virtual environments, for instance, the *distributed interactive virtual environment* (DIVE) (which is still in use since 1991). DIVE allows a number of users to interact with each other in a virtual world [24]. The central feature in DIVE is the shared, distributed database where all interactions occur through this common medium.

Another environment that became popular during the 1990s was the Common Object Request Broker Architecture (CORBA) [25]. CORBA introduced new interoperability improvements since it was independent of the programming language used. On the other hand, CORBA use had sharply declined in new projects since 2000. Some reflect this for being very complicated to developers or by the process CORBA standard was created (e.g., the process did not require a reference implementation for a standard before being adopted). Further, Web services became popular in the 2000s as an alternative approach to achieve interoperability among heterogeneous applications (which also contributed to CORBA's decline).

Web services standards were fully finalized in 2000. However, TCP/IP, HTTP, and XML (which are the major Web services standards) had matured since the 1990s. These standards have opened the way for Simple Object Access Protocol (SOAP) version 1.0, which was developed in 1999 by Userland Software and Microsoft [26]. SOAP provided a common language (based on XML) to interface different applications. A major breakthrough came when IBM backed up the SOAP proposal in early 2000 and joined the effort for producing the SOAP standard version 1.1 [27]. It was followed in the same year by the definition of the standards version 1.0 of the Web Services Description Language (WSDL), which is used to describe exposed services [28]. The final boost for making Web services popular came when five major companies (Sun, IBM, Oracle, HP, and Microsoft) announced their support for Web services in their products in 2000. It did not take long for the distributed simulation community to take advantage of Web services technology. Web services are now being used even to wrap the HLA interfaces to overcome its interoperability shortcomings, or to perform pure distributed simulation across the WAN/Internet. Web services presented the service-oriented architecture (SOA) concept, which means services are deployed in interoperable units in order to be consumed by other applications. The U.S. DoD Global Information Grid (GIG) is based on SOA to interoperate the DoD heterogeneous systems. At the time of this writing, Web service is the technology of choice for interoperating heterogeneous systems across the Internet [29].

## SYNCHRONIZATION ALGORITHMS FOR PARALLEL AND DISTRIBUTED SIMULATION

Parallel/distributed simulations are typically composed of a number of sequential simulations where each is responsible for a part of the entire model. In parallel and distributed simulations, the execution of a system is subdivided in smaller, simpler parts that run on different processors or nodes. Each of these subparts is a sequential simulation, which is usually referred to as a logical processor (LP). These LPs interact with each other using message passing to notify each other of a simulation event. In other words, LPs use messages to coordinate the entire simulation [1]. The main purpose of synchronization algorithms is to produce the same results as if the simulation was performed sequentially in a single processor. The second purpose is to optimize the simulation speed by executing the simulation as fast as possible.

In order to reduce the execution times, the parallel simulator tries to execute events received on different LPs concurrently (in order to exploit parallelism). Nevertheless, this might cause errors in a simulation. Consider the scenario presented in Figure 11.1, in which two LPs are processing different events. Consider that the LPs receive two events: $E_{200}$ is received by $LP_2$ (with time stamp 200), and event $E_{300}$ is received by $LP_1$ (with time stamp

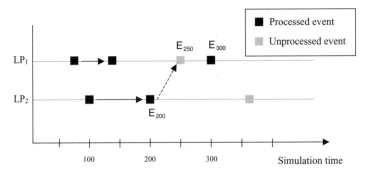

**Figure 11.1** Causality error in a distributed simulation.

300). Suppose that $LP_2$ has no events before time 200, and $LP_1$ has no events before time 300. It thus seems reasonable to process $E_{200}$ and $E_{300}$. Suppose that, when we execute $E_{200}$, it generates a new event, $E_{250}$ (with time stamp 250), which must be sent to $LP_1$. When $LP_1$ receives the event $E_{250}$, it was already processing (or had processed) the event $E_{300}$ with time stamp 300. As we can see, we receive an event from the past in the future (an event that requires immediate attention, and might affect the results of processing event $E_{300}$). This is called a *causality error*.

The *local causality constraint* guarantees the conditions under which one cannot have causality errors: If each LP processes events and messages in nondecreasing time stamp order, causality errors cannot occur [1]. This brings us to a fundamental issue in synchronization algorithms: Should we avoid or deal with local causality constraints? Based on these ideas, two kinds of algorithms were defined. *Conservative* (pessimistic) algorithms avoid local causality errors by carefully executing safe events, not permitting local causality errors. On the other hand, *optimistic* algorithms allow causality errors to occur, but fix them when detected. It is difficult to decide which type is better than the other one, because simulation is application dependent. In fact, the support for both types of algorithms may exist within one system.

## Conservative Algorithms

*Conservative algorithms* were introduced in the late 1970s by Chandy and Misra and Bryant [30,31]. This approach always satisfies local causality constraint via ensuring safe time stamp-ordered processing of simulation events within each LP [30,32]. Figure 11.2 shows the data structures used: input and output queues on each LP, and a local virtual time (LVT) representing the time of the last processed event. For instance, LP-B uses two input queues: one from LP-A and one from a different LP (not showed in the figure). At this point, it has processed an event at time 4 and has advanced the LVT = 4. Its output queue is connected to LP-A.

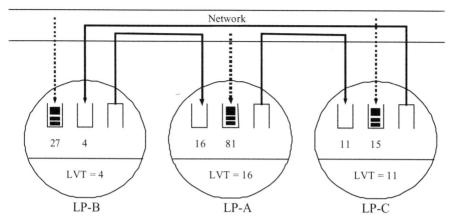

**Figure 11.2** Deadlock situation where each LP is waiting for an event from another LP.

For instance, LP-B has received an event with time stamp = 27; thus, we know that it will never receive an event with a smaller time stamp from the same LP. If at that point it receives an event with time stamp 4 from LP-C, LP-B can safely process it (in fact, it can process any event from LP-C earlier than time 27, as we know that we will not receive an event with an earlier time stamp). However, LP-B must be blocked once all the unprocessed events from LP-C are processed. If one of the input queues is empty (as in the figure), the LP must be blocked. We cannot guarantee the processing of other events (for instance, although we have plenty of events in the second queue, as the first one is empty, and the associated time stamp is 4, the LP cannot continue: if we receive, for instance, a new event with time stamp 5 from LP-C, this will cause a causality error). As we can see, the simulation can enter into a deadlock when a cycle of empty queues is developed where each process in the simulation is blocked (as shown in the figure).

A solution to break the deadlock in Figure 11.2 is to have each LP broadcasting the lower bound on its time stamp to all other relevant LPs. This can be accomplished by having LPs sending each other "null" messages with its time stamp. In this case, when an LP processes an event, it sends other LPs a null message, allowing other LPs decide the safe events to process. For instance, in our example, LP-A will inform LP-C that the earliest time stamp for a future event will be 16. Therefore, is it now safe for LP-C to process the next event in the input queue (with time stamp 15), which breaks the hold-and-wait cycle (thus preventing deadlock to occur). These are the basic ideas behind the Chandy/Misra/Bryant algorithm [1,30,31].

Further, runtime performance in conservative algorithms depends on an application property called *lookahead*, which is the time distance between two LPs. The lookahead value can ensure an LP to process events in the future safely. Let us suppose that LP-A and LP-B in Figure 11.2 represent the time taken to traverse two cities by car (which takes 3 units of simulation time). In this case, if LP-A is at simulation time 16, we know the smallest time stamp it will send to LP-B is 19, so LP-B can safely process events with that time stamp or lower. The lookahead is an important value because it determines the degree of parallelism in the simulation, and it affects the number of exchanged null messages. Naturally, the lookahead value is very difficult to extract in complex applications. Further, null messages could harshly degrade system performance [1]. Therefore, an LP can advance and process events safely once it realizes the lower time stamp bound and the lookahead information for all other relevant LPs. As a result, many algorithms were proposed during the late 1980s and 1990s to arm each LP with this information as efficiently as possible. For example, the barrier algorithms execute the simulation by cycling between phases. Once an LP reaches the barrier (i.e., a wall clock time), it is blocked until all other LPs get the chance to reach the barrier. In this case, an LP knows that all of its events are safe to process when it executes the barrier primitive (e.g., semaphore). Of course, the algorithms need to deal with the messages remaining in the network (called transient messages) before LPs cross the barrier. Examples of such algorithms are bounded lag, synchronous protocol, and a barrier technique, which deals with the transit messages problem [33–35]. Different algorithms are discussed in detail in *Parallel and Distribution Simulation Systems* [1].

The above-described algorithms still form the basis of recent conservative distributed simulation. For example, the distributed CD++ (DCD++) [36] is using a conservative approach similar to the barrier algorithms, as shown in Figure 11.3. DCD++ is a distributed simulation extension of the CD++ toolkit

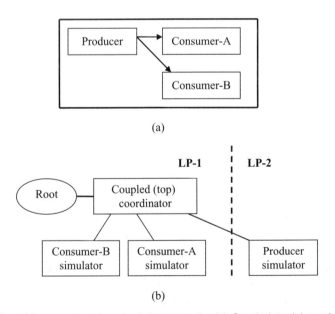

**Figure 11.3** DCD++ conservative simulation example. (a) Coupled model consists of three atomic models. (b) Model hierarchy during simulation split between two LPs.

[37], which is based on the DEVS formalism [38]. Figure 11.3 shows a DEVS coupled model that consists of three atomic models. An atomic model forms an indivisible block. A coupled model is a model that consists of one or more coupled/atomic models. The Producer model in Figure 11.3(a) has one output port linked with the input port of two consumer models. Suppose that this model hierarchy is partitioned between two LPs, as shown in Figure 11.3(b).

In this case, each LP (which is a component running in DCD++) has its own unprocessed event queue, and the simulation is cycling between phases. In this case, the Root Coordinator starts a phase by passing a simulation message to the topmost Coordinator in the hierarchy. This message is propagated downward in the hierarchy. In return, a DONE message is propagated upward in the hierarchy until it reaches the Root Coordinator. Each model processor uses this DONE message to insert the time of its next change (i.e., an output message to another model, or an internal event message) before passing it to its parent coordinator. A coordinator always passes to its parent the least time change received from its children. Once the Root Coordinator receives a DONE message, it advances the clock and starts a new phase safely without worrying about any lingering transit messages in the network. Further, each coordinator in the hierarchy knows which child will participate in the next simulation phase. Furthermore, each LP can safely process any event exchanged within a phase since an event is generated at the time it is supposed to be executed by the receiver model. In this approach, the barrier is represented by the arrival of the DONE message at the Root Coordinator. However, the Root Coordinator does not need to contact any of the LPs because they are already synchronized.

The lookahead value is the most important parameter in conservative algorithms. Therefore, lookahead extraction has been studied intensively by researchers. Recently, the effort has focused on determining the lookahead value dynamically at runtime (instead of static estimation). This is done by collecting lookahead information from the models as much as possible [39–42].

## Optimistic Algorithms

Conservative algorithms avoid violating LPs' local causality constraints, while optimistic algorithms allow such violations to occur but provide techniques to undo any computation errors. Jefferson's time warp mechanism remains the most well-known optimistic algorithm [43]. The simulation is executed via a number of *time warp processors* (TWLPs) interacting with each other via exchanging time-stamped event messages. Each TWLP maintains its LVT and advances "optimistically" without explicit synchronization with other processors. On the other hand, a causality error is detected if a TWLP receives a message from another processor with a time stamp in the past (i.e., with a time stamp less than the LVT), as shown in Figure 11.4. Such messages are called *straggler* messages.

To fix the detected error, the TWLP must roll back to the event before the straggler message time stamp; hence, undo all performed computation.

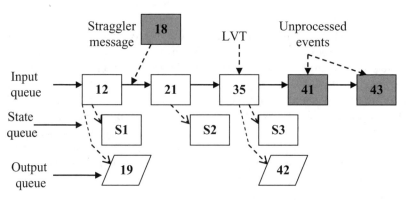

**Figure 11.4** TWLP internal processing.

Therefore, three types of information are needed to be able to roll back computation:

(1) An input queue to hold all incoming events from other LPs. This is necessary because the TWLP will have to reprocess those events in case of rollback. The events in this queue are stored according to their received time stamp.

(2) A state queue to save the TWLP states that might roll back. This is necessary because the simulation state usually changes upon processing an event. Thus, to undo an event processing affect, the prior state of its processing must be restored. For example, as shown in Figure 11.4, the TWLP must roll back events 21 and 35, upon event 18 (i.e., with time stamp 18) arrival. Thus, the simulation must be restored to state S1, the state that resulted from processing event 12. Afterward, the processor can process event 18 and reprocess events 21 and 35.

(3) An output queue to hold the output messages sent to other processors. These messages are sorted according to their sending time stamps. This is necessary because part of undoing an event computation is to undo other events scheduled by this event on the other processors. Such messages are called *antimessages* and they may cause a rollback at its destination, triggering other antimessages, resulting in a cascade of rollbacks in the simulation system. Upon event 18 arrival, in Figure 11.4, all antimessages resulted from events 21 and 35 are triggered. In this example antimessage 42 is sent. When antimessage meets its counterpart positive message, they annihilate each other. Suppose the shown processor in Figure 11.4 receives an antimessage for event 43. In this case, unprocessed event 43 is destroyed without any further actions. On the other hand, if an antimessage is received for event 21, the simulation must be rolled back to state S1, LVT is set to 12, and antimessage 42 must be sent to the appropriate processor.

The time warp computation requires a great deal of memory throughout the simulation execution. Therefore, a TWLP must have a guarantee that rollback will not occur before a certain virtual time. In this case, a TWLP must not receive a positive/negative message before a specific virtual time, called *global virtual time* (GVT). This allows TWLP to reclaim memory via releasing unneeded data such as saved previous simulation states, and events in the input/output (I/O) queues with time stamp less than the GVT. Further, the GVT can be used to ensure committing certain operations that cannot be rolled back such as I/O operations. GVT serves as the lower floor for the simulation virtual time. Thus, as the GVT never decreases and the simulation must not roll back below the GVT, all events processed before the GVT can be safely committed (and their memory can be reclaimed) [44]. Releasing memory for information older than GVT is performed via a mechanism called *fossil collection*. How often the GVT is computed is a trade-off. The more often the computation, it allows better space utilization, but it also imposes a higher communication overhead [1,45,46]. For example, the pGVT (passive global virtual time) algorithm allows users to set the frequency of GVT computation at compile time [47]. The GVT computation algorithm described in Bauer et al. uses clock synchronization to have each processor start computation at the same time instant [48]. Each processor should have a highly accurate clock to be able to use this algorithm.

The purpose of computing the GVT is to release memory, since the simulation is guaranteed to not roll back below it. The fossil collection manager cleans up all of the objects in the state/input/output queues with time stamp less than the GVT. Many techniques have been used to optimize this mechanism: the infrequent state saving technique (which avoids saving the modified state variables for each event), the one antimessage rollback technique (which avoids sending multiple antimessages to the same LP), or the antimessage with the earliest time stamp (only sent to that LP since it suffices to cause the required rollback). Lazy cancellation is a technique that analyzes if the result of the new computed message is the same as the previous one. In this case, an antimessage is not sent [49].

## DISTRIBUTED SIMULATION MIDDLEWARE

The main purpose of a distributed simulation middleware is to interoperate different simulation components and between different standards. Integrating new simulation components should be easy, fast, and effortless. To achieve this, certain prerequisite conditions must be met [4]:

(1) The middleware Application Programming Interface (API) should be easy to understand.
(2) It should follow widely accepted standards.
(3) It should be fast to integrate with new simulation software.

```
module BankAccount
{
    interface account {

        readonly attribute float balance;
        readonly attribute string name;

        void deposit (in float amount);
        void withdraw (in float amount);
    };

    interface accountManager {
        exception reject {string reason;};

        account createAccount (in string name) raises (reject);
        void deleteAccount (in account acc);
    };
};
```

Figure 11.5   CORBA IDL example.

(4) It should be interoperable with other middleware and be independent of diverse platforms.

In the following sections, we will discuss some of the features of existing simulation middleware.

## CORBA

As discussed earlier, CORBA is an open standard for distributed object computing defined by the Object Management Group (OMG) [25,50].

CORBA object services are defined using the Interface Definition Language (IDL), which can then be compiled into a programming language stubs such as C, C++, or Java (IDL syntax is similar to other programming languages, as shown in Fig. 11.5). Clients in CORBA invoke methods in remote objects using a style similar to remote procedure calls (RPCs), using IDL stubs, as shown in Fig. 11.6. The method call may return another CORBA handle where the client can invoke methods of the returned objects.

Figure 11.6 shows a picture of CORBA architecture: CORBA IDL stubs and skeletons glue operations between the client and server sides. The object request broker (ORB) layer provides a communication mechanism for transferring client requests to target object implementations on the server side. The ORB interface provides a library of routines (such as translating strings to object references and vice versa). The ORB layer uses the object adapter with routing client requests to objects and with objects activation.

Building distributed simulations using CORBA is straightforward, since CORBA enables application objects to be distributed across a network. Therefore, the issue becomes identifying distributed object interfaces and defining them in IDL; hence, a C++/Java local operation call becomes an RPC (hidden by CORBA). Therefore, to support distributed simulation using

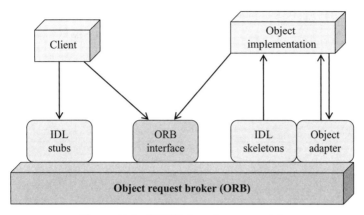

**Figure 11.6** CORBA 2.x reference model.

CORBA, you just need to translate your existing C++/Java simulation interfaces into CORBA IDL definition.

The work by Zeigler and Doohwan is an example of implementing a distributed DEVS simulation using CORBA [51]. For instance, a DEVS Simulator IDL interface (to the Coordinator, presented in Fig. 11.3) could be defined as follows (tN is the global next event time):

```
Module Simulator{
    Interface toCoordinator
    {
        boolean start();
        double tN?();
        double set_Global_and_Sendoutput (in double tN);
        boolean appIyDeltaFunc(in message);
    };
};
```

The *Simulator* module above is initialized via the method *start*. The *Simulator* module receives its *tN* via the method *set_Global_and_Sendoutput*, and in response, it returns its output. The above IDL code can then be compiled into specific Java/C++ code. For instance, in the Zeigler and Doohwan article, a DEVS coordinator IDL (to the simulator) was defined as follows [51]:

```
Module Coordinator{
    Interface toSimulator
    {
        boolean register
            (in Simulator::toCoordinator SimObjRef);
        boolean startSimulation();
        boolean stopSimulation();
    };
};
```

**Table 11.1 General HLA terminology**

| Term | Description |
| --- | --- |
| Attribute | Data field of an object |
| Federate | HLA simulation processor |
| Federation | Multiple federates interacting via RTI |
| Interaction | Event (message) sent between federates |
| Object | Collection of data sent between federates |
| Parameter | Data field of an interaction |

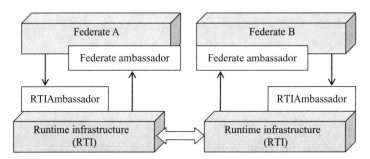

**Figure 11.7** HLA interaction overview.

The above IDL description shows that the simulator is given an object reference for the *toCoordinator* interface (via register method). As a result, simulators and coordinators can now invoke each other methods (across the network).

## HLA

As discussed earlier, the HLA was developed to provide a general architecture for simulation interoperability and reuse [5,22,23,52]. Table 11.1 shows the common terminology used in HLA.

Figure 11.7 shows the overall HLA simulation interaction architecture. The figure shows HLA simulation entities (called *federates*). Multiple federates (called a *federation*) interact with each other using the *runtime infrastructure* (RTI), which implements the HLA standards. Federates use the *RTIAmbassador* method to invoke RTI services, while the RTI uses the *FederateAmbassador* method to pass information to a federate in a callback function style. A callback function is a function passed to another function in the form of a reference (e.g., a C++ function pointer) to be invoked later via its reference. For example, in Figure 11.7, when the federate A sends an interaction (via *RTIAmbassador*) to the federate B, the RTI invokes a function in federate B via that function reference.

The HLA consists of three parts: the *object model template* (OMT) (to document exchanged shared data), the *HLA interface specification* (to define RTI/federates interfaces), and the *HLA rules* (to describe federate obligations and interactions with the RTI) [22,23].

The OMT provides a standard for documenting HLA object model information. This ensures detailed documentation (in a common format) for all visible objects and interactions managed by federates. Therefore, the data transmitted can be interpreted correctly by receivers to achieve the federation's objectives. The OMT consists of the following documents:

(1) The *federation object model* (FOM), which describes the shared object's attributes and interactions for the whole federation (several federates connected via the RTI).
(2) The *simulation object model* (SOM), which describes the shared object, attributes, and interactions for a single federate. The SOM documents specific information for a single simulation.

The HLA interface specification standardized the API between federates and RTI services [23]. The specification defines RTI services and the required callback functions that must be supported by the federates. Many contemporary RTI implementations conform to the IEEE 1516 and HLA 1.3 API specifications such as Pitch pRTI™ (C++/Java), CAE RTI (C++), MÄK High Performance RTI (C++/Java), and poRTIco (C++). However, the RTI implementation itself is not part of the standards. Therefore, interoperability between different RTI implementations should not be assumed, since HLA standards do not define the RTI network protocol. In this sense, the standards assume homogeneous RTI implementations in a federation. However, federation should be able to replace RTI implementations since APIs are standardized (but relinking and compiling are required). Unfortunately, this is not always the case.

The RTI services are grouped as follows:

(1) *Federation management*: services to create and destroy federation executions.
(2) *Declaration management*: federates must declare exactly what objects (or object attributes) they can produce or consume. The RTI uses this information to tell the producing federates to continue/stop sending certain updates.
(3) *Object management*: basic object functions, for instance, deletion/ updates of objects.
(4) *Ownership management*: services that allow federates to exchange object attributes ownership between themselves.
(5) *Time management*: these services are categorized in two groups:
    • *Transportation*: services to ensure events delivery reliability and events ordering.

- *Time advance*: services to ensure logical time advancement correctly. For example, a conservative federate uses the *time advance request* service with a parameter $t$ to request time to advance to $t$. The RTI then responds via invoking the *time granted* callback function.
(6) *Data distribution management*: it controls filters for data transmission (data routing) and reception of data volume between federates.

The HLA has been widely used to connect HLA-compliant simulations via RTI middleware. However, it presents some shortcomings:

(1) No standards exist to interoperate different RTI implementations. Therefore, interoperability should not be assumed among RTIs provided by different vendors. Further, standards are too heavy and no load balancing as part of the standards [4].
(2) The system does not scale well when many simulations are connected to the same RTI. This is because RTI middleware acts as a bus that manages all activities related to connected simulations in a session.
(3) HLA only covers syntactic (not semantic) interoperability [4].
(4) Interfacing simulations with RTIs can vary from a standard to another. It is a strong selling point for commercial RTIs that you can use your old HLA 1.3 federates with HLA 1516 RTI implementations [5].
(5) HLA API specifications are tied to programming languages. Some interoperability issues need to be resolved when federates are developed with different programming languages.
(6) Firewalls usually block RTI underlying communication when used on WAN/Internet networks.

A new WSDL API has been added to the HLA IEEE 1516-2007 standard, allowing HLA compliant simulation to be connected via the Internet using SOAP-based Web services. Some examples of existing HLA-based simulation tools using Web services include Boukerche et al. [53], Möller and Dahlin [54], and Zhu et al. [55]. As shown in Figure 11.8, the Web service provider RTI

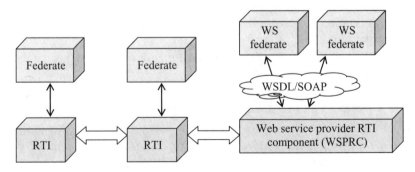

**Figure 11.8** Interfacing RTI with Web services.

component (WSPRC) is an RTI with one or more Web service ports, allowing HLA to overcome some of its interoperability problems. Therefore, this solution uses Web service interoperability in the WAN/Internet region while maintaining the standard HLA architecture locally. The WSPRC and Web Services (WS) federate APIs are described in WSDL where a standard federate and standard RTI API is described in actual programming languages. For instance, the Pitch pRTI version 4.0 supports Web services.

WS-based solutions solved interoperability issues at the federate level. However, this solution still does not solve interoperation of different WSPRC implementations, since the standard does not cover this part. Further, it does not provide a scalable solution, since many simulation components are still managed by a single component.

### SOAP-Based Web Services Middleware

*SOAP-based Web services* (or big Web services) provide a standard means of interoperating between different heterogeneous software applications, residing on a range of different platforms mainly for software reuse and sharing. At present, it is the leading technology for interoperating remote applications (including distributed simulations) across WAN/Internet networks. For example, a new WSDL API has been added to the HLA IEEE 1516-2007 standard, allowing HLA-compliant simulations to be connected via the Internet using SOAP-based Web services. Efforts in Boukerche et al., Möller and Dahlin, and Zhu et al. are examples of HLA-based simulation using Web services [53–55]. Further, the DEVS community is moving toward standardizing interoperability among different DEVS implementations using SOAP-based Web services [56].

The SOAP-based WS programming style is similar to RPCs, as depicted in Figure 11.9. The server exposes a group of *services* that are accessible via *ports*. Each service can actually be seen as an RPC, with semantics described via the

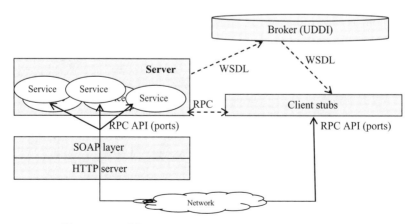

**Figure 11.9** SOAP-based Web service architecture overview.

procedure parameters. Ports can be viewed as a class exposing its functionality with a number of operations, forming an API accessible to the clients. On other hand, *clients* need to access those services, and they do so using procedure *stubs* at their end. The stubs are local and allow the client to invoke services as if they were local procedure calls.

Client programmers need to construct service stubs with their software at compile time. The clients consume a service at runtime by invoking its stub. In a WS-based architecture, this invocation is in turn converted into an XML SOAP message (which describes the RPC). This SOAP message is wrapped into an HTTP message and sent to the server port, using an appropriate port Uniform Resource Identifier (URI). Once the message is received at the server, an HTTP server located into the same machine passes the message to the SOAP layer (also called SOAP engine; it usually runs inside the HTTP server as Java programs called *servlets*). The SOAP layer parses the SOAP message and converts it into an RPC, which is applied to the appropriate port (which activates the right service). In turn, the server returns results to the clients in the same way.

Service providers need to publish the services available (using WSDL documents), in order to enable clients to discover and use the services. One way of doing so is via a broker called Universal Description, Discovery, and Integration (UDDI). UDDI is a directory for storing information about Web services and is based on the World Wide Web Consortium (W3C) and Internet Engineering Task Force (IETF) standards.

To achieve interoperability, services need to be described in WSDL and published so that clients can construct their RPC stubs correctly [57]. Further, XML SOAP messages ensure a common language between the client and the server regardless of their dissimilarities.

To demonstrate the role of SOAP and WSDL in an example, suppose that a simulation Web service exposes a port that contains a number of simulation services. Suppose further that the *stopSimulation* service (which takes an integer parameter with the simulation session number, and returns true or false indicating the success or the failure of the operation) is used to abort a simulation, as shown below:

```
boolean stopSimulation(int in0);   // method prototype
...
result = stopSimulation(1000);     // method call
```

From the client's viewpoint, the *stopSimulation* service is invoked similarly to any other procedure (using the SOAP-engine API). The responsibility of the SOAP engine (e.g., AXIS server) is to convert this procedure call into XML SOAP message as shown in Figure 11.10.

The SOAP message in Figure 11.10 will then be transmitted in the body of an HTTP message using the HTTP POST method. It is easy to see how an RPC is constructed in this SOAP message. The *stopSimulation* RPC is mapped to

```
1  <?xml version="1.0" encoding="UTF-8"?>
2  <SOAP-ENV:Envelope xmlns:xsd="http://www.w3.org/2001/XMLSchema"
3     xmlns:SOAP-ENV="http://schemas.xmlsoap.org/soap/envelope/"
4     xmlns:xsi="http://www.w3.org/2001/XMLSchema-instance">
5   <SOAP-ENV:Body>
6     <ns1:stopSimulation xmlns:ns1="http://WS-Port-URI/">
7       <in0 xsi:type="xsd:int">1000</in0>
8     </ns1:stopSimulation>
9   </SOAP-ENV:Body>
10 </SOAP-ENV:Envelope>
```

**Figure 11.10**  SOAP message request example.

```
1  <?xml version="1.0" encoding="UTF-8"?>
2  <SOAP-ENV:Envelope xmlns:xsd="http://www.w3.org/2001/XMLSchema"
3     xmlns:SOAP-ENV="http://schemas.xmlsoap.org/soap/envelope/"
4     xmlns:xsi="http://www.w3.org/2001/XMLSchema-instance">
5   <SOAP-ENV:Body>
6     <ns1:stopSimulationResponse xmlns:ns1="http://WS-Port-URI/">
7  SOAP-ENV:encodingStyle="http://schemas.xmlsoap.org/soap/encoding/">
8             <return xsi:type="xsd:boolean">true</return>
9     </ns1:stopSimulationResponse>
10  </SOAP-ENV:Body>
11 </SOAP-ENV:Envelope>
```

**Figure 11.11**  SOAP message response example.

lines 6–8 in Figure 11.10. Line #6 indicates invocation service *stopSimulation* on the Web service port with URI (http://WS-Port-URI/). URIs are WS port addresses (which correspond, for instance, to CORBA object references). Line #7 indicates that this service takes one integer parameter (i.e., simulation session number) with value 1000. Figure 11.11 shows a possible response to the client as a SOAP message, responding with the *stopSimulation* return value.

The above-explained example shows how SOAP is used to achieve interoperability. Because the participant parties have agreed on a common standard to describe the RPC (in this case SOAP), it becomes straightforward for software to convert an RPC (from any programming language) to a SOAP message (and vice versa).

Interoperability cannot only be achieved with SOAP messages because RPCs are programming procedures; hence, they need to be compiled with the clients' software. For example, a programmer writing a Java client needs to know that the *stopSimulation* service method looks exactly as *boolean stopSimulation (int in0)*. Here is where WSDL helps in achieving interoperability for SOAP-based WS. Web service providers need to describe their services in a WSDL document (and publish them using UDDI) so that clients can use it to generate services stubs.

The major elements of any WSDL document are the type, message, port Type, binding, port, and service elements. Some of these elements (type, message, and portType) are used to describe the functional behavior of the

```
1  <wsdl:message name="stopSimulationRequest">
2    <wsdl:part name="in0" type="xsd:int"/>
3  </wsdl:message>
4
5  <wsdl:message name="stopSimulationResponse">
6    <wsdl:part name="stopSimulationReturn" type="xsd:boolean"/>
7  </wsdl:message>
8
9  <wsdl:portType name="CDppPortType">
10  <wsdl:operation name="stopSimulation" parameterOrder="in0">
11    <wsdl:input message="impl:stopSimulationRequest"
12           name="stopSimulationRequest"/>
13    <wsdl:output message="impl:stopSimulationResponse"
14           name="stopSimulationResponse"/>
15  </wsdl:operation>
16
17 </wsdl:portType>
18
19 <wsdl:binding name="CDppPortTypeSoapBinding"
20               type="impl:CDppPortType">
21    <wsdlsoap:binding style="rpc"
22         transport="http://schemas.xmlsoap.org/soap/http"/>
23    <wsdl:operation name="stopSimulation">
24       <wsdlsoap:operation soapAction=""/>
25       <wsdl:input name="stopSimulationRequest">
26          <wsdlsoap:body encodingStyle="http://.../"
27               namespace="http://..." use="encoded"/>
28       </wsdl:input>
29
30       <wsdl:output name="stopSimulationResponse">
31          <wsdlsoap:body encodingStyle="http://.../"
32               namespace="http://..." use="encoded"/>
33       </wsdl:output>
34    </wsdl:operation>
35 </wsdl:binding>
```

**Figure 11.12**  Excerpt of a WSDL document example.

Web service in terms of the functionality it offers. On the other hand, binding, port, and service define the operational aspects of the service, in terms of the protocol used to transport SOAP messages and the URL of the service. The former is referred to as abstract service definition, and the latter is known as concrete service definition.

To carry on with our previous example, the simulation service provider should describe the *stopSimulation* service (along with other provided services) in a WSDL document. Figure 11.12 shows an excerpt of the WSDL description for the *boolean stopSimulation (int in0)* service.

Lines 1–7 show the messages used by the Web service to send the request and to handle the response. The *stopSimulation* operation uses an input message called *stopSimulationRequest* (which is an *int*eger), and an output message called *stopSimulationResponse* (a Boolean value). Lines 9–17 show the *portType* definition, which is used by operations accessing the Web service. It defines *CDppPortType* as the name of the port and *stopSimulation* as the name of the exposed operation by this port. As discussed earlier, ports define connection points to a Web service. If we want to relate this with a traditional program, *CDppPortType* defines a class where *stopSimulation* is a method with *stopSimulationRequest* as the input parameter and *stopSimulationResponse* as the return parameter. Lines 19–35 show the *binding* of the Web service, which defines the message format and ports protocol details. The <wsdlsoap:binding> element has two attributes: *style* and *transport*. In this example, the style attribute uses the RPC style, and the transport attribute defines the SOAP protocol to apply. The <wsdl:operation> element defines each operation the port exposes. In this case, operation *stopSimulation* is the only one. The SOAP I/O encoding style for operation *stopSimulation* is defined in lines 25–33.

As we can see, it is a great deal of work to describe one RPC. However, mature tools are one of the main advantages of SOAP-based WS. The WSDL document is usually converted to programming language stubs and vice versa with a click of a button (or with a simple shell command).

Using SOAP and WSDL, interoperability is achieved at the machine level regardless of their differences such as programming languages and operating systems. However, interoperability at the human level is still needed. For example, a programmer still needs to know that the integer input parameter to service *stopSimulation* means the simulation session number (even if that programmer was able to compile and invoke the service). This gets worse when a service procedure is complex with many input parameters. Therefore, in practice a text description can be helpful for client programmers. It is possible to add comments to WSDL document like any other XML documents (and WSDL without comments is worse than programming code without them). However, WSDL documents are typically generated by tools (and they need to move comments between WSDL and programming code stubs).

In addition, we need a standardized protocol when using Web services to interoperate various remote applications (such as interoperation of different simulations to perform distributed simulation). This is because Web services provide interoperability to overcome differences between machines rather than to overcome the differences between various application functionalities. Therefore, standards are still needed to accomplish simulations among different simulators successfully. As part of this effort, the DEVS simulation community is in the progress of developing standards to interface different DEVS (and non-DEVS) implementations using SOAP-based Web services (is an example of such proposals [56]).

In contemporary WS-based distributed simulations (e.g., Wainer et al. and Mittal et al.), simulation components act as both client and server at the same time [58,59]. In this case, a simulator becomes the client when it wants to send a simulation message to a remote simulator (which the later becomes the server), as shown in Figure 11.13.

SOAP-based distributed simulations share in common that synchronized messages are exchanged in RPC style where contents are usually passed as input parameters to the RPC (so it becomes similar to invoking a local call

**394** AN INTRODUCTION TO DISTRIBUTED SIMULATION

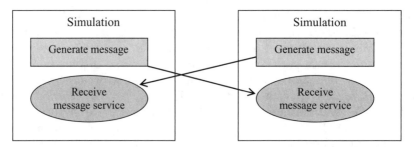

**Figure 11.13** Distributed simulation using SOAP-based WS.

procedure). Further, those RPCs are based on internal software implementation, which makes interfacing standards not easy to achieve among existing systems. This is because each system has already defined its RPC interfaces.

To summarize the major drawback points with SOAP-based Web services:

(1) Heterogeneous interface by exposing few URIs (ports) with many operations. Building programming stubs correctly (i.e., compiled without errors) is not enough to interface two different simulators quickly and efficiently. One possible solution is one of the participant parties has to wrap their simulator API with the simulator API to be able to interact with it. Another possible solution is to combine both simulator APIs and expose a new set of APIs, assuming this solution works. What happens if many vendor simulators need to interface with each other? It becomes a complex process. In fact, exposing heterogeneous programming procedures of a simulator and expecting it to interoperate with another simulator that also exposes heterogeneous procedures quickly and efficiently is a naive assumption.

(2) It uses an RPC style, which is suitable for closed communities that need to coordinate new changes among each other. In fact, those APIs (services) are programming procedures, which means that they reflect the internal implementation. Therefore, different vendors, for example, have to hold many meetings before they reach an agreement on defining those stubs, because they are tied into their internal implementation; hence, it affects a great deal the internal design and implementation of the simulation package. However, suppose that those different simulator vendors came to an agreement of standardizing the same exposed API, and assume that some changes are required during development or in the future. How easy is it to change those standardized APIs? A new coordination among different teams becomes inevitable to redefine new services.

(3) To use SOAP-based services requires building services stubs at compile time. This can cause more complexity in future advancements if the

simulation components can join/leave the simulation at runtime. For example, in "Ad hoc distributed simulations," the authors present the ad hoc distribution simulation, where the number of LPs is not known in advance and can be changed during runtime [60].

## RESTful Web Services Middleware

The *representational state transfer* (REST) provides interoperability by imitating the World Wide Web (WWW) style and principles [61]. RESTful Web services are gaining increased attention with the advent of Web 2.0 and the concept of mashup (i.e., grouping various services from different providers and present them as a bundle) because of its simplicity [62]. REST exposes services as "resources" (which are named with unique URIs similar to Web sites) and manipulated with uniform interface, usually HTTP methods: GET (to read a resource), PUT (to create/update a resource), POST (to append to a resource), and DELETE (to remove a resource). For example, a client applies the HTTP GET method to a resource's URI in order to retrieve that resource representation (e.g., this is what happens when you browse a Web site). Further, a client can transfer data by applying HTTP methods PUT or POST to a URI. REST applications need to be designed as resource oriented to get the benefits of this approach (see "RESTful Web services" for design guidelines [63]). REST is sometimes confused with HTTP, since HTTP perfectly matches REST principles. However, REST is an approach that devotes principles such as standardized uniform interface, universal addressing schemes, and resource-oriented design. REST has been used in many applications such as Yahoo, Flicker, and Amazon S3. It is also used in distributed systems such as National Aeronautics and Space Administration (NASA) SensorWeb (which uses REST to support interoperability across Sensor Web systems that can be used for disaster management) [64]. Another example of using REST to achieve plug-and-play interoperability heterogeneous sensor and actuator networks is described in Stirbu [65]. Example of REST usage in business process management (BPM) is described in Kumaran et al., which focuses on different methods and tools to automate, manage, and optimize business processes [66]. REST has also been used for modeling and managing mobile commerce spaces [67].

REST architecture separates the software interface from internal implementation; hence, services can be exposed while software internal implementation is hidden from consumers, and providers need to conform to the service agreement, which comes in the form of messages (e.g., XML). This type of design is a recipe for a plug-and-play (or at least semiautomatic) interoperability, as a consumer may search, locate, and consume a service at runtime (this is why Web 2.0 applications have expanded beyond regular computer machines to cell phones or any other device connected to the Internet). In contrast, other RPC-style form of interfacing require a programmer to build the interface stubs and recompile the application software before being able to use the intended service. This is clearly not the way to reach a plug-and-play interoperability. Distributed simulation can benefit of this capability toward future challenges (see the study by Strassburger et al. [4]) such as having middleware that have a plug-and-play (semiautomatic) interoperability, and accessed by any device from anywhere. Indeed, interoperating two independent developed simulators where each one of them exposes heterogeneous defined set of RPCs is not a trivial task to do. In fact, RPCs are often tied to internal implementation, and semantics are described in programming parameters. To add to the situation complexity, many simulators expose many RPCs of many objects (or ports). One has to question if this task is worth the cost, particularly if we need to add more independent developed simulators and models. The bottom line is that those simulators are software packages; hence, they interface with their APIs. Therefore, the API design matters when connecting diverse software together. To achieve plug-and-play interoperability, simulators need to have uniform interface, and semantics need to be described in the form of messages such as XML.

Based on these ideas, we designed RESTful-CD++, the first existing distributed simulation middleware, based on REST [2]. The RESTful-CD++'s main purpose is to expose services as URIs. Therefore, RESTful-CD++ routes a received request to its appropriate destination resource and apply the required HTTP method on that resource. This makes the RESTful-CD++ independent of a simulation formalism or a simulation engine. CD++ is selected to be the first simulation engine to be supported by the RESTful-CD++ middleware.

In this case, as shown in Figure 11.14, the simulation manager component is constructed to manage the CD++ distributed simulation such as the geographic existence of model partitions, as shown in Figure 11.3. The simulation manager is seen externally as a URI (e.g., similar to Web site URIs). On the other hand, it is a component that manages a distributed simulation LP instance; in our case, an LP is a CD++ simulation engine. Therefore, LPs exchange XML simulation messages among each other according to their wrapped URIs (using the HTTP POST method). The RESTful-CD++ middleware API is expressed as URI template, as shown in Figure 11.15, that can be created at

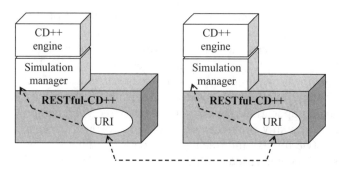

**Figure 11.14** RESTful-CD++ distributed simulation session.

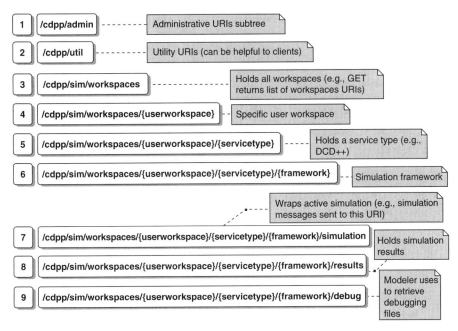

**Figure 11.15** RESTful-CD++ URIs template (APIs).

runtime. Variables (written within braces {}) in the URI template are assigned at runtime by clients, allowing modelers to create and name their resources (URIs) as needed. Consequently, the RESTful-CD++ exposes its APIs as regular Web site URIs that can be mashed up with other Web 2.0 applications (e.g., to introduce real systems in the simulation loop). In addition, it is capable of consuming services from SOAP-based Web services (Fig. 11.15).

## CONCLUSION

Distributed simulation deals with executing simulations on multiple processors connected via communication networks. Distributed simulation can be used to achieve model reuse via interoperation of heterogeneous simulation components, reducing execution time, connecting geographically distributed simulation components, avoiding people/equipment relocations and information hiding—including the protection of intellectual property rights. These simulations are typically composed of a number of sequential simulations where each is responsible for a part of the entire model.

The main purpose of a distributed simulation middleware is to interoperate different simulation components and between different standards. Integrating new simulation components should be easy, fast, and effortless. A number of middlewares have been used to achieve interoperability among different simulation components such as CORBA, HLA, and SOAP-based/REST-based

Web services. HLA is the used middleware in the military sector where various simulation components are plugged into the RTI, which manages the entire simulation activities. On the other hand, SOAP-based and CORBA expose services as RPC style via ports/objects, where semantic is described in the parameters of those RPCs. REST-based WS, instead, separate interface from internal implementation via exposing standardized uniform interface and describing semantics in the form of messages (e.g., XML). REST can provide a new means of achieving a plug-and-play (automatic/semiautomatic) distributed simulation interoperability over the Internet and can introduce real systems in the simulation loop (e.g., Web 2.0 mashup applications). This approach has the potential of highly influencing the field, as it would make the use of distributed simulation software more attractive for the industry (as one can reuse existing applications and integrate them with a wide variety of e-commerce and business software applications already existing on the Web).

## REFERENCES

[1] Fujimoto RM. *Parallel and Distribution Simulation Systems*. New York: John Wiley & Sons; 2000.

[2] Boer C, Bruin A, Verbraeck A. A survey on distributed simulation in industry. *Journal of Simulation*, 3(1):3–16; 2009.

[3] Boer C, Bruin A, Verbraeck A. Distributed simulation in industry—A survey, part 3—The HLA standard in industry. Proceedings of Winter Simulation Conference (WSC 2008). Miami, FL, 2008.

[4] Strassburger S, Schulze T, Fujimoto R. Future trends in distributed simulation and distributed virtual environments: Results of a peer study. Proceedings of Winter Simulation Conference (WSC 2008). Miami, FL, 2008.

[5] IEEE-1516-2000. Standard for modeling and simulation (M&S) high level architecture (HLA)—Frameworks and rules. 2000.

[6] Gan BP, Liu L, Jain S, Turner SJ, Cai WT, Hsu WJ. Distributed supply chain simulation across the enterprise boundaries. Proceedings of Winter Simulation Conference (WSC 2000). Orlando, FL, December 2000.

[7] Wainer G, Zeigler B, Nutaro J, Kim T. DEVS Standardization Study Group Final Report. Available at http://www.sce.carleton.ca/faculty/wainer/standard. Accessed March 2009.

[8] Lenoir T, Lowood H. Theaters of Wars: The Military—Entertainment Complex. Available at http://www.stanford.edu/class/sts145/Library/Lenoir-Lowood_TheatersOfWar.pdf. Accessed March 2009.

[9] Calvin J, Dickens A, Gaines B, Metzger P, Miller D, Owen D. The SIMNET virtual world architecture. Proceedings of Virtual Reality Annual International Symposium (IEEE VRAIS 1993). Seattle, WA, 1993.

[10] Taha HA. Simulation with SIMNET II. Proceedings of Winter Simulation Conference (WSC 1991). Phoenix, AZ, December 1991.

[11] Taha HA. Introduction to SIMNET v2.0. Proceedings of Winter Simulation Conference (WSC 1988). San Diego, CA, December 1988.

[12] Fitzsimmons EA, Fletcher JD. Beyond DoD: Non-defense training and education applications of DIS. *Proceedings of the IEEE*, 83(8):1179–1187; 1995.
[13] Hofer RC, Loper ML. DIS today. *Proceedings of the IEEE*, 83(8):1124–1137; 1995.
[14] Pullen JM, Wood DC. Networking technology and DIS. *Proceedings of the IEEE*, 83(8):1156–1167; 1995.
[15] IEEE-1278.1-1995. Standard for distributed interactive simulation—Application protocols. 1995.
[16] IEEE-1278.2-1995. Standard for distributed interactive simulation—Communication services and profiles. 1995.
[17] IEEE 1278.3-1996. Recommended practice for distributed interactive simulation—Exercise management and feedback. 1996.
[18] IEEE 1278.4-1997. Recommended practice for distributed interactive—Verification validation & accreditation. 1997.
[19] Anita A, Gordon M, David S. Aggregate Level Simulation Protocol (ALSP) 1993 Confederation Annual Report. The MITRE Corporation. 1993. Available at http://ms.ie.org/alsp/biblio/93_annual_report/93_annual_report_pr.html. Accessed March 2009.
[20] Fischer M. Aggregate Level Simulation Protocol (ALSP)—Future training with distributed interactive simulations. U.S. Army Simulation, Training and Instrumentation Command. International Training Equipment Conference. The Hague, Netherlands, 1995.
[21] Babineau WE, Barry PS, Furness CS. Automated testing within the Joint Training Confederation (JTC). Proceedings of the Fall Simulation Interoperability Workshop. Orlando, FL, September 1998.
[22] IEEE-1516.1-2000. Standard for modeling and simulation (M&S) high level architecture (HLA)—Federate interface specification. 2000.
[23] IEEE-1516.2-2000. Standard for modeling and simulation (M&S) high level architecture (HLA)—Object model template (OMT) specification. 2000.
[24] Frécon E, Stenius M. DIVE: A scalable network architecture for distributed virtual environments. *Distributed Systems Engineering Journal*, 5(3):91–100; 1998.
[25] Henning M, Vinoski S. *Advanced CORBA Programming with C++*. Reading, MA: Addison-Wesley; 1999.
[26] Kakivaya G, Layman AS, Thatte S, Winer D. SOAP: Simple Object Access Protocol. Version 1.0. 1999. Available at http://www.scripting.com/misc/soap1.txt. Accessed March 2009.
[27] Box D, Ehnebuske D, Kakivaya G, Layman A, Mendelsohn N, Nielsen H, Thatte S, Winer D. Simple Object Access Protocol (SOAP) 1.1. May 2000. Available at http://www.w3.org/TR/2000/NOTE-SOAP-20000508/. Accessed March 2009.
[28] Christensen E, Curbera F, Meredith G, Weerawarana S. Web Service Description Language (WSDL) 1.1. March 2001. Available at http://www.w3.org/TR/wsdl. Accessed March 2009.
[29] Zeigler B, Hammods P. *Modeling and Simulation-Based Data Engineering: Pragmatics into Ontologies for Net-Centric Information Exchange*. Burlington, MA: Academic Press; 2007.

[30] Chandy KM, Misra J. Distributed simulation: A case study in design and verification of distributed programs. *IEEE Transactions on Software Engineering*, SE-5(5):440–452; 1979.

[31] Bryant RE. Simulation of packet communication architecture computer systems. *Technical Report LCS, TR-188*. Cambridge, MA: Massachusetts Institute of Technology; 1977.

[32] Chandy KM, Misra J. Asynchronous distributed simulation via a sequence of parallel computations. *Communications of the ACM*, 24(4):198–205; 1981.

[33] Lubachevsky B. Efficient distributed event-driven simulations of multiple-loop networks. Proceedings of the 1988 ACM SIGMETRICS Conference on Measurement and Modeling of Computer Systems. Santa Fe, NM, 1988.

[34] Nicol D. The cost of conservative synchronization in parallel discrete event simulation. *Journal of the ACM*, 40(2):304–333; 1993.

[35] Nicol D. Noncommittal barrier synchronization. *Parallel Computing*, 21(4):529–549; 1995.

[36] Al-Zoubi K, Wainer G. Using REST Web services architecture for distributed simulation. Proceedings of Principles of Advanced and Distributed Simulation PADS 2009. Lake Placid, NY, 2009.

[37] Wainer G. *Discrete-Event Modeling and Simulation: A Practitioner's Approach*. Boca Raton, FL: CRC Press and Taylor & Francis Group; 2009.

[38] Zeigler B, Kim T, Praehofer H. *Theory of Modeling and Simulation*. 2nd ed. San Diego, CA: Academic Press; 2000.

[39] Chung M, Kyung C. Improving lookahead in parallel multiprocessor simulation using dynamic execution path prediction. Proceedings of Principles of Advanced and Distributed Simulation (PADS 2006). Singapore, May 2006.

[40] Zacharewicz G, Giambiasi N, Frydman C. Improving the lookahead computation in G-DEVS/HLA environment. Proceedings of Distributed Simulation and Real-Time Applications (DS-RT 2005). Montreal, QC, Canada, 2005.

[41] Liu J, Nicol DM. Lookahead revisited in wireless network simulations. Proceedings of Principles of Advanced and Distributed Simulation (PADS 2002). Washington, DC, 2002.

[42] Meyer R, Bagrodia L. Path lookahead: A data flow view of PDES models. Proceedings of Principles of Advanced and Distributed Simulation (PADS 1999). Atlanta, GA, 1999.

[43] Jefferson DR. Virtual time. *ACM Transactions on Programming Languages and Systems*, 7(3):405–425; 1985.

[44] Frey P, Radhakrishnan R, Carter HW, Wilsey PA, Alexander P. A formal specification and verification framework for time warp-based parallel simulation. *IEEE Transactions on Software Engineering*, 28(1):58–78; 2002.

[45] Samadi B. Distributed simulation, algorithms and performance analysis. PhD Thesis, Computer Science Department, University of California, Los Angeles, CA; 1985.

[46] Mattern F. Efficient algorithms for distributed snapshots and global virtual time approximation. *Journal of Parallel and Distributed Computing*, 18(4):423–434; 1993.

[47] D'Souza LM, Fan X, Wisey PA. pGVT: An algorithm for accurate GVT estimation. Proceedings of Principles of Advanced and Distributed Simulation (PADS 1994). Edinburgh, Scotland, July 1994.

[48] Bauer D, Yaun G, Carothers CD, Yuksel M, Kalyanaraman S. Seven-o'clock: A new distributed GVT algorithm using network atomic operations. Proceedings of Principles of Advanced and Distributed Simulation (PADS 2005). Monterey, CA, 2005.

[49] Lubachevsky B, Weiss A, Schwartz A. An analysis of rollback-based simulation. *ACM Transactions on Modeling and Computer Simulation (TOMACS)*, 1(2):154–193; 1991.

[50] Object Management Group (OMG). Available at http://www.omg.org/. Accessed February 2009.

[51] Zeigler BP, Doohwan K. Distributed supply chain simulation in a DEVS/CORBA execution environment. Proceedings of Winter Simulation Conference (WSC 1999). Phoenix, AZ, 1999.

[52] Khul F, Weatherly R, Dahmann J. *Creating Computer Simulation Systems: An Introduction to High Level Architecture*. Upper Saddle River, NJ: Prentice Hall; 1999.

[53] Boukerche A, Iwasaki FM, Araujo R, Pizzolato EB. Web-based distributed simulations visualization and control with HLA and Web services. Proceedings of Distributed Simulation and Real-Time Applications (DS-RT 2008). Vancouver, BC, Canada, 2008.

[54] Möller B, Dahlin C. A first look at the HLA evolved Web service API. Proceedings of 2006 Euro Simulation Interoperability Workshop, Simulation Interoperability Standards Organization. Stockholm, Sweden, 2006.

[55] Zhu H, Li G, Zheng L. Introducing Web services in HLA-based simulation application. Proceedings of IEEE 7th World Congress on Intelligent Control and Automation (WCICA 2008). Chongqing, China, June 2008.

[56] Al-Zoubi K, Wainer G. Interfacing and Coordination for a DEVS Simulation Protocol Standard. Proceedings of Distributed Simulation and Real-Time Applications (DS-RT 2008). Vancouver, BC, Canada, 2008.

[57] Christensen E, Curbera F, Meredith G, Weerawarana S. Web Services Description Language (WSDL) 1.0. Available at http://xml.coverpages.org/wsdl20000929.html. Accessed March 2009.

[58] Wainer G, Madhoun R, Al-Zoubi K. Distributed simulation of DEVS and cell-DEVS models in CD++ using Web services. *Simulation Modelling Practice and Theory*, 16(9):1266–1292; 2008.

[59] Mittal S, Risco-Martín JL, Zeigler BP. DEVS-based simulation Web services for net-centric T&E. Proceedings of the 2007 Summer Computer Simulation Conference. San Diego, CA, 2007.

[60] Fujimoto R, Hunter M, Sirichoke J, Palekar M, Kim H, Suh W. Ad hoc distributed simulations. Proceedings of Principles of Advanced and Distributed Simulation (PADS 2007). San Diego, CA, June 2007.

[61] Fielding RT. Architectural Styles and the Design of Network-based Software Architectures. PhD Thesis, University of California, Irvine, CA, 2000.

[62] O'Reilly T. What is Web 2.0. Available at http://www.oreillynet.com/pub/a/oreilly/tim/news/2005/09/30/what-is-web-20.html. Accessed May 2009.

[63] Richardson L, Ruby S. *RESTful Web Services*. 1st ed. Sebastopol, CA: O'Reilly Media; 2007.

[64] Cappelaere P, Frye S, Mandl D. Flow-enablement of the NASA SensorWeb using RESTful (and secure) workflows. 2009 IEEE Aerospace Conference. Big Sky, MT, March 2009.

[65] Stirbu V. Towards a RESTful plug and play experience in the Web of things. IEEE International Conference on Semantic Computing (ICSC 2008). Santa Clara, CA, August 2008.

[66] Kumaran S, Rong L, Dhoolia P, Heath T, Nandi P, Pinel F. A RESTful architecture for service-oriented business process execution. IEEE International Conference on e-Business Engineering (ICEBE '08). Xi'an, China, October 2008.

[67] McFaddin S, Coffman D, Han JH, Jang HK, Kim JH, Lee JK, Lee MC, Moon YS, Narayanaswami C, Paik YS, Park JW, Soroker D. Modeling and managing mobile commerce spaces using RESTful data services. 9th IEEE International Conference on Mobile Data Management (MDM '08). Beijing, China, 2008.

# 12

# INTEROPERABILITY AND COMPOSABILITY

Andreas Tolk

For many modeling and simulation (M&S) developers, questions regarding the future interoperability and composability of their solution are not the main concern during design and development. They design their M&S system or application to solve a special problem and provide a solution. There is nothing wrong with this perception. However, there are many reasons why it is preferable to design interoperability and composability from the early phases on, for example, by using open standards for the communication of information or by using standardized interfaces to common services. The main driving factor for this is the desire to enable the *reuse of existing solutions*. Why should we invest in rewriting a solution that already exists?

The second aspect is that of *modularity*, in particular when dealing with complex systems. While it may be possible to use and evaluate small models as a whole, large and complex system rapidly become too big to be handled in one block. Development and testing for such systems should be conducted in modules; however, these modules need to be interoperable and composable to allow bringing them back into a common system.

The aspect of *reducing costs* is also playing a significant role. The idea is to reduce development cost by reducing reliable solutions and avoiding to "reinvent the wheel" in new models. However, again this assumes that the components can be identified, selected, composed, and orchestrated.

---

*Modeling and Simulation Fundamentals: Theoretical Underpinnings and Practical Domains*,
Edited by John A. Sokolowski and Catherine M. Banks
Copyright © 2010 John Wiley & Sons, Inc.

In addition, the growing connectivity of real-world problems is reflected in the requirement to *compose cross-domain solutions* as well. Examples are, among others, the evaluation of interdependencies between the transportation systems and possible energy support in the domain of homeland security, or the analysis and support of common operations of several nations with several branches of their armed forces hand in hand with nonmilitary and often even nongovernment organizations for the organizations like the North Atlantic Treaty Organization (NATO) or the European Defense Agency (EDA). Other examples from medical simulation can easily be derived for biologic and medical simulations, where similar problems are observed when composing models on the enzyme, cell, or organism level with each other. The common challenge of these compositions is that such joint operations are more than just the parallel execution of part solutions. Synergistic effects need to be taken into consideration, as the whole new operation is likely to be more than just the sum of its part solution.

The growing connectivity requiring interoperable and composable parts is also reflected in the ideas of *service-oriented architectures* (SOAs) and system of systems. In both cases, solutions are composed on-the-fly reusing preexisting services that provide the required functionality. While engineers traditionally conduct the evaluation and adaptation of existing solutions to make them fit for a new environment, these engineering tasks now need to be conducted by machines, such as intelligent agents. This requires that all information needed to allow for

- the identification of applicable services
- the selection of the best subset for the given task
- the composition of these services to produce the solution
- the orchestration of their execution

must be provided in machine-readable form as annotations. Consequently, services and systems must be annotated with information on their interoperability and composability characteristics to allow and enable their composition on-the-fly.

Finally, interoperability and composability challenges are not limited to M&S applications and services. Many M&S application areas as defined earlier in this book require the interoperation of M&S systems and operational infrastructure, such as traffic information systems and evacuation models, or military command and control systems and combat simulation systems.

This chapter will focus on the technical challenges of interoperability and composability, current proposed standardized solutions, and ongoing related research. It will not deal with business models supporting the idea. It will also leave out security aspects (you do not want your opponent or competition to use your best tools for his solutions) and questions of intellectual property out. These are valuable research fields on their own.

To deal with the selected subset of challenges, the chapter is structured as follows:

(1) We will start with an overview of currently used *definitions for interoperability and composability*. In this section, we will also include ongoing discussions of models for composability and interoperability that can guide the M&S professional. This section will introduce the levels of conceptual interoperability model (LCIM).
(2) The next section will give an overview of *current standards* supporting interoperability and composability, including the Institute of Electrical and Electronics Engineers (IEEE) standards for distributed interactive simulation (DIS) and the high-level architecture (HLA) and the Simulation Interoperability Standards Organization (SISO) standard for base object models (BOMs). We will also have a look at solutions in support of Web-based solutions, in particular contributions proposed in support of the semantic Web. The section ends with a comparison of these solutions using the LCIM.
(3) The last section will look into ongoing research on *engineering methods* in support of interoperability and composability. The methods of data engineering, process engineering, and constraint engineering build the center of this section, showing how they contribute to solutions in the frame of the LCIM.

Students and scholars of the topics of interoperability and composability are highly encouraged to use this chapter as a first step toward understanding M&S fundamentals: theoretical underpinnings and practical domains. These topics, interoperability and composability, have implications for nearly all domains captured in this book, in particular for distributed simulation development and validation, verification, and accreditation (VV&A). However, it also implies new views on conceptual modeling beyond established needs as well as the need for extended annotations of M&S services in SOAs. It also implies the need for new M&S standards as current solutions are too focused on implementation. We will focus on these issues in this chapter.

## DEFINING INTEROPERABILITY AND COMPOSABILITY

It is good practice to start discussions on the need for unambiguous definitions with respective definitions of terms that are used. We will start with the more traditional definitions used by IEEE and other organizations before looking at ongoing research on layered models of interoperations that are applied to improve the community understanding of what interoperability and composability are and how they can be reached.

## Selected Interoperability Definitions

IEEE defines *interoperability* as the *ability of two or more systems or components to exchange information and to use the information that has been exchanged* [1]. This simple definition has already a number of implications:

(1) Interoperability is defined between *two or more systems*. As such, it includes peer-to-peer solutions as well as hub solutions.
(2) Interoperability allows systems to *exchange information*. This means that systems must be able to produce the required information as well as to consume the provided information. In particular, when information is encapsulated, this may be challenging, which explains that it is necessary to take interoperability requirements into account early enough, so that the design does not hide information from accessibility.
(3) Interoperability allows systems to *use the information* in the receiving system. This implies some common understanding that is shared between sender and receiver. If a system "just listens" to provided information but ignores everything it cannot use, this is not an interoperable solution.

Other organizations, like the Open Group, are stricter in their definition. The Open Group [2] defines interoperability not only as an exchange of information, but they add the ability of systems to provide and receive services from other systems and to use the services so interchanged to enable them to operate effectively together. The notion of actively using other services and taking action based on the received information is a new element in this view.

The U.S. Department of Defense (DoD) adds the component of efficiency to the collaboration and defines interoperability as *the ability of systems, units, or forces to provide data, information, materiel, and services to and accept the same from other systems, units, or forces, and to use the data, information, materiel, and services so exchanged to enable them to operate effectively together* [3]. The same directive furthermore states that joint concepts and integrated architectures shall be used to characterize the interoperations of interest.

Thus, interoperability is understood as the ability of systems to effectively collaborate together on the implementation level to reach a general common objective. To this end, they exchange information that both sides understand well enough to make use of it in the receiving system. Interoperability is a characteristic of a group of systems.

## Selected Composability Definitions

In the M&S community, the term *composability* is also used to address similar issues. Petty and Weisel [4] documented various definitions. Examples for definitions of composability are the following.

Harkrider and Lunceford defined composability as *the ability to create, configure, initialize, test, and validate an exercise by logically assembling a unique simulation execution from a pool of reusable system components in order to meet a specific set of objectives* [5]. They introduced the aspect of logically assembling and, as such, emphasized the necessity for a common basis for the conceptual models that describe the underlying logic.

Pratt et al. approached the challenge of composability from the common architecture perspective. They defined it as *the ability to build new things from existing pieces* [6]. These existing pieces, however, are components of a common architecture, or at least can be captured in a common architecture framework.

Kasputis and Ng emphasized the simulation view. They defined composability as *the ability to compose models/modules across a variety of application domains, levels of resolution, and time scales* [7].

In their work, Petty and Weisel recommended the following definition: *Composability is the capability to select and assemble simulation components in various combinations into valid simulation systems to satisfy specific user requirements* [4]. They also observed that composability deals with the composition of M&S applications using components that exist in the community (e.g., in a common repository). The composition is driven by requirements defining the intended use of the desired composition. Their definition became a common basis of composability research within the community. Composability therefore resides in the models, dealing with the conceptualizations and how they can support a set of requirements.

In comparison, interoperability is seen as the broader, technical principle of interacting systems based on information exchange, while composability deals with the selection and composition of preexisting domain solutions to fulfill user requirements. This idea to distinguish between interoperability of implementation or simulation systems and composability of conceptualizations or simulation models is also the result of current layered approaches.

## Toward a Layered Model of Interoperation

Several researchers introduced layered models to better understand the theoretical underpinnings of interoperation, not only in M&S. Computer science has a tradition of using layered models to better understand the concepts underlying successful interoperation on the implementation level. One of the better known examples is the International Organization for Standards (ISO)/Open System Interconnect (OSI) reference model that introduced seven layers of interconnection, each with well-defined protocols and responsibilities [8]. This section uses the idea of introducing a reference model with well-defined layers of interoperation to better deal with challenges of interoperability of simulation systems and composability of simulation models.

Dahmann introduced the idea of distinguishing between substantive and technical interoperability [9]. In her presentation, technical interoperability ensures connectivity and distributed computation, while substantive interoperability ensures the effective collaboration of the simulation systems contributing to the common goal.

Petty built on this idea in his lectures and short courses [10]. He explicitly distinguished between the implemented model representing substantive interoperability and layers for protocols, the communication layers, and hardware representing technical interoperability.

Tolk and Muguira introduced the first version of a layered model for substantive interoperability, which was very data-centric [11]. In this first model, they distinguished among system specific data, documented data, aligned static data, aligned dynamic data, and harmonized data. These categories describe gradual improvements of interoperability and composition. While system specific data result in independent systems with proprietary interfaces, documented data allow for ad hoc peer-to-peer federations. If these data follow a common model, they are statically aligned and allow for easier collaboration. If their use in the systems is also understood, the data are dynamically aligned as well, and systems can be integrated. Finally, when assumptions and constraints regarding the data and their use are captured as well, the data are harmonized, allowing a unified view.

Using the responding articles of Hoffmann [12], Page et al. [13], and Tolk et al. [14], the LCIM was improved into its current form, which was successfully applied in various application domains, which are not limited to M&S applications. The main improvement was to adapt the names of the layers of interoperation to the terms known from the computer linguistic spectrum regarding the increasing level of understanding based on the information exchanged [12]. In addition, Page et al. [13] proposed to clearly distinguish between the three governing concepts of interoperation:

(1) *Integratability* contends with the physical/technical realms of connections between systems, which include hardware and firmware, protocols, networks, and so on.
(2) *Interoperability* contends with the software and implementation details of interoperations; this includes exchange of data elements via interfaces, the use of middleware, mapping to common information exchange models, and so on.
(3) *Composability* contends with the alignment of issues on the modeling level. The underlying models are purposeful abstractions of reality used for the conceptualization being implemented by the resulting systems.

In summary, *successful interoperation of solutions requires integratability of infrastructures, interoperability of systems, and composability of models.* Successful standards for interoperable solutions must address all three categories.

## The LCIM

The current version of the LCIM was first published in *Theory of Modeling and Simulation* [15]. In this and the following papers, the LCIM exposes the following six layers of interoperation:

(1) The *technical layer* deals with infrastructure and network challenges, enabling systems to exchange carriers of information. This is the domain of integratability.
(2) The *syntactic layer* deals with challenges to interpret and structure the information to form symbols within protocols. This layer belongs to the domain of interoperability.
(3) The *semantic layer* provides a common understanding of the information exchange. On this level, the pieces of information that can be composed to objects, messages, and other higher structures are identified. It represents the aligned static data.
(4) The *pragmatic layer* recognizes the patterns in which data are organized for the information exchange, which are in particular the inputs and outputs of procedures and methods to be called. This is the context in which data are exchanged as applicable information. These groups are often referred to as (business) objects. It represents the aligned dynamic data.
(5) The *dynamic layer* recognizes various system states, including the possibility for agile and adaptive systems. The same business object exchanged with different systems can trigger very different state changes. It is also possible that the same information sent to the same system at different times can trigger different responses.
(6) Finally, assumptions, constraints, and simplifications need to be captured. This happens in the *conceptual layer*. This layer represents the harmonized data.

The following figure shows the LCIM in connection with the three interoperation categories as defined in Page et al. [13] (Fig. 12.1). The figure adds an additional basis level 0 in which no interoperation takes place and where no interoperability has been established.

The LCIM is unique regarding the dynamic and conceptual level. The viewpoint of the LCIM is to distinguish clearly between the three interoperation categories—integratability, interoperability, and composability—and their related concepts within infrastructures, simulations, and models of the systems or services.

## Alternative Layered Views

Although the LCIM has been successfully applied in various domains, alternative layered models exist that are of interest and at a similar maturity level

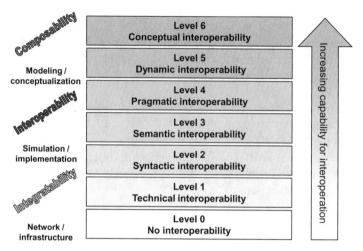

Figure 12.1 Levels of conceptual interoperability model.

[14]. Of particular interest is the following model that finds application in the net-centric environment.

Zeigler et al. proposed the following architecture for M&S that also comprises six layers [15]. They define these layers as follows:

(1) The *network layer* contains the infrastructure including the computer and network.
(2) The *execution layer* comprises the software used to implement the simulation. This includes protocols, databases, and so on.
(3) The *modeling layer* captures the formalism for the model behavior.
(4) The *design and search layer* supports the design of systems based on architectural constraints, comparable to the ideas captured in Pratt et al. [6] and mentioned earlier in this chapter.
(5) The *decision layer* applies the capability to search, select, and execute large model sets in support of what-if analyses.
(6) The *collaboration layer* allows experts—or intelligent agents in support of experts—to introduce viewpoints and individual perspectives to achieve the overall goal.

The LCIM maps well to the network, execution, and modeling layer that deal with infrastructure, simulation, and model. The upper three layers are metalayers that capture the intended and current use of the model, including architectural constraints, which are not dealt with by the LCIM. Using this architecture for M&S, Zeigler and Hammonds defined syntax, semantics, and pragmatics as linguistic levels in a slightly different way [16]. They are defined as follows (see also Fig. 12.2):

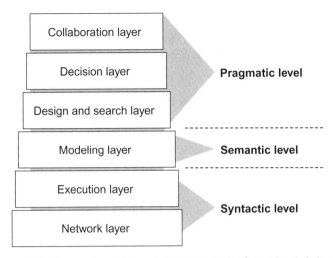

**Figure 12.2** Association between architecture for M&S and linguistic levels.

(1) *Syntax* focuses on structure and adherence to the rules that govern that structure, such as Extensible Markup Language (XML).
(2) *Semantics* consists of low-level and high-level parts. Low-level semantics focus on definition of attributes and terms; high-level semantics focus on the combined meaning of multiple terms.
(3) *Pragmatics* deals with the use of data in relation to its structure and the context of the application (why is the system applied).

These definitions are different from the similar terms introduced in the LCIM. In particular, the pragmatics as defined by Zeigler and Hammonds represent the context of the application; in the LCIM, pragmatics is the context of data exchange within the application.

Zeigler and Hammonds associated these linguistic levels with the architecture for M&S [16]. The difference in the definition of pragmatics becomes obvious, as the intended use capture in the linguistic definition of the term is mapped to the metalayers of the architecture for M&S. In summary, the LCIM focuses on interoperation challenges between models (composability), implementing simulation systems (interoperability), and underlying infrastructure (integratability). These questions are addressed in the syntactic and semantic layer of Zeigler's model; the questions addressed by his pragmatic level are outside the scope of the LCIM, as they deal with the use of the systems.

Both viewpoints are valid and offer a different perspective of the challenges of interoperability and composability. While the LCIM is unique in defining the dynamic and the agility of systems in the dynamic layer as well as the assumption, constraints, and simplifications in the conceptual layer, the approach of Zeigler and Hammonds introduces the intended and current use of linguistic pragmatics as an additional challenge.

In the next section, we will use the LCIM to evaluate the contribution of current interoperability standard solutions. The interested reader is referred to the Further Readings section for more information on alternative approaches.

## CURRENT INTEROPERABILITY STANDARD SOLUTIONS

At this point, the reader should know that interoperation of systems requires support regarding integratability, interoperability, and composability for the infrastructures, the simulations, and the models. Within this section, we will look at the interoperability standard solutions that are applied in the M&S domain. This selection of current solutions is neither complete nor exclusive. Furthermore, this chapter can only give short- and high-level introductions that by no means can replace the in-depth study of these standards.

### IEEE 1278: DIS

Military training is one of the earliest adaptors of distributed simulation applications. The use of training simulators in the armed forces became standard procedure in the education every soldier goes through when he learns to handle his equipment. However, soon after simulation was applied routinely in the armed forces, the requirement for team and group training came up. This required sharing information between the simulators. The objective was to train soldiers in their simulators as a team; that is, the tank or flight simulators had to be represented in the situation displays of the other soldiers. To this end, the Close Combat Tactical Trainer (CCTT) of the U.S. Army was integrated into a SIMulator NETworking (SIMNET) system. SIMNET was developed within a research project out of the Defense Advanced Research Projects Agency (DARPA) and resulted in a network system that allowed tank simulators located in the United States, Germany, and Korea to exchange information about location, activities, and interactions with enemy forces allowing for a whole unit of tank crews to simulate training simultaneously.

The success of SIMNET let to the standardization efforts for the IEEE 1278 Standard for DIS [17]. DIS was built on the foundation of SIMNET. The standard consists of five parts that were released and updated in the following documents:

(1) IEEE 1278-1993—Standard for DIS—Application protocols
(2) IEEE 1278.1-1995—Standard for DIS—Application protocols
(3) IEEE 1278.1-1995—Standard for DIS—Application protocols—Errata (May 1998)
(4) IEEE 1278.1A-1998—Standard for DIS—Application protocols
(5) IEEE-1278.2-1995—Standard for DIS—Communication Services and Profiles

## CURRENT INTEROPERABILITY STANDARD SOLUTIONS 413

(6) IEEE 1278.3-1996—Recommended Practice for DIS—Exercise Management and Feedback
(7) IEEE 1278.4-1997—Recommended Practice for DIS—Verification, Validation, and Accreditation
(8) IEEE 1278.5-XXXX—Fidelity Description Requirements (unpublished).

Under the lead of the SISO, the DIS community is currently working on another update of this standard.

In this section, the application protocols defined in the first part of the standard are of particular interest. Following the introductory remarks, it becomes obvious that the interoperability achieved by applying the IEEE 1278 standard is narrowly focused, as the standard is rooted in the exchange of well-defined *protocol data units* (PDUs) with a shared understanding of PDUs and their defining attributes.

The DIS community defined and standardized PDUs for all sorts of possible events that could happen during such a military training. Whenever a preconceived event happens—such as one tank firing at another, two system colliding, artillery ammunition being used to shoot into special area, a report being transmitted using radio, a jammer being used to suppress the use of communication or detection devices, and more—the appropriate PDU is selected from the list of available PDUs and used to transmit the standardized information describing this event. Within a PDU, syntax and semantics are merged into the information exchange specification. To a certain degree, even the pragmatics are part of this standard PDUs, as the intent is standardized for some interactions as well: When two systems are in a duel, the shooting systems determines if a shot hits the target or not, while the victim determines what effect the hit produces. The following table shows the structure of each PDU.

As described before, DIS is used for military simulators that represent weapon systems. The objects that DIS can represent are categorized in IEEE 1278 as platforms, ammunition, life forms, environmental cultural features, supplies, radios, sensors, and others. DIS also supports the notion of an "expendable" object that allows user-specific representations, but this object is not standardized. The current DIS version defines 67 different PDU types, such as fire, collisions, service calls, and so on. Each PDU is defined as shown in Table 12.1.

The general characteristics of DIS are the absence of any central management; all simulations remain autonomous and are just interconnected by information exchange via PDUs; each simulator has an autonomous perception of the situation; cause–effect responsibilities are distributed for the PDUs to minimize the data traffic. There is no time management or data distribution management. The PDUs are transmitted in a ring or on a bus and each simulator uses PDUs that are directed at one of his entities.

**Table 12.1  PDU structure**

| | | |
|---|---|---|
| PDU header | Protocol version | 8 bit enumeration |
| | Exercise ID | 8 bit unsigned integer |
| | PDU type | 8 bit enumeration |
| | Protocol family | 8 bit enumeration |
| | Time stamp | 32 bit unsigned integer |
| | Length | 16 bit unsigned integer |
| | Padding | 16 bit unsigned integer |
| Originating entity ID | Site | 16 bit unsigned integer |
| | Application | 16 bit unsigned integer |
| | Entity | 16 bit unsigned integer |
| Receiving entity ID | Site | 16 bit unsigned integer |
| | Application | 16 bit unsigned integer |
| | Entity | 16 bit unsigned integer |

DIS is clean and efficient for its purpose. The main advantage of DIS—that everything is defined in detail for the application domain the standard was defined for—is also its main disadvantage: DIS is only applicable for military training on the chosen modeling level. Whenever a new information exchange requirement is identified, the standard needs to be extended. If a new cause–effect responsibility has to be modeled, the standard needs to be extended.

Nonetheless, DIS has been and is still very successful in the military modeling domain, and other M&S application areas are considering comparable domain-specific solutions utilizing well-defined information exchange specifications with well-defined cause–effect responsibilities.

The interested reader is referred to David Neyland's book on virtual combat, listed in the Further Readings section, for more information on DIS and its applications. Information on current developments is accessible via the Web sites of SISO.

## IEEE 1516: HLA

In comparison with DIS, the IEEE 1516 *Standard for M&S HLA* is much more flexible. The declared objective of HLA is to define a software architecture that allows the federation of simulation systems. The driving assumptions behind HLA are that (1) no single simulation system can satisfy the need of all users and (2) no single simulation system developer is an expert across all domains; the logical approach is to compose existing simulation systems into a federation to satisfy the user's requirements with the best simulation systems available.

Different from DIS, this is not limited to a special M&S application area. As a software architecture, HLA supports all M&S application domains, all M&S formalisms, discrete and continuous simulation, and so on. This flexibility comes with the price that much more detail must be aligned between the simulation system providers for their federation.

The IEEE 1516 standard consists of five parts that were released and updated in the following documents:

(1) IEEE 1516-2000—Standard for M&S HLA—Framework and Rules;
(2) IEEE 1516.1-2000—Standard for M&S HLA—Federate Interface Specification;
(3) IEEE 1516.1-2000 Errata (October 2003);
(4) IEEE 1516.2-2000—Standard for M&S HLA—Object Model Template (OMT) Specification;
(5) IEEE 1516.3-2003—Recommended Practice for HLA Federation Development and Execution Process (FEDEP);
(6) IEEE 1516.4-2007—Recommended Practice for Verification, Validation, and Accreditation of a Federation an Overlay to the HLA FEDEP.

Under the lead of SISO, the HLA community is currently working on another update of this standard called *HLA-evolved*, which includes dynamic link capabilities, extended XML support, increased fault tolerance, Web capabilities, and more. In addition, particularly in the United States, many HLA federations are built using the predecessor of IEEE 1516, the U.S. DoD HLA Specification 1.3 NG, which has a slight difference. In this section, we will use the current version of IEEE 1516, and the *federate interface specification* and the *OMT specification* are of particular interest.

***Federate Interface Specification*** One of the main differences between IEEE 1278 DIS and IEEE 1516 HLA is the fact that HLA defines a common *runtime infrastructure* (RTI) with standardized interfaces and well-defined functional categories as part of the standard. The participating systems—called federates—agree to exclusively exchange data during runtime via the RTI. This facilitates management and synchronization of a federation. The functions provided by the RTI to the federation are divided into the six management areas shown in the following figure (Fig. 12.3).

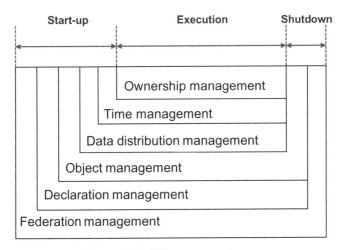

**Figure 12.3** RTI management areas.

The six RTI management areas can be defined as follows:

(1) The purpose of the *federation management* is to determine the federation. Federates join and leave the federation using the functions defined in this group.
(2) The purpose of the *declaration management* is to identify which federate can publish and/or subscribe to which information exchange elements. This defines the type of information that can be shared.
(3) The purpose of *object management* is managing the instances of shareable objects that are actually shared in the federation. Sending and receiving and updating belong into this group.
(5) The purpose of *data distribution management* is to ensure the efficiency of information exchange. By adding additional filters, this group ensures that only data of interest are broadcasted.
(6) The purpose of *time management* is the synchronization of federates.
(7) The purpose of *ownership management* is to enable the transfer of responsibility for instances or attributes between federates.

The figure shows in which order the functions belonging to the management areas are normally invoked (join the federation, declare publish and subscribe capabilities, update objects and send interactions, leave federation). In contrast to the PDUs of DIS, the types and the structure of information are not standardized in HLA, but the objects can be agreed upon between all federates.

IEEE 1516 defines the interface from the federate to the RTI as well as the interface from the RTI to the federate and their interplay. For example, if federate A updates one of his objects O, he calls the update function of the RTI with the object parameters. The RTI calls the receive functions of all other federates that subscribed to the object type of object O, if not excluded by additional filters. All allowed sequences of functions and callbacks are standardized in the federate interface specification.

The functions comprised in these groups use the objects defined in the OMT. The possibility to define its own objects makes the IEEE 1516 HLA flexible. The requirement to use the functions defined in the RTI makes it stable.

**OMT Specification** As mentioned earlier, IEEE 1516 HLA only standardizes the form in which the data to be exchanged are documented. To this end, the OMT is used to define all parameters that are needed to call functions or callbacks provided by the RTI or the federate. As mentioned before, the OMT specification prescribes the format and syntax for describing information but does not standardize the domain-specific data that will appear in the object model. However, IEEE 1516 HLA requires that all shareable information is captured in an object model following the OMT specification. This HLA object models may be used to describe an individual federate (which is called a simulation object model [SOM], defining the information exchange capabilities of this federate) or to describe the information shared between the federates within a federation (which is called a federation object model [FOM]) In addition, management information for the federation is also captured in this schema (which is called the management object model [MOM]).

IEEE 1516 distinguishes between *objects* and *interactions* as information exchange types:

- *Objects exist over time.* They need to be created, updated, and destroyed. They are described by their attributes. Different attributes can be owned by different federates, and their ownership can change over time.
- *Interactions are events.* They are just created and sent to interested federates as a single event. They are described by parameters.

Besides this distinction, objects and interactions are equivalent regarding their information. It is often a modeling decision if something is an object or an interaction. If, for example, one object in federate fires a missile at an object in another federate, the developer can model the missile as an object that updates its position and state (to allow for interceptions), or the developer can decide to just model the impact as an unavoidable event in the form of an interaction.

To specify the exchangeable information as needed to support the federate interface specification, 14 tables are needed. The interested reader is referred to the standard IEEE 1516 and the book of Kuhl et al. in the Further Readings section for details [18]:

(1) The *object model identification table* enumerates information on the developer of the model, contact information, version, and so on. This is meta-information needed for administration purposes and to facilitate the identification of an object model for potential reuse. The structure is pretty easy. The table has two columns to specify the category to be described (such as name, organization, address, and version) and the content describing the category.

(2) The *object class structure table* describes all objects in the form of an object hierarchy of subclasses and superclasses. All objects are subclasses from the generic object called HLAobjectRoot. Following the idea of inheritance as known from object-oriented languages, subclasses are specializations of their superclasses and therefore inherit the attributes of their superclasses. Their specialization is modeled by additional attributes. It should be pointed out that this table does not specify any attributes. This is done in another table.

(3) The *interaction class structure table* comprises the equivalent information for interactions.

(4) The *attribute table* specifies the attributes and their characteristics that are used to model objects. Each class of objects is characterized by a

fixed set of attribute types. This attribute table describes all object attributes represented in a federation. The first column specifies the object class that is specified by the attribute, the second column specifies the attribute itself, the third column the data type. Valid data types are described in the data type table. The remaining columns define characteristics needed for the management functions of the RTI, such as updates, ownership, transportation, and so on.

(5) The *parameter table* comprises the equivalent information for parameters characterizing interactions.

(6) The *dimension table* comprises information needed for data distribution management that can be referred to in the attribute and parameter tables.

(7) The *time representation table* comprises information needed for time management that can be referred to in the attribute and parameter tables.

(8) The *user-supplied tag table* comprises user-defined tags that can be referred to in the attribute and parameter tables. These flags can be evaluated in well-defined RTI functions and callback functions defined in the federate interface specification.

(9) The *synchronization table* comprises the description needed for synchronization points. These points provide support for the mechanism for federates to synchronize their activities.

(10) The *transportation type table* comprises information needed for the transportation of data that can be referred to in the attribute and parameter tables.

(11) The *switches table* comprises switches that can be used by federates. As the RTI performs actions on behalf of federates, switches are used to configure if these actions are enabled or disabled based on the current circumstances. These actions comprise automatically soliciting updates of instance attribute values when an object is newly discovered and advising federates when certain events occur.

(12) The *data type tables* are actually a group of several tables for the various data type categories. HLA 1516 defines the following tables:
- simple data type table
- enumerated data type table
- fixed record data type table
- array data type table
- variant record data type table

Each attribute data type or parameter data type needs to be defined in one of these tables before it can be used.

(13) The *notes table* may comprise additional descriptive information for any entry within any of the other tables. The notes can be referred to in the other tables.

(14) The *FOM/SOM lexicon* comprises semantic information necessary to better understand the information by defining all tags used. This list of terms and definitions is the controlled vocabulary used within the federation.

To summarize this section without going into detail, the information exchange within IEEE 1516 HLA is based on objects characterized by attributes and interactions characterized by parameters. Allowable data types for attributes and parameters must be defined by the federation developer. The RTI specifies management function areas for data management, data distribution management, time management, and more. These functions use the characteristics specified for attributes and parameters in additional tables.

The IEEE 1516 HLA significantly increased the flexibility of simulation federation definitions. Instead of being limited to predefined information exchange groups, the developer can specify the objects and interactions and can even support different time model philosophies. However, the IEEE 1516 HLA does not support semantic transparency of federates regarding procedures and processes. As such, the IEEE 1516 HLA standard is limited to the specification of data to be used for information exchange and the definition of synchronization points.

The interested reader is referred to Kuhl et al. on creating computer simulation systems, listed in the Further Readings section, for more information on HLA and its applications. Information on current developments is accessible via the Web sites of SISO.

## SISO-STD-003-2006: BOMs

The SISO Standard 003-2006 on BOMs is currently the most recent contribution to M&S interoperability standards [19]. BOMs utilize the IEEE 1516 HLA structures for the definition of objects and interactions. In addition, BOMs add the idea of patterns of interplay to the annotations standardized in support of reuse and composition of services. The main characteristics of BOMs can be summarized as follows:

(1) A BOM is a standards-based approach for describing a *reusable component or piece part* of a simulation or system.
(2) BOMs are unique in that they are able to represent or support *discrete patterns of interplay among conceptual entities* within a simulation environment.
(3) BOMs are intended to serve as *building blocks* for supporting composability.
(4) This includes the *composition of HLA object models*, federate capabilities, and/or federation agreements regardless of the hardware platform, operating system, or programming language required of a participating federate.

Two documents define BOMs and their use, which can be downloaded and used by everyone being interested in this for free [19]; in contrast to DIS and HLA, BOM is not standardized by IEEE:

(1) SISO-STD-003-2006—BOM Template Specification
(2) SISO-STD-003.1-2006—Guide for BOM Use and Implementation

SISO-STD-003-2006 BOM uses the XML to capture model metadata, aspects of the *conceptual model*, the class structures of an object model, which are to be used by a federation for representing the conceptual model aspects, and the mapping that exists between that conceptual model and object model. The following figure shows the BOM template (Fig. 12.4).

The categories of the BOM template—in the standard referred to as sections of the template—are clearly motivated by IEEE 1516 HLA, as the following enumeration shows as well.

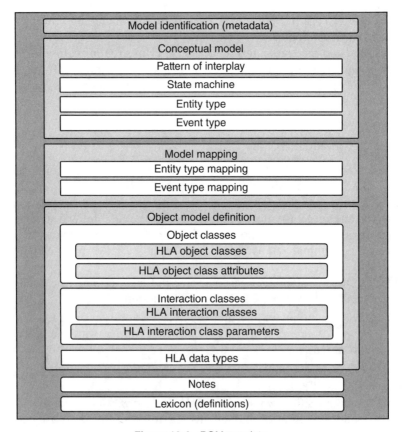

**Figure 12.4** BOM template.

(1) Essential metadata needed so that the BOM can be described, discovered, and properly reused is captured in the *model identification section*. This section is very similar to the object model identification table used to describe the federation as defined in the last section, but the BOM version is machine readable and comprises more information.

(2) Conceptual entities and the events, which occur among those entities as well as the states attainable by those entities, are described in the *conceptual model section*. It should be pointed out that the conceptual model referred to here is close to the computer science view: BOM uses state machines and other methods to capture a principal view of what is modeled inside, without giving away the implementation details. This approach is comparable to use Unified Modeling Language (UML) artifacts to describe a component.

(3) Mapping of conceptual entities and events to object model, object, and interaction classes result in the *model mapping section*. This section maps the implementation-independent conceptual ideas to implementation-driven details, as they are defined in the next section of the BOM template.

(4) The conceptual entities and events are represented by object classes, interaction classes, and data types used to perform the behavior described in the conceptual model. These implementation details are captured in the *object model definition section*. In the example of IEEE1516, these are HLA objects and interactions, specified by attributes and parameters as defined in the HLA OMT, but other implementations are foreseen for future extensions of the BOM already.

(5) Finally, notes and definitions supporting any of the above-mentioned elements are captured including a lexicon with definitions of all used terms in the *notes section* and the *lexicon (definitions) section*. Again, the lexicon defines the controlled vocabularies that are used within the component defined by the BOM.

Of particular interest is the conceptual model section, as this is the main new contribution of BOM. It introduces the notion of patterns of interplay as a specific type of pattern characterized by a sequence of events involving one or more conceptual entities [20]. Conceptual entities are represented in the participating federates. Each event can result in a state change of the conceptual entity. The conceptual model section captures the entity types for the conceptual entities, the event types for the events by or between the entities, the state machine to capture state changes of the entities, and the patterns of interplay showing entities and events.

Although this view on conceptual modeling is not sufficient to ensure interoperability, it is a significant step into the direction of semantic transparency of federates and components.

BOM is not yet sufficiently covered in textbooks. However, the standard is easy to read, and BOM tutorials are available for free on the Internet.

### Web Services and the Semantic Web

The last set of standards, *Web services* and the *semantic Web*, differs from DIS, HLA, and BOM. These standardization efforts are driven by the M&S community. However, the interest to use applicable standards from other domains in support of interoperable M&S solutions instead of using M&S-specific solutions is increasingly observable in current discussions on interoperability and composability. Of particular interest here is the semantic Web as it searches for solutions for composable services and semantic consistency as well. This section only deals with a limited selected subset of means to show future trends and current applicability of solutions [21,22].

***XML*** The backbone of semantic Web methods is XML [23]. XML resulted from improvements—mainly simplifications allowing the easier application—of the Standard Generalized Markup Language (SGML). SGML itself was developed out of an IBM project, in the 1960s, for inserting tags that could be used to describe data and evolved into the ISO Standard 8879. Different features or sections of a document could be marked as serving a particular function. One of the most successful descendants of SGML, prior to the development of XML, is the Hypertext Markup Language (HTML). HTML is the language that currently makes most of the World Wide Web (WWW) documents possible. Since its introduction in 1998, XML has become widely and almost universally adopted by all levels of data and system modelers and developers.

XML is important to the development of portable solutions because it is extensible—the markup tags are self-describing—universally readable—it is based on Unicode—and highly portable—it can be transferred via almost any medium, and its self-contained and embedded nature make it a perfect partner to the data it is describing.

However, XML is not a Rosetta stone. Its application must be accompanied by management effort to support the semantic consistency of tags being used, the mapping of different tags, and more.

***Resource Description Framework (RDF) and RDF Schema*** The World Wide Web Consortium (W3C) developed the RDF as a standard framework to capture ontologies. By definition, RDF is a standardized method for describing resources. In RDF, a description is made of three parts, namely the "subject" (what you are describing) and the "object" (the definition) joined together by the "predicate" (the link between the two). By linking the subject to the object, the predicate gives meaning to the relationship. The set of subject, object, and predicate is commonly referred to as an RDF triple. RDF triples mostly rely on universal resource identifiers (URIs) to provide a physical address for each member of the RDF triple. A URI, although originally envisioned as being quite useful, is not used universally within RDF. Other possibilities include simple terms, literals, and probably in the future extensible resource identifiers (XRIs). The application of RDF resulted soon in the creation of an XML-based RDF schema commonly known as the RDF/XML schema, or RDFS. For interoperability and composability, RDF offers a standardized way to define subjects using standardized Web-based means across federations as well as in heterogeneous environments with systems from various domains.

RDFS describes resources using URIs, simple terms, and literals. Literals are simple terms that may be typed. Literals include a reference to a description of their type. Literal types in RDF are similar to those found within programming languages—integer, string, and so on. RDFS follows the basic RDF triple structure, but models predicate as a property, and the object as a value for that property. This allows modeling how objects, or property values, can be related to subjects. Also, property values (or objects) can be treated as subjects themselves. A machine can infer that certain properties are transitive.

Relying on RDFS as an enhancement to XML for data interchange has the potential to increase the level interoperation between systems. If the RDF structures are well formed and complete in helping to describe the semantic meaning of the data being interchanged, and if each system is capable of making use of those RDF structures, then it becomes possible to evaluate semantic alignment between RDF/RDFS described solutions.

**Web Ontology Language (OWL) and Application to Services** The purpose of the OWL is similar to that of RDFS: to provide an XML-based vocabulary to express classes, properties, and relationships among classes [24]. However, RDFS does this at a very rudimentary level and is not rich enough to reflect the complex nature of many systems.

DARPA targeted overcoming these shortcomings with the development of the DARPA Agent Markup Language (DAML), an RDFS-based language that makes it possible to describe systems at a higher level of detail. DARPA later combined DAML with the European Community's ontology interface layer (OIL) to create DAML+OIL, a full-fledged ontology modeling language. A revision of DAML+OIL, lead by the W3C resulted in the creation of OWL, a new standard for expressing the ontology of a system. Some of OWL's main capabilities include the following:

(1) *Defining property characteristics*: RDFS defines a property in terms of its range (possible values), its domain (class it belongs to), and as a "sub-Property-Of" to narrow its meaning. OWL makes it possible to describe the nature of properties by defining them as symmetric, transitive, functional, inverse, or inverse functional.

(2) *Object property versus data type properties*: In owl, as opposed to RDFS, object properties and data type properties are members of two disjoint classes. Object properties are used to relate resources to one another, while data properties link a resource to a literal (number, string, etc.) or a built-in XML schema data type.

(3) *Property restriction*: OWL classes have a higher level of expressiveness than RDFS classes from which they are inherited. OWL classes allow restrictions on global properties. A property can have all of its values belonging to a certain class, at least one value coming from a certain class or simply have a specific value. OWL also allows restrictions on the cardinality of properties, by simply specifying cardinality, minimum cardinality, or maximum cardinality.

(4) *Class relationships*: OWL classes have most of the properties of sets in set theory (union, disjoint, complement, etc.).

In summary, OWL increases the power of inference that systems can make about one another. OWL provides a powerful framework for expressing ontologies.

OWL for Services (OWL-S) is the application of OWL to describe, in particular, Web services in a much more detailed fashion than the current Web Service Description Language (WSDL) [24]. WSDL is a W3C standard that describes a protocol-and-encoding-independent mechanism for Web service providers to describe the means of interacting with service, but the definition of information exchange is limited to the power provided by XML. WSDL defines operations and messages and provides means to describe how these are bound to implementation using the Simple Object Access Protocol (SOAP) or other executable protocols.

OWL-S adds meaning to the description. OWL-S models a service using three components that are captured in the following list:

(1) The *service profile* provides a concise description of the capabilities implemented by the service. The service profile specifies what the service does. This allows clients and search agents to determine whether the service fulfills their needs.

(2) The *service model* describes the behavior and state changes of a service. The service model specifies how the service works. The service model specifies the inputs, outputs, preconditions, and effects (IOPE) of the service.

(3) The *service grounding* defines how to make use of a service. The service grounding specifies how to access the service. Because WSDL is suitable to express the grounding of a service, such as formats and protocols, OWL-S applies these ideas as well.

The combination of OWL methods for the service profile and the service model and WSDL method for the service grounding results in the best of both worlds. OWL-S provides a semantic description of services, while WSDL specifies how to access the services. Potential clients can use the service profile to discover the service, use the service model to understand the behavior of the service at the abstract level, and finally use WSDL to identify the protocols to bind and interact with the service at the implementation level.

As such, OWL-S enables a new level of interoperation for services. It makes it possible to automatically discover and invoke Web services. It also supports service composition and interoperation and allows more complex tasks to be performed in an automated fashion. This means that interoperability and composability that standardize means are provided to describe IOPE in machine readable form, similar to using BOM templates.

**Summarizing Observations on Standardization Efforts and Alternatives** DIS and HLA are well-established IEEE standards that have been successfully applied in worldwide distributed simulation experimentation. BOM is a SISO standard that introduces the idea of conceptual models of representing entities, events, and patterns of interplay in support of reusability and composability.

XML, RDF/RDFS, and OWL/OWL-S provide means that can be applied in support of interoperability and composability of M&S applications. In particular, when M&S applications need to interoperate with operational systems that do not follow M&S interoperability standards, this knowledge becomes important.

There are other possibilities that may be considered for special application domains, among these is the *Test and Training Enabling Architecture* (TENA) [25]. TENA is not a standard as those described here but an integrated approach to develop distributed simulation for military testing and training, including the integration of live system on test ranges. TENA supports the integration of HLA- and DIS-based systems, but is neither HLA nor DIS based. Following the TENA philosophy, interoperability requires a common architecture, which is TENA, an ability to meaningfully communicate, which requires a common language provided by the TENA Object Model and a common communication mechanism, which is the TENA Middleware and Logical Range Data Archive. In addition, a common context in the form of a common understanding of the environment, a common understanding of time as provided by TENA Middleware, and a common technical process as provided by TENA processes is needed. The specialization of test and training in the military domain is the strength as well as the weakness of TENA, as test and training is well supported, but the transition to other domains requires significant changes to the object model, the data archives, and even the middleware.

Another branch not evaluated in this chapter but worthy of being considered is the use of the *Discrete-Event System Specification* (DEVS) as documented in Zeigler et al. [15]. DEVS is a formalism rooted in systems theory.

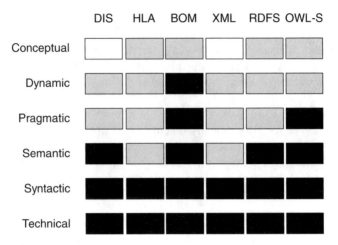

**Figure 12.5** LCIM applied to interoperability standards.

Using this formalism consistently improves reuse and composability but goes beyond the scope of this chapter.

### Contributions and Gaps of Current Solutions

Using the LCIM defined earlier in this chapter, a comparison between these different contributions becomes possible. The following figure shows the six levels of interoperation (technical, syntactic, semantic, pragmatic, dynamic, and conceptual) and the evaluated standardization efforts (DIS, HLA, BOM, XML, RDFS, OWL-S) to show the degree of support of integratability, interoperability, and composability as defined in the LCIM can be provided. The dark gray fields indicate that this level of interoperation is well covered by the respective approach. The light gray fields indicate that the standard comprises means to cope with challenges of this level, but not sufficiently to allow for unambiguous solution.

The justifications for the shades shown in Figure 12.5 are as follows:

(1) DIS uses established infrastructures like Ethernet or Token Rings and established Web protocols to ensure the technical connectivity. The definition of PDUs ensure unambiguous syntax and semantic. However, which PDUs can be sent to whom in which context (pragmatic) and how the system reacts (dynamics) are not part of the standard, although there is room to capture agreements. The conceptual assumptions and constraints, however, are not covered.

(2) HLA has means to support all levels of interoperation, but only the lower levels are standardized. As DIS, the technical level is supported by using established protocols. The HLA OMT unambiguously defines

the syntax. However, as no common data model is given (on purpose, in order to be able to support all M&S application domains and areas), every level from the semantic level must be agreed to and documented in the artifacts provided by the HLA. Unfortunately, these artifacts are not mandatory, so that the upper layers are not supported unambiguously.

(3) BOM evaluated the lessons learned from DIS and HLA and introduced an extension of necessary artifacts in the form of the BOM template that requires to identify the conceptual entities and events (semantics), resulting state changes in the entities (dynamic), and the patterns of interplay (pragmatic) for the higher levels of interoperation. The mapping connects these to the syntactically consistent representations in the services or components represented by the BOM. However, the capturing of assumptions and constraints is limited to the documentation what is modeled, which is not sufficient to support the conceptual level of the LCIM.

(4) To evaluate XML in this context is a little bit unfair, as XML has been designed to describe information exchange between information technology systems in general. It uses established infrastructures and defines the syntax unambiguously. It also provides the frame for higher levels (as RDFS, OWL-S, and also HLA and BOM and some extensions on DIS utilize XML).

(5) RDFS adds the level of semantics to XML but stays on the same level as DIS: unambiguous definition of objects to be exchanged (or shared) between applications. To a certain degree, RDFS allows to capture the context as well, but not enough to fully support the pragmatic level.

(6) OWL-S explicitly models IOPE and, by doing so, unambiguously defines syntax, semantics, and pragmatics. If the effects were unambiguously mapped to inputs of other services to model state changes as well, it would also support dynamics, but OWL-S misses the common conceptual representations that are used in BOM to reach this level.

This evaluation shows the possible descriptive and prescriptive use of the LCIM, which has been introduced and expanded in Tolk et al. [26] and Wang et al. [27]. What is currently still lacking is a set of metrics for each level that can be used to evaluate if a system supports this level or not. Currently, this evaluation is conducted by subject matter expertise using their personal judgment, which leaves too much room for interpretation and discussion. Current discussions show, for example, that some M&S experts explain the opinion that BOMs already support the conceptual level, while others state that the statecharts used in BOM are not even sufficient to support the dynamic level. A similar discussion is possible for the expressiveness of OWL-S. Additional research and agreements are necessary.

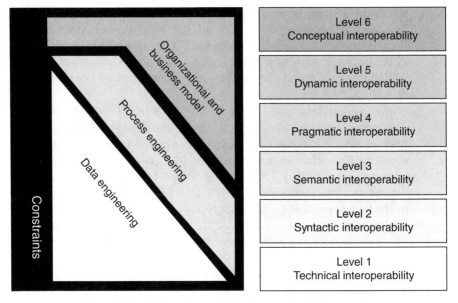

**Figure 12.6** Engineering methods in support of interoperation.

## ENGINEERING METHODS SUPPORTING INTEROPERATION AND COMPOSITION

The elusiveness of the conceptual level of interoperability drives current research. When the LCIM was first presented in 2003, many experts doubted that reaching level 6 will ever be possible. However, several current research projects that will be discussed in this section are working on contribution. Figure 12.6 was introduced to motivate a general layered approach of interoperation in complex systems [28].

Interoperability and composability are not values per se. They are required to support the composition of components that provide a part solution to a problem into a coherent and consistent solution. It is therefore necessary to understand the *organizational and business model* first before any questions regarding the applicability of solutions can be answered. This model identifies and defines the necessary organizations, entities, and capabilities as well as the business processes that need to be supported. It can be argued that this model builds the conceptual model of the operation that needs to be supported. Although this model itself is a simplification of reality, it needs to be complete regarding the required functionality, the conceptual providers and their capability, their relations, and the common processes. This allows that the organizational and business model becomes the blueprint against which the solutions are measured.

Based on this blueprint, process engineering and data engineering are conducted as engineering methods that mutually support each other. *Data engineering* evaluates the *information exchange requirements* (what data need to be exchanged to support the business) from the operational view and compares this with the *information exchange capability* (what data can be produced by the data source) and the information exchange need (what data can be consumed by the data target). *Process engineering* evaluates if the processes that transform and utilize these data can be synchronized, or even orchestrated in support of the business processes.

Data engineering is conducted in four main phases that in practical applications are often iterative and even circular. The four phases are:

(1) *Data administration* answering the questions where the data are, and how the data can be obtained (including business and security concerns).
(2) *Data management* doing the planning, organizing, and managing of data including definition and standardization of the meaning of data as of their relations.
(3) *Data alignment* evaluating information exchange requirement, capabilities, and needs to ensure obtainability and exchangeability of all relevant data.
(4) *Data mediation* ensuring the lossless mediation between different viewpoints.

Introducing a common reference model capturing the information exchange requirements derived from the organizational and business model including dependencies between information elements extends data engineering to *model-based data engineering* [29].

Process engineering is conducted in four similar phases as well. Process engineering is conducted to synchronize, align, and orchestrate processes of the participating systems in support of the common business process. The four phases are:

(1) *Process administration* identifying the relevant processes including their source and their operational contexts.
(2) *Process identification* conducting the organizing and managing of the processes and their specifications.
(3) *Process alignment* defining attributes of the processes, such as determining where in the life of the system the process will occur and what effect it has internally on the system.
(4) *Process transformation* involves performing the necessary transformation to one or more attributes of some of the processes in question to support the required synchronization, alignment, and orchestration.

To intrude into the elusiveness of conceptual interoperability, *constraint engineering* becomes necessary. King described the foundation for constraint engineering [30]. Using ontological means to capture the organizational and business model including the assumptions, constraints, and simplifications making up the conceptual model, he outlined a process for capturing and aligning assertions. Constraint engineering is also conducted in several phases, namely:

- *Capture assertions* as propositions (assumptions, constraints, implemented considerations, and competencies) for the model and organizational environment that are known within the scope or that are otherwise important.
- *Encode propositions* in a knowledge representation language.
- *Compare assertion lists* requiring a multilevel strategy that is described in detail in King [30].

In the first applications, we were able to show that it is possible that services can be conceptually misaligned without showing evidence thereof in their implementation. In other words, services showed no evidence of creating a problem even if their source code were open; however, they exposed problems on the conceptual layer as their assertions resulted in conflicts. While on the implementation level nothing spoke against composing them, the resulting composition were conceptually faulty and would create conceptually wrong answers. Assumptions and constraints leading to simplifications of *what is not modeled* in these evaluated services resulted in conceptual conflicts. Capturing just a conceptual model of *what is implemented* using methods, such as those supported by BOM or generally UML, is therefore not sufficient. *Constraints and assumptions of what is and what is not implemented* must be modeled explicitly, showing the need for constraint engineering in support of conceptual interoperability.

## CONCLUSION

This chapter introduced the student and scholar to the concepts needed to understand the challenges of integratability, interoperability, and composability. It motivated why it is necessary to distinguish between interoperability of simulation systems focusing on aspects of their implementation and composability of simulation models focusing on aspects of their conceptualization. The LCIM was introduced to systematically evaluate in prescriptive and descriptive applications the various layers of interoperation. The LCIM also identifies artifacts needed to annotate systems and service, allowing identifying applicable potential solutions, selecting the best candidates, composing the selected candidates to provide the solution, and orchestrating their execution.

Following these theoretical concepts, current interoperability standardization efforts were introduced: IEEE 1278 DIS [17], IEEE 1516 HLA [18], and SISO-STD-003-2006 BOM [19]. In addition to these M&S-specific standards, Web-based standards were described as well: XML, RDF/RDFS, and

OWL/OWL-S. When applying the LCIM descriptively, the elusiveness of the conceptual level becomes apparent.

Current research evaluates engineering methods in support of enabling interoperation in complex systems: data, process, and constraint engineering. This is an ongoing research that contributes to the next generation of interoperability standards.

Although this chapter focuses on the technical challenges of interoperability and composability, technical solutions do only support interoperability as a point solution in time. To ensure interoperability and composability over the lifetime of a system or a federation, management processes are needed that are integrated into project management as well as into strategic project management [31]. What management processes are needed and which artifacts need to be produced to support them are topics of ongoing research.

## REFERENCES

[1] Institute of Electrical and Electronics Engineers. *IEEE Standard Computer Dictionary: A Compilation of IEEE Standard Computer Glossaries.* New York: IEEE Press; 1990.

[2] The Open Group. The Open Group Architecture Framework Version 8.1.1, Enterprise Edition: Part IV: Resource Base—Glossary. Berkshire, UK, 2006, available at http://www.opengroup.org/architecture/togaf8-doc/arch/chap36.html (accessed December 2009).

[3] U.S. Department of Defense (DoD) Directive 5000.01. The Defense Acquisition System. Certified as current as of November 20, 2007 (former DoDD 5000.1, October 23, 2004).

[4] Petty MD, Weisel EW. A composability lexicon. Proceedings of the Spring Simulation Interoperability Workshop. Orlando, FL, March 30–April 4, 2003, pp. 181–187.

[5] Harkrider SM, Lunceford WH. Modeling and simulation composability. Proceedings of the Interservice/Industry Training, Simulation and Education Conference. Orlando, FL, November 29–December 2, 1999.

[6] Pratt DR, Ragusa LC, von der Lippe S. Composability as an architecture driver. Proceedings of the Interservice/Industry Training, Simulation and Education Conference. Orlando, FL, November 29–December 2, 1999.

[7] Kasputis S, Ng HC. Composable simulations. Proceedings of the Winter Simulation Conference. Orlando, FL, December 10–13, 2000, pp. 1577–1584.

[8] International Organization for Standardization (ISO)/International Electrotechnical Commission (IEC) 10731:1994. *Information Technology—Open Systems Interconnection—Basic Reference Model—Conventions for the Definition of OSI Services.* ISO Press; 1994.

[9] Dahmann JS. High level architecture interoperability challenges. *Presentation at the NATO Modeling & Simulation Conference.* Norfolk, VA: NATO RTA Publications; October 25–29, 1999.

[10] Petty MD. Interoperability and composability. In Modeling & Simulation Curriculum of Old Dominion University. Old Dominion University.

[11] Tolk A, Muguira JA. The levels of conceptual interoperability model (LCIM). Proceedings of the Simulation Interoperability Workshop. Orlando, FL, September 14–19, 2003.

[12] Hofmann M. Challenges of model interoperation in military simulations. *Simulation*, 80:659–667; 2004.

[13] Page EH, Briggs R, Tufarolo JA. Toward a family of maturity models for the simulation interconnection problem. Proceedings of the Simulation Interoperability Workshop. Arlington, VA, April 18–23, 2004.

[14] Tolk A, Turnitsa CD, Diallo SY. Implied ontological representation within the levels of conceptual interoperability model. *International Journal of Intelligent Decision Technologies*, 2(1):3–19; 2008.

[15] Zeigler BP, Kim TG, Praehofer H. *Theory of Modeling and Simulation*. 2nd ed. New York: Academic Press; 2000.

[16] Zeigler BP, Hammonds PE. *Model and Simulation-Based Data Engineering*. New York: Elsevier Science & Technology Books-Academic Press; 2007.

[17] Institute of Electrical and Electronics Engineers. *IEEE 1278 Standard for Distributed Interactive Simulation*. New York: IEEE Publication.

[18] Institute of Electrical and Electronics Engineers. *IEEE 1516 Standard for Modeling and Simulation High Level Architecture*. New York: IEEE Publication.

[19] Simulation Interoperability Standards Organizations. SISO-STD-003-2006 Base Object Model (BOM) Template Specification; SISO-STD-003.1-2006 Guide for BOM Use and Implementation. Available at http://www.sisostds.org. Accessed May 15, 2009.

[20] Hou B, Yao Y, Wang B. Mapping from BOM conceptual model definition to PDES models for enhancing interoperability. Proceedings of the 7th International Conference on System Simulation and Scientific Computing. Venice, Italy, November 21–23, 2008, pp. 349–354.

[21] Daconta MC, Obrst LJ, Smith KT. *The Semantic Web: The Future of XML, Web Services, and Knowledge Management*. Indianapolis, IN: John Wiley; 2003.

[22] Tolk A. What comes after the semantic Web—PADS implications for the dynamic Web. Proceedings of the 20th Workshop on Principles of Advanced and Distributed Simulation. Singapore, May 24–26, 2006, pp. 55–62.

[23] Harold ER, Means WS. *XML in a Nutshell: A Desktop Quick Reference*. 3rd ed. Sebastopol, CA: O'Reilly; 2004.

[24] Alesso HP, Smith CF. *Developing Semantic Web Services*. Wellesley, MA: AK Peters; 2004.

[25] Noseworthy JR. The Test and Training Enabling Architecture (TENA) supporting the decentralized development of distributed applications and LVC simulations. Proceedings of the 12th IEEE/ACM International Symposium on Distributed Simulation and Real-Time Applications. Vancouver, Canada, October 27–29, 2008, pp. 259–268.

[26] Tolk A, Turnitsa CD, Diallo SY. Implied ontological representation within the levels of conceptual interoperability model. *Journal of Systemics, Cybernetics and Informatics*, 5(5):65–74; 2007.

[27] Wang W, Tolk A, Wang W. The levels of conceptual interoperability model: Applying systems engineering principles to M&S. Proceedings of the Spring Simulation Multiconference. San Diego, CA, March 22–27, 2009, pp. 375–384.

[28] Tolk A, Diallo SY, King RD, Turnitsa CD. A layered approach to composition and interoperation in complex systems. In *Complex Systems in Knowledge-Based Environments: Theory, Models and Applications*. Vol. 168. Tolk A, Jain LC (Eds.). Secaucus, NJ: Springer SCI; 2009, pp. 41–74.

[29] Tolk A, Diallo SY. Model-based data engineering for Web services. *IEEE Internet Computing*, 9(4):65–70; 2005.

[30] King RD. On the Role of Assertions for Conceptual Modeling as Enablers of Composable Simulation Solutions. PhD Thesis, Engineering, Old Dominion University, Norfolk, VA; 2009.

[31] Tolk A, Landaeta RE, Kewley RH, Litwin TT. Utilizing strategic project management processes and the NATO code of best practice to improve management of experimentation events. Proceedings of the International Command and Control Research and Technology Symposium. Washington, DC, June 15–17, 2009.

## FURTHER READINGS

Davis PK, Anderson RH. *Improving the Composability of Department of Defense Models and Simulation*. Santa Monica, CA: National Defense Research Institute (US), Rand Corporation; 2003. Available at http://www.rand.org/pubs/monographs/MG101/index.html.

Kuhl F, Dahmann J, Weatherly R. *Creating Computer Simulation Systems: An Introduction to the High Level Architecture*. Upper Saddle River, NJ: Prentice Hall; 2000.

Neyland DL. *Virtual Combat: A Guide to Distributed Interactive Simulation*. New York: Stackpole Books; 1997.

Page EH. Theory and practice for simulation interconnection: interoperability and composability in defense simulation. In *Handbook of Dynamic Systems Modeling*, Fishwick PA (Ed.). Boca Raton, FL: CRC/Taylor & Francis Group; 2007.

# INDEX

Aggregate modeling, 19
Analysis, 1, 21
  data, 345
  regression, 349
  sensitivity, 142, 346
Analysis and operations research, 10
Applications, 20
ARENA simulation, 87–92

Central limit theorem, 139
Computational technologies, 8
Computer simulation, 4
Confidence interval estimate, 50
  statistical estimation, 50
Continuous systems, 99
  simulating, 110
  system class, 100
CountrySim, 229, 255–265

Data/information technologies, 8
Development technologies, 7
Direct3D, 210
Discrete event simulation, 57
  queue discipline, 61
  queuing, 57, 156
  queuing model system, 60
Distributed simulation, 373–402
  commercial off the shelf (COTS), 374
  conservative algorithms, 378
  CORBA, 384
  high level architecture (HLA), 374, 386
  middleware, 383
  optimistic algorithms, 381
  RESTful web services middleware, 395
  SIMulator NETworking (SIMNET), 375
  SOAP-based web services middleware, 389
  synchronization algorithms, 377
Distribution functions, 35
  empirical, 41
  exponential, 36
  normal, 36
  theoretical, 44
  triangular, 35
  uniform, 35
Domains, 22

Euler's method, 111

Fidelity, 13
Flash, 216

Google Earth, 217–220
Google Maps, 218–222

Human behavior modeling, 11, 271–324
  artificial intelligence, 278
  finite state machines, 287
  fuzzy logic, 277
  human behavior, 271
  lumped versus distributed, 275

*Modeling and Simulation Fundamentals: Theoretical Underpinnings and Practical Domains*,
Edited by John A. Sokolowski and Catherine M. Banks
Copyright © 2010 John Wiley & Sons, Inc.

Human behavior modeling (*continued*)
  pattern recognition, 296
  physical level, 273
  strategic level, 274
Human-computer interfacing (HCI), 11, 308
Human factors, 11, 305
  augmented cognition, 306
Human information processing, 307
Hybrid modeling, 19
Hypothesis testing, 350

International Council of Systems Engineering (INCOSE), 2
Interoperability and composability, 403–433
  cross domain solutions, 404
  federate interface specification, 415
  IEEE 1278: DIS, 412
  IEEE 1516: HLA, 414
  integratability, 408
  layered model, 407
  levels of conceptual interoperability model (LCIM), 409–411
  modularity, 403
  OMT specification, 416
  resource description framework, 422
  resource description schema, 422
  SISO-STD-003–2006: BOMs, 419
  Web Ontology Language (OWL), 423
  XML, 422

Java 3D, 213

MATLAB, 215
Maya, 215
Model, 1, 3
  conceptual, 327
  constraint, 160
  executable, 327
  functional, 158
  mathematical, 3
  notional, 2
  physical, 2
  spatial, 161
  stochastic, 10
Model examples, 104
  Predator–prey, 104

Model types, 16
  agent-based modeling, 18, 233
  data-based modeling, 17
  finite element modeling (FEM), 17, 163
  formal modeling, 165
  game theory, 236
  input data modeling, 39
  multimodels, 164
  other modeling, 19
  physics-based modeling, 16
  semiformal modeling, 168
Modeling technologies, 7
Modeling and simulation (M&S), 1
  analysis, 165
  analyze phase, 7
  characteristics and descriptors, 12
  code phase, 7
  execute phase, 7
  life cycle (figure), 9
  model phase, 7
  strategy, 101

OODA loop (observe, orient, decide, act), 245
OpenGL, 209
OpenSceneGraph, 211
Operational Research (OR) Methods, 174
Oscillators, 107
  critically damped solution, 109
  overdamped solution, 109
  underdamped solution, 108
Output data analysis, 48

Probability and statistics, 9, 10, 26
  density function, 32
  distribution function, 32
  set, 26
  space, 28
Project management, 11

Random number, 38
Random process, 10, 21
Random variables, 10, 31
  state variables, 61, 102
Random variates, 37
Resolution, 13

Runge–Kutta methods, 112
  adaptive time step, 116

Scale, 13
Simulation, 1, 4, 328
  constructive, 20
  continuous, 2, 12
  deterministic, 21
  discrete, 2, 12, 25
  error estimation, 133
  exercise, 6
  hand, 67
  implementation, 118
  live, 19
  Monte Carlo simulation, 12, 131–145
  nonterminating, 54
  paradigms, 12
  run, 6
  stochastic, 21
  terminating, 54
  trial, 6
  virtual, 19
Social sciences, 227, 229
  cognitive modeling, 228
  economic institution modeling, 228
  ethnographic modeling, 228
  political strategy modeling, 228
  social agent systems, 228
System, 2
  dynamic systems, 12
Systems modeling, 147–180
  conceptual models, 149
  declarative models, 152
  Markov chains, 154
  methodologies and tools, 148
  types, 147

Time-advance mechanism, 62

XNA Game Studio, 212

Use cases, 168

Verification and validation, 8, 173
  verification and validation process, 15
Verification, Validation, and
    Accreditation (VV&A), 325–372
  cause-effect graphing, 345
  comparison testing, 348
  dynamic methods, 346
  face validation, 341
  inductive assertions, 353
  inspection, 341
  methods, 340
  performing, 333
  predicate calculus, 353
  predictive validation, 346
  referent, 327
  risks, bounds, model credibility, 336
  simuland, 327
  static V&V methods, 345
  Turing test, 343
Visualization, 1, 11, 181–226
  3D representations, 184
  computer animations, 11
  computer graphics, 181
  computer visualization, 11
  digital images, 202
  lighting and shading, 194
  reflection models, 197
  shading models, 201
  synthetic camera and projections, 189
  texture mapping, 202, 205